Springer Proceedings in Physics

Volume 301

Indexed by Scopus

The series Springer Proceedings in Physics, founded in 1984, is devoted to timely reports of state-of-the-art developments in physics and related sciences. Typically based on material presented at conferences, workshops and similar scientific meetings, volumes published in this series will constitute a comprehensive up to date source of reference on a field or subfield of relevance in contemporary physics. Proposals must include the following:

- Name, place and date of the scientific meeting
- A link to the committees (local organization, international advisors etc.)
- Scientific description of the meeting
- List of invited/plenary speakers
- An estimate of the planned proceedings book parameters (number of pages/ articles, requested number of bulk copies, submission deadline).

Please contact:

For Americas and Europe: Dr. Zachary Evenson; zachary.evenson@springer.com

For Asia, Australia and New Zealand: Dr. Loyola DSilva; loyola.dsilva@springer.com

Mohamed Belhaq
Editor

Advances in Nonlinear Dynamics and Control of Mechanical and Physical Systems

Selected Articles from CSNDD 2023; 15–17 May; Marrakesh, Morocco

Springer

Editor
Mohamed Belhaq
University Hassan II—Casablanca
Casablanca, Morocco

ISSN 0930-8989 ISSN 1867-4941 (electronic)
Springer Proceedings in Physics
ISBN 978-981-99-7957-8 ISBN 978-981-99-7958-5 (eBook)
https://doi.org/10.1007/978-981-99-7958-5

This Springer imprint is published by the registered company Springer Nature Singapore Pte Ltd.
The registered company address is: 152 Beach Road, #21-01/04 Gateway East, Singapore 189721, Singapore

Paper in this product is recyclable.

Preface

This book presents the peer-reviewed post-proceedings of the 5th International Conference on Structural Nonlinear Dynamics and Diagnosis (CSNDD 2023). It highlights recent advances in nonlinear dynamics and control with applications in mechanics and physics. The book addresses the state of the art and presents contributions by outstanding scientists in the most active current lines of research in the field of structural nonlinear dynamics and diagnosis. The book covers various aspects of applications, including: micro and nano-electromechanical systems, atomic force microscopy, energy harvesting, time-delayed feedback control, quasi-periodic oscillations and synchronization, anti-resonance, parametric resonance and frequency modulation, solitons in macro-mechanical damped resonators, analytical and semi-analytical methods, control and analysis of switching systems, structural health monitoring, nonlinear vibrations of structures, multi-scale dynamics and nonlinear eigenvalue problem, dynamic modeling and fault diagnosis, constrained multi-catenary systems, conservative chaotic system, hysteretic structures, and nonlinear PDEs.

Casablanca, Morocco | Mohamed Belhaq

Contents

Contributors

Karam Allali Laboratory of Mathematics Computer Science and Applications, Faculty of Sciences and Technologies, Hassan II University of Casablanca, Mohammedia, Morocco

Rabie Aloui R &D Department, Design Office, Safran Seats, Safran Group, Tunisia

Noureddine Atalla Groupe d'Acoustique de l, Université de Sherbrooke, Université de Sherbrooke, Québec, Canada

Taoufik Bakri TNO Sustainable Urban Mobility and Safety, The Hague, The Netherlands

J. M. Balthazar Federal University of Technology - Paraná, R. Doutor Washington Subtil Chueire 330-Jardim Carvalho, Ponta Grossa - PR, Brazil;
São Paulo State University, av. Eng. Luís Edmundo Carrijo Coube, 14-01 - Vargem Limpa, Bauru - SP, Brazil;
Universidade Tecnológica Federal do Paraná - Campus Ponta Grossa, Ponta Grossa, PR, Brazil;
School of Engineering, São Paulo State University, Bauru, SP, Brazil;
Federal University of Technology - Paraná - UTFPR, Ponta Grossa, Brazil;
College of Egineering - Unesp, Bauru, Brazil

Arthur Barbosa University of Franche-Comté, Besançon, France

Mohamed Belhaq Faculty of Sciences Ain Chock, University Hassan II of Casablanca, Casablanca, Morocco;
Department of Physics, Faculty of Sciences Ain Chock, Hassan II University of Casablanca, Casablanca, Morocco

Luis Benadero Universitat Politècnica de Catalunya, Barcelona, Spain

Noureddine Bouhaddi University of Franche-Comté, Besançon, France

Átila M. Bueno Polytechnic School of the University of São Paulo, So Paulo, SP, Brazil

Mnaouar Chouchane Mechanical Engineering Laboratory, ENIM Monastir, Tunisia

Angel Cid-Pastor Universitat Rovira i Virgili, Tarragona, Spain

Dongze Cui Vibroacoustics & Complex Media Research Group, École Centrale de Lyon, Écully, France;
Groupe d'Acoustique de l, Université de Sherbrooke, Université de Sherbrooke, Québec, Canada

Hilson H. Daum Universidade Tecnológica Federal do Paraná - Campus Guarapuava, Guarapuava, PR, Brazil

Abdelali El Aroudi Universitat Rovira i Virgili, Tarragona, Spain

Peter Hagedorn Technical University of Darmstadt, Dynamics and Vibrations Group, FNB, Darmstadt, Germany

Mustapha Hamdi Faculty of Sciences and Technology, University Abdelmalek Essaâdi of Tétouan, Al Hoceima, Morocco

Morten H. Hansen Department of Mechanical and Electrical Engineering, University of Southern Denmark, Sønderborg, Denmark

J. L. Huang Department of Applied Mechanics and Engineering, Sun Yat-sen University, Guangzhou, P.R. China

Mohamed Ichchou Vibroacoustics & Complex Media Research Group, École Centrale de Lyon, Écully, France

Najib Kacem University of Franche-Comté, Besançon, France

Olesya I. Kanishcheva Digital Technologies Department, Voronezh State University, Voronezh, Russia

Markus Kantner Weierstrass Institute for Applied Analysis and Stochastics, Berlin, Germany

Evgeny A. Karpov Digital Technologies Department, Voronezh State University, Voronezh, Russia

Ilham Kirrou ENSA-Agadir, University Ibn Zohr, Agadir, Morocco

Zacharias Kraus Technical University of Darmstadt, Dynamics and Vibrations Group, FNB, Darmstadt, Germany

G. A. Kurina Voronezh State University, Universitetskaya pl., 1, Voronezh, Russia; Federal Research Center "Computer Science and Control" of Russian Academy of Sciences, Moscow, Russia

Walid Larbi Structural Mechanics and Coupled Systems Laboratory, CNAM, Paris, France

Zeljan Lozina University of Split, Faculty of Electrical Engineering, Mechanical Engineering and Naval Architecture, Split, Croatia

Kuntal Mandal Universitat Rovira i Virgili, Tarragona, Spain

Luis Martínez-Salamero Universitat Rovira i Virgili, Tarragona, Spain

Peter A. Meleshenko Digital Technologies Department, Department of Geometry and Functional Theory, Voronezh State University, Voronezh, Russia

Lutz Mertenskötter Weierstrass Institute for Applied Analysis and Stochastics, Berlin, Germany

Omar D. Mohammed Mechanical Engineering Department, Prince Mohammad Bin Fahd University PMU, Khobar, Saudi Arabia

Airton Nabarrete Aeronautics Institute of Technology - ITA, São José Dos Campos, Brazil

Jorge Luis Nabarrete Engineering consultant, São Bernardo do Campo, Brazil

M. A. Ribeiro Federal University of Technology - Paraná, R. Doutor Washington Subtil Chueire 330-Jardim Carvalho, Ponta Grossa - PR, Brazil;
Universidade Tecnológica Federal do Paraná - Campus Ponta Grossa, Ponta Grossa, PR, Brazil

Max Sebastiá-Rullo Universitat Rovira i Virgili, Tarragona, Spain

Damir Sedlar University of Split, Faculty of Electrical Engineering, Mechanical Engineering and Naval Architecture, Split, Croatia

Lucas G. C. Silveira Department of Mechanical and Electrical Engineering, University of Southern Denmark, Sønderborg, Denmark

Mikhail E. Semenov Digital Technologies Department, Voronezh State University, Voronezh, Russia;
Department of Geometry and Functional Theory, Voronezh State University, Voronezh, Russia;
Geophysical Survey of Russia Academy of Science, Obninsk, Russia;
Applied Mathematics and Mechanics Department, Voronezh State Technical University, Voronezh, Russia

A. M. Tusset Federal University of Technology - Paraná, R. Doutor Washington Subtil Chueire 330-Jardim Carvalho, Ponta Grossa - PR, Brazil;
Universidade Tecnológica Federal do Paraná - Campus Ponta Grossa, Ponta Grossa, PR, Brazil

Hugo Valderrama-Blavi Universitat Rovira i Virgili, Tarragona, Spain

Ferdinand Verhulst Mathematisch Instituut, University of Utrecht, Utrecht, The Netherlands

B. X. Zhang Department of Applied Mechanics and Engineering, Sun Yat-sen University, Guangzhou, P.R. China

W. D. Zhu Department of Applied Mechanics and Engineering, Sun Yat-sen University, Guangzhou, P.R. China

Abdel-Malek Zine Institute Camille Jordan–CNRS, École Centrale de Lyon, Écully, France

Emergence and Approximation of Tori

Taoufik Bakri and Ferdinand Verhulst

Abstract We will consider isolated quasi-periodic solutions and families of quasi-periodic solutions. A remarkable aspect of the system Sprott A and a few related systems is the observed presence of families of quasi-periodic solutions organised on invariant tori, a phenomenon known earlier in the context of conservative systems and the so-called KAM-theorem. We describe how the tori bifurcation phenomenon in dissipative systems has been linked to time-reversal and canards. This will be illustrated for the systems NE8 and NE9. Numerical bifurcation techniques in combination with averaging are used to study the presence of periodic and quasi-periodic solutions and possibly invariant manifolds. An interesting mechanical application will be the interaction of a parametrically excited oscillator and an oscillator with self-excitation where Hopf and Neimark–Sacker bifurcations occur naturally. This illustrates the strength of combining analytical and numerical techniques.

1 Introduction

To study applications in mechanics and other fields, in general in dynamical systems, it is useful to focus on basic ingredients of the systems. One considers for instance periodic solutions, integral manifolds, chaos, energy exchange of oscillators and many other phenomena. We will pay attention to quasi-periodic solutions that can be isolated or are organised on a torus.

In the case of dissipative systems the theory of quasi-per iodic solutions and tori was developed after the results for conservative systems, see for instance [2]. Dissipative systems that have a torus containing quasi-periodic solutions can be

T. Bakri (✉)
TNO Sustainable Urban Mobility and Safety, PO Box 96800, 2509 JE The Hague, The Netherlands
e-mail: taoufik.bakri@tno.nl

F. Verhulst
Mathematisch Instituut, University of Utrecht, PO Box 80.010, 3508 TA Utrecht, The Netherlands
e-mail: f.verhulst@uu.nl

M. Belhaq (ed.), *Advances in Nonlinear Dynamics and Control of Mechanical and Physical Systems*, Springer Proceedings in Physics 301,
https://doi.org/10.1007/978-981-99-7958-5_1

perturbed. If the torus is normally hyperbolic the torus will persist in the perturbed system under rather general conditions, see for references [19] or [20]. In practice it is difficult to check whether a system contains a normally hyperbolic torus and most research for dissipative systems start with perturbing coupled harmonic equations, see [4, 5] for references. An exception is the Sprott A system, see [7], where certain symmetries enable us to establish families of tori. We will briefly discuss 2 related systems NE8 and NE9, the analysis is complicated as the systems contain canards.

In subsequent sections we will also study a quasi-periodically perturbed Van der Pol-equation and a system describing interactions between parametric excitation and self-excitation. The model is inspired by the work of the Czechian scientist A. Tondl, see [6]. In the case of interacting oscillators we have more dimensions and parameters and so rich dynamics behaviour; this motivates us to combine asymptotic analysis with the tools AUTO [3] and MATCONT [12].

2 The Jafari–Sprott–Golpayegani Project

A number of chaotic three-dimensional systems, in fact 17 autonomous systems with linear and quadratic terms only and one parameter (a), have been listed and studied in [9], see also [17]; all these systems are dissipative, i.e. the three-dimensional phase-flow is not volume-preserving. The systems are numbered NE1, …, NE17 with one of them, called Sprott A (or NE1). A systematic study of these 17 systems is very instructive as three-dimensional systems generally show much more complexity in bifurcations and chaos than two-dimensional ones as the Jordan separation theorem for 2 spatial dimensions is very restrictive. The 17 systems are relatively simple, linear and quadratic terms and have only one parameter. The evidence for chaos in [9] is mainly numerical but it is an interesting start. Strong evidence in [13, 14] has shown that in the Sprott A system families of tori arise. It has recently been demonstrated in [7] that these families are caused by certain symmetry characteristics of the equations.

In Sect. 3 we will discuss the NE8 and NE9 that have common features but show different dynamics qualitatively.

2.1 An Intermezzo on Canards

Symmetries are an essential ingredient of understanding the dynamics of the Sprott A system. Another interesting aspect is the presence of slow-fast systems that arise for certain values of the parameter and for certain initial values. In the Sprott A system this involves the presence of canards that complicate the dynamics of quasi-periodic solutions. We give a short description of the canard phenomenon, see for an extensive discussion and references [19]. In general canards may be present in slow-fast systems of the form

$$\dot{x} = f(x, y), \ \varepsilon\dot{y} = g(x, y).$$

The dynamics of $x(t)$ will in general be relatively slow with respect to the dynamics of $y(t)$ as $\dot{y}$ will be proportional to $1/\varepsilon$. However, this will not be true in a neighbourhood of space where $g(x, y) = 0$. Suppose that we can solve the equation $g(x, y) = 0$ that can be an algebraic or transcendental equation or a set of such equations (as we did not specify the dimensions of x and y).

Suppose the solution of $g(x, y) = 0$ is $y = y(x)$ that represents a curve or more general a manifold in x, y-space. We will call this a slow manifold as the dynamics on or near this manifold will have a time-behaviour comparable to that of the $x(t)$ component. To understand the canard phenomenon 2 elements are important. First, that the slow manifold under fairly general conditions really exists as a solution of the system and may be stable for a finite interval of time; it has been shown in [19] that in the case of linear attraction the solutions in a neighbourhood of the slow manifold will become exponentially close proportionally to $e^{-ct/\varepsilon}$ with c a positive constant dependent on the system. Secondly, if at some point in (x, y)-space the slow manifold changes stability, it takes some time to leave the slow manifold because of the extreme exponential closeness. We call this delay of instability and the dynamical phenomenon a 'canard'. An illustration by a simple example is given in Fig. 1 (left). The equation used here is

$$\dot{x} = 1, \ \varepsilon\dot{y} = xy. \tag{1}$$

We start at $x(0) = -0.5, y(0) = 0.83$. As we took $\varepsilon = 0.01$ the initial $\dot{y}(0)$ will be -42.5 and $y(t)$ will move quickly to the x-axis. We can solve the problem by explicitly producing:

$$x(t) = x(0) + t, \ y(t) = y(0)e^{\frac{1}{\varepsilon}(x(0)+t/2)t}.$$

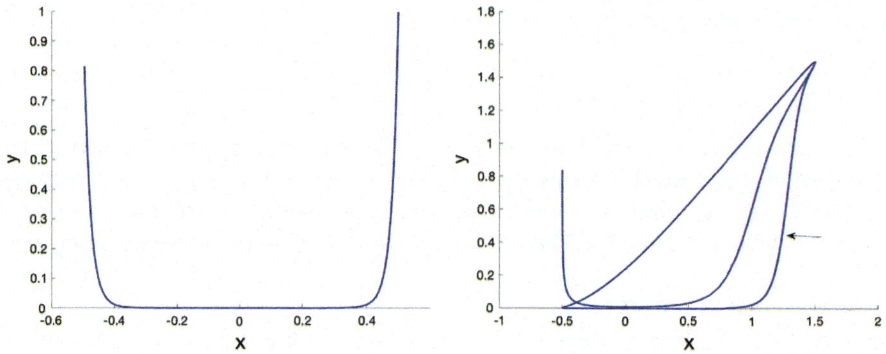

Fig. 1 Left, the canard behaviour demonstrated for (1), $\varepsilon = 0.01$. Right, a solution of (2) with $\varepsilon = 0.05$ where the solution tends to a periodic solution with canards indicated by an arrow. The amplitude depends on the initial conditions

Starting with $x(0)$ negative and $y(0)$ positive, the x-axis ($y = 0$) will be the approximate slow manifold. The exponential attraction with timelike variable t/ε has as a consequence that the solution approaches the x-axis exponentially close. It was shown in [19] that this exponential closeness is a general feature of stable slow manifolds in slow-fast systems. The solution $x(t)$ passes zero after which the slow manifold becomes unstable. However, this instability is delayed because the solution has become extremely close to the slow manifold in its stable stage. This causes the canard behaviour.

A slightly more complicated system where the solutions remain bounded and tend to a periodic solution with a canard is the two-dimensional system:

$$\dot{x} = \sin(t), \ \varepsilon\dot{y} = xy - y^2. \tag{2}$$

We choose again $x(0) = -0.5$, $y(0) = 0.83$. In this case we have 2 slow manifolds that intersect: $y = 0$ and $y = x$. We have again canard behaviour when the solutions follow the x-axis (approximate slow manifold) and pass $x = 0$. The slow manifold $y = x$ becomes stable and, after a canard, the solutions return following $y = x$. The solutions approach the closed curve indicated by an arrow. Numerical experiments suggest that when varying the initial conditions we have a family of periodic solutions.

2.2 *The Sprott A System*

An interesting aspect of the system Sprott A (system NE1) is the observed presence of families of invariant tori, known in conservative systems but, remarkably, also in a dissipative systems with a small parameter. This aspect was studied in more detail for Sprott A in [13, 14] who correctly observe that we have a kind of KAM tori, see for KAM-theory [2]; their evidence is numerical. For completeness we formulate the equations for the system Sprott A:

$$\dot{x} = y, \ \dot{y} = -x - yz, \ \dot{z} = y^2 - a, \tag{3}$$

with parameter $a > 0$. The Sprott A system has a physical meaning as it is a special case of the Nosé-Hoover oscillator. It was shown in [7] that in this dissipative system an infinite number of tori are present, a geometric structure that is caused by symmetries of the system. This is based on the following time-reversal symmetry:

Time-reversal of Sprott A
If $(x(t), y(t), z(t))$ *is a solution of system* (3) *then by putting* $\bar{x} = x, \bar{y} = -y, \bar{z} = -z$ *and reversing time* $\tau = -t$ *then* $(\bar{x}(\tau), \bar{y}(\tau), \bar{z}(\tau))$ *is a solution.*
Such time-reversal symmetry is called R-symmetry in [10].

The background of reversal symmetry is discussed with other physical examples in [15], see for the theory also [8, 10].

In addition, for system Sprott A we can identify in [7] many periodic solutions on the tori. Lazar's frequency condition [11] can be used to demonstrate the process of chaos in the nearly invisible gaps between the tori.

3 Systems NE8 and NE9

A novel result in [7] is that we have completed the theoretical picture for Sprott A by linking the tori bifurcation phenomenon to time- reversal and canards. A interesting question is whether such results apply to other examples of the Jafari–Sprott–Golpayegani project. Candidates are the systems NE8 and NE9 that have similar equations.

For the Sprott A system unbounded solutions can only be found on the z-axis. A novel aspect is that for system NE9 this is different; we find in [7] 'rings' of initial values leading to unbounded solutions. Scaling near infinity and using again geometric singular perturbation theory provides insight in this dynamics.

More information on systems NE1, NE8 and NE9 can be found in [1, 7]. We discuss here some aspects in detail.

3.1 System NE9

Consider the system NE9 (Fig. 2):

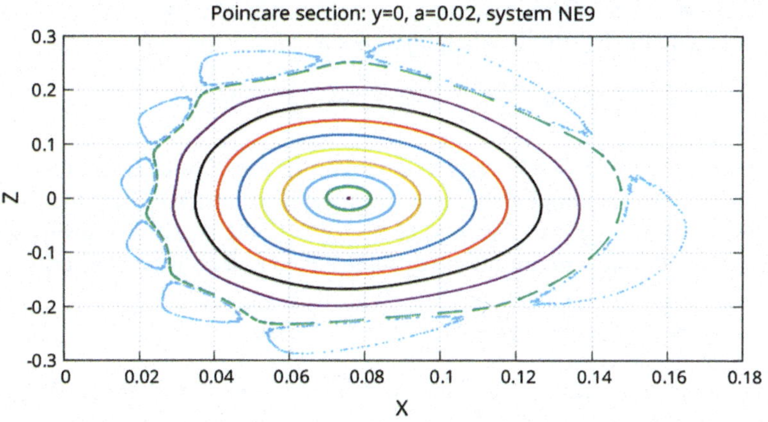

Fig. 2 Poincaré map in the plane $y = 0$ of the Sprott NE9 system near the origin of phase-space for $a = 0.02$. The behaviour near the z-axis shows complex dynamics for various initial conditions

$$\dot{x} = y,\ \dot{y} = -x - yz,\ \dot{z} = -xz + 7x^2 - a, \tag{4}$$

with $a > 0$. Near the origin we rescale

$$x = \varepsilon\bar{x},\ y = \varepsilon\bar{y},\ z = \varepsilon\bar{z},\ a = \varepsilon^2 a_0,$$

and drop the bars. Applying first order averaging with polar coordinates $x = r\cos(t + \phi)$, $y = -r\sin(t + \phi)$, see [16] or [20] for the technique, we find the system:

$$\dot{r} = -\frac{1}{2}\varepsilon rz, \dot{\phi} = 0, \dot{z} = \varepsilon\left(\frac{7}{2}r^2 - a_0\right).$$

A critical point (equilibrium) of the averaged vector field is found for

$$r_0 = \sqrt{\frac{2a_0}{7}}, z = 0.$$

We conclude with the second Bogoliubov theorem, see [20], that in an $O(\varepsilon)$ neighbourhood of r_0 and $z = 0$ a periodic solution exists; the solution is neutrally stable (purely imaginary eigenvalues). System NE9 satisfies the time-reversal condition formulated in Sect. 2.2 so we conclude that a family of invariant tori exists near the origin with the periodic solution near the origin as an organising centre. This is confirmed by the numerics.

The numerics also suggests canard behaviour near the z-axis. This complicates the description of periodic solutions and tori near the z-axis.

It was shown in [7] that system NE9 contains sets of unbounded solutions not containing the z-axis. We intend to avoid this phenomenon here. To study the dynamics near the z-axis consider the case $a = \varepsilon a_0$ and rescale $x = \varepsilon\bar{x}$, $y = \varepsilon\bar{y}$; leaving out the bars the equations valid in a neighbourhood of the z-axis are

$$\dot{x} = y,\ \dot{y} = -x - yz,\ \dot{z} = \varepsilon(-xz + 7\varepsilon x^2 - a_0).$$

The system describes relatively fast motion of the x, y components and slow motion of $z(t)$. Exceptional in this dynamics are the orbits on or near the slow manifold described by $y = -x - yz = 0$ or $x = y = 0$, the z-axis. For orbits spatially removed from the slow manifold the dynamics is to first order in ε described by

$$\dot{x} = y,\ \dot{y} = -x - yz(0),$$

with harmonic solutions (x, y), damped if $z(0) > 0$ and excited if $z(0) < 0$. The symmetry of the equations will cause alternating damping and excitation separated by motion along the slow manifold. This is shown in Fig. 3 for 5000 timesteps. The slow manifold is normally hyperbolic outside a neighbourhood of $z = 0$ and shows repeated canard behaviour when passing $z = 0$.

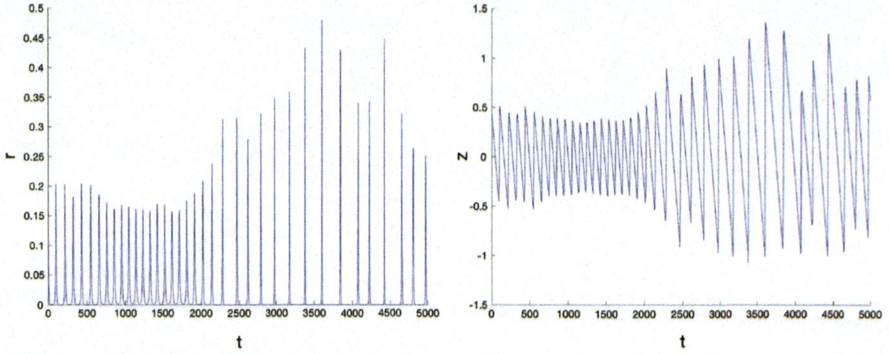

Fig. 3 Dynamics of system NE9 in the case $a = 0.01, x(0) = 0.1, y(0) = 0, z(0) = 0.4$. Left $r(t) = \sqrt(x^2 + y^2)$ passing regularly near the slow manifold where $z(t) \approx 0$. Right the zigzag behaviour of $z(t)$ causing alternatively damping and excitation

3.2 System NE8

Consider system NE8 of [9]:

$$\dot{x} = y, \ \dot{y} = -x - yz, \ \dot{z} = xy + 0.5x^2 - a, \tag{5}$$

with $a_0 > 0$. Again we will look for $0 < a \ll 1$ for periodic solutions near the origin of phase-space. As before we rescale

$$x = \varepsilon\bar{x}, = \varepsilon\bar{y}z = \varepsilon\bar{z}, a = \varepsilon^2 a_0,$$

and drop the bars. Applying first order averaging with polar coordinates $x = r\cos(t + \phi)$, $y = -r\sin(t + \phi)$, we find the system:

$$\dot{r} = -\frac{1}{2}\varepsilon r z, \dot{\phi} = 0, \dot{z} = \varepsilon\left(\frac{1}{4}r^2 - a_0\right).$$

We conclude with the second Bogoliubov theorem that the critical point $r = \sqrt{2a_0}, z = 0$ corresponds with a nearby periodic solution of the system. Again we find neutral stability from the eigenvalues of the critical point. We have to compute a second-order averaging calculation to find out if the neutral stability is persistent as is the case for systems Sprott A and NE9. Interestingly we find to $O(\varepsilon^2)$ that the periodic solution is attracting. Note at this point that system NE8 does not satisfy the reversal-time property formulated in Sect. 2.2. We do not expect families of tori and the dynamics is expected to be completely different. In Fig. 4 we illustrate the attracting behaviour of the flow

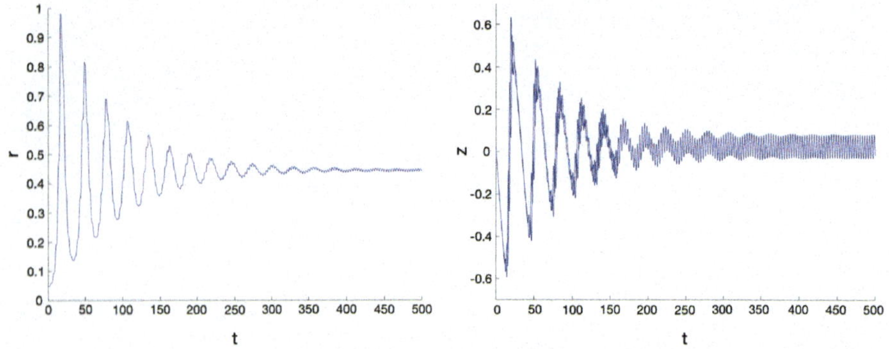

Fig. 4 In the case NE8 an orbit is shown attracted to a periodic solution near the origin of phase-space. Starting point is $x(0) = 0.05$, $y(0) = 0$, $z(0) = 0.5$ with $a = 0.05$. Left $r(t)$, right $z(t)$ during 500 timesteps

4 A Van der Pol-Equation with Quasi-periodic Forcing

Consider the Van der Pol-equation with a cubic (Duffing) term added and quasi-periodic forcing:

$$\ddot{x} + x - \varepsilon a x^3 = \varepsilon \dot{x}(1 - x^2) + f(t), \tag{6}$$

For $f(t)$ we have a quasi-periodic function:

$$f(t) = \sum_{n=1}^{N} c_n \cos \omega_n t, \tag{7}$$

with natural number N, the frequencies $\omega_n, n = 1, \ldots, N$ are irrational numbers, incommensurable and not ε-close to 1. We intend to determine and approximate for ε small a quasi-periodic solution of the equation. We will take as an example the case

$$f(t) = c \cos qt,$$

with q irrational, for instance $q = \sqrt{2}$ or $\sqrt{3}$. The analysis for 1 term in expression (7) will be typical for a more general treatment.

If $\varepsilon = 0$ a particular solution $\eta(t)$ of (6) is

$$\eta(t) = \frac{c}{1 - q^2} \cos qt, \tag{8}$$

To obtain variational equations we sketch the general procedure for n-dimensional ODEs in variable y with external forcing. Consider the equation:

$$\dot{y} = Ay + f(t) + \varepsilon F(t, y).$$

A is a constant $(n \times n)$ matrix. Putting $\varepsilon = 0$ we have a particular solution $\eta(t)$, so

$$\dot{\eta} = A\eta + f(t).$$

The equation $\dot{y} = Ay$ has fundamental matrix solution $\Phi(t)$. We transform $y \mapsto z$

$$y(t) = \Phi(t)z + \eta(t).$$

Substitution in the equations produces the variational (slowly varying) equation:

$$\dot{z} = \Phi^{-1}(t)F(t, \Phi(t)z + \eta(t)).$$

More details on variational equations can be found in [20].

4.1 Averaging Based on Variational Equations

In the case of the forced Van der Pol-equation (6) the fundamental matrix $\Phi(t)$ consists of trigonometric functions. Transforming $x, \dot{x} \mapsto r, \phi$ we put

$$x = r\cos(t + \phi) + \eta(t), \ \dot{x} = -r\sin(t + \phi) + \dot{\eta}(t).$$

to find for r, ϕ:

$$\begin{cases} \dot{r} &= -\varepsilon\sin(t+\phi)[(-r\sin(t+\phi) + \dot{\eta}(t))(1 - (r\cos(t+\phi) + \eta(t))^2) + \\ & a(r\cos(t+\phi) + \eta(t))^3], \\ \dot{\phi} &= -\frac{\varepsilon}{r}\cos(t+\phi)[(-r\sin(t+\phi) + \dot{\eta}(t))(1 - (r\cos(t+\phi) + \eta(t))^2) + \\ & a(r\cos(t+\phi) + \eta(t))^3]. \end{cases} \tag{9}$$

With $f(t) = c\cos\sqrt{3}\,t$ and corresponding $\eta(t)$ from (8) the righthand side is quasi-periodic in t. We have in this case (Fig. 5):

$$\eta(t) = -\frac{c}{2}\cos\sqrt{3}\,t. \tag{10}$$

Averaging over t produces the system:

$$\dot{r} = \frac{\varepsilon}{2}r\left(1 - \frac{1}{4}r^2 - \frac{c^2}{8}\right),\ \dot{\phi} = -\varepsilon\frac{3a}{8}(r^2 + 2c^2). \tag{11}$$

The forcing causes an important change of the standard Van der Pol-equation. With this choice of the forcing frequency ($\omega^2 = 3$) the approximate solution with constant r vanishes if $1 - c^2/8$ becomes negative so if $|c| \geq 2\sqrt{2}$. The self-excitation is

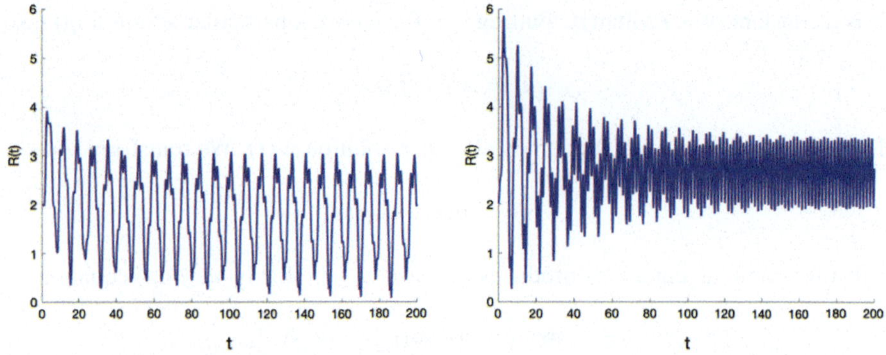

Fig. 5 Two timeseries for $R(t) = \sqrt{x^2(t) + \dot{x}^2(t)}$ of the forced Van der Pol-equation (6) with $f(t) = c\cos\sqrt{3}t$. In both cases $x(0) = 2$, $\dot{x}(0) = 0$, $\varepsilon = 0.05$, $a = 0.5$. Left the case $c = 2$ where $r(t)$ tends to $\sqrt{2}$ and is modulated by the quasi-periodic forcing. Right the case $c = 4$ where $r(t) \to 0$ and the motion $x(t)$ tends to the forced solution

destroyed by the external excitation. In this case we have that the solutions tend to the periodic solution $x(t) = -\frac{c}{2}\cos\sqrt{3}t$.

If $|c| < 2\sqrt{2}$ we have that the quasi-periodic expression

$$x(t) = 2\sqrt{1 - c^2/8}\,\cos\left(t - \varepsilon\frac{3a}{8}\left(4 + \frac{3}{2}c^2\right)t\right) - \frac{c}{2}\cos\sqrt{3}t$$

is an $O(\varepsilon)$ approximation valid on the timescale $1/\varepsilon$.

4.2 An Iteration Approach

We have obtained periodic and quasi-periodic solutions by considering the averaged system (11). Another, but related method is the Poincaré–Lindstedt method for periodic solutions where one uses iteration of a corresponding integral equation while applying periodicity conditions. The iteration method was generalised in [20] by iteration, expansion with respect to the small parameter and applying secularity conditions instead of periodicity conditions. As the Poincaré expansions produce convergent approximations on the timescale $O(1)$ the method does also apply to quasi-periodic solutions and families of them as these solutions can be characterised on such a short timescale.

For (6) we have formulated the variational equations (9). For quasi-periodic solutions we have no restriction on the phase but for $r(t)$ we propose secularity conditions. The quasi-periodic case presupposes $r(t) > 0$, we expand following Poincaré $r(t) = r_0 + \varepsilon r_1(t) + \varepsilon^2 \ldots$ (with r_0 a positive constant) and for the phase $\phi(t) = \phi_0 + \varepsilon\phi_1(t) + \varepsilon^2 \ldots$. We use $\eta(t)$ from (10). Substituting the expansion in the equation for $\dot{r}$ in system (9) we derive to first order in ε the integral for $r_1(t)$:

$$\begin{cases} r_1(t) &= -\int_0^t [\sin(s+\phi_0)(-r_0\sin(s+\phi_0)+\dot{\eta}(s))(1-(r_0\cos(s+\phi_0) \\ & \quad +\eta(s))^2)+a(r_0\cos(s+\phi_0)+\eta(s))^3]ds. \end{cases} \tag{12}$$

We find the secularity condition:

$$r_0\left(1-\frac{1}{4}r_0^2-\frac{1}{8}c^2\right)=0, \tag{13}$$

leading to the requirement:

$$r_0 = 2\sqrt{1-c^2/8},\ |c| < 2\sqrt{2}. \tag{14}$$

The initial value of the phase ϕ_0 plays no part in the secularity condition. Substituting in the integral for $r_1(t)$ expression (12) we find a quasi-periodic expression. We can compute $\phi_1(t)$ by substituting r_0 in the equation for $\phi(t)$ of system (9) and integrating. We find

$$\phi_1(t) = -\frac{3a}{8}(r_0^2+2c^2)t + Q(t). \tag{15}$$

$Q(t)$ is a quasi-periodic function. The expansions are convergent on the timescale $O(1)$ but the results are stronger. From the averaging results in Sect. 4.1 we have that $r_0+\varepsilon r_1(t)$ approximates $r(t)$ with error $O(\varepsilon^2)$ on the timescale $1/\varepsilon$. According to [16], see also [20], the approximation for the amplitude is valid for all time.

The phase $\phi(t)$ is not bounded, the approximation $\varepsilon\phi_1(t)$ has error $O(\varepsilon)$ on the timescale $1/\varepsilon$. We can improve the approximations by substituting for $r(t)$, $\phi(t)$ the expressions $r_0+\varepsilon r_1(t)$, $\phi_0+\varepsilon\phi_1(t)$ in system (9) and integrating to obtain second-order approximations in ε for the quasi-periodic solutions.

5 Interaction of Self-excited and Parametric Excitation

Consider a modification of the system formulated in [6]:

$$\begin{cases} \ddot{x}+\varepsilon\kappa\dot{x}+(1+\varepsilon\cos 2t)x+\varepsilon cy &= 0, \\ \ddot{y}-\varepsilon\mu(1-y^2)\dot{y}+y+\varepsilon a y^3+\varepsilon cx &= 0. \end{cases} \tag{16}$$

The system formulation is inspired by the work of A. Tondl, see [6, 18]. We have weak linear interaction between the 2 oscillators governed by parameter c, with additional parameters κ, μ, a. General position periodic solutions are candidates to produce tori by bifurcation. Without interaction ($c = 0$), the equation for $x(t)$ will have periodic solutions forced by the parametric $\cos 2t$ term. The equation for

$y(t)$ without interaction will have periodic solutions with a period branching off the harmonic solutions, the period will be close to 2π. The parameter a is introduced as the Van der Pol-equation has a 2π-periodic solution in an $O(\varepsilon)$ approximation, if $a = 0$ one needs higher order approximations as a change of the period 2π arises at $O(\varepsilon^2)$. The case $a = 0$ was studied in [6], choosing $a \neq 0$ produces many changes.

5.1 *Periodic Solutions and Tori by Averaging*

We use the transformations $x, \dot{x} \mapsto r, \phi$ and $y, \dot{y} \mapsto R, \psi$:

$$x(t) = r(t)\cos(t + \phi(t)), \dot{x}(t) = -r(t)\sin(t + \phi(t)),$$

$$y(t) = R(t)\cos(t + \psi(t)), \dot{y}(t) = -R(t)\sin(t + \psi(t)).$$

Transforming produces slowly variational equations for the *positive* amplitudes r, R and phases ϕ, ψ. We leave out the variational equations but see system (34) for the amplitudes. After averaging over time we find

$$\begin{cases} \dot{r} &= \frac{\varepsilon}{2}(-\kappa r + \frac{1}{2} r \sin 2\phi + cR\sin\chi), \\ \dot{\phi} &= \frac{\varepsilon}{2r}(\frac{1}{2} r\cos 2\phi + cR\cos\chi), \\ \dot{R} &= \frac{\varepsilon}{2}(\mu R(1 - \frac{1}{4}R^2) - cr\sin\chi), \\ \dot{\psi} &= \varepsilon\frac{3a}{8}R^2 + \frac{\varepsilon}{2}c\frac{r}{R}\cos\chi, \end{cases} \tag{17}$$

with $\chi = \phi - \psi$. The solutions of system (17) produce $O(\varepsilon)$ approximations of the solutions of system (16) valid on the timescale $1/\varepsilon$. Note that if $\sin\chi = 0$ satisfies the equations the amplitudes r and R decouple in the averaged system; see also Sect. 5.5.

The zeros of the righthand sides of system (17) with $r, R \neq 0$ correspond with general position periodic solutions of system (16). If $\cos\chi = 0$ we have for positive R that $a = 0$. This case was studied in [6]. The detuning $a \neq 0$ of the Van der Pol Oscillator complicates the analysis and introduces new phenomena.

The four equations of system (17) yield the following relations:

$$\begin{cases} \cos\chi = & \dfrac{-3aR^3}{4cr}; \\ \sin\chi = & \dfrac{\mu R(1 - R^2/4)}{cr}; \\ \cos 2\phi = & \dfrac{3aR^4}{2r^2}; \\ \sin 2\phi = & 2\kappa - 2\mu\left(\dfrac{R}{r}\right)^2(1 - R^2/4). \end{cases} \tag{18}$$

Requiring that $\sin^2\chi + \cos^2\chi = 1$ yields the following relationship between r and R.

$$r = R\sqrt{\frac{9a^2 + 12\mu^2(1 - R^2/4)^2}{16c^2}} \tag{19}$$

Requiring the same for the angle 2ϕ and using (19) yields after some simplifications the expression for R as function of the parameters only. We find

$$R = \sqrt{Z + 4}; \tag{20}$$

with

$$Z_{1,2} = \frac{-B \pm \sqrt{B^2 - 4AC}}{2A}; \tag{21}$$

$$A = 36a^2\kappa^2 - 9a^2 + 4\kappa^2\mu^2 - \mu^2; \tag{22}$$

$$B = 288a^2\kappa^2 - 72a^2 + 32c^2\kappa\mu; \tag{23}$$

$$C = 576a^2\kappa^2 - 144a^2 + 64c^4 \tag{24}$$

A necessary and sufficient condition for the existence of nontrivial equilibria is $B^2 - 4AC \geq 0$ and ($Z_1 > -4$ or $Z_2 > -4$). Note that the sign of a and c plays no role here. Note also that taking $a = 0$ yields exactly the same nontrivial equilibria as in [6]. See Fig. 6 for the case $a = 0$.

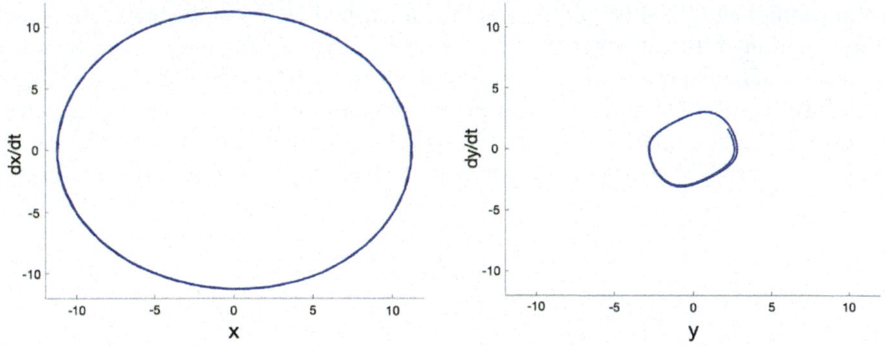

Fig. 6 Projections of a solution of system (16) if $a = 0$ with parameters $\kappa = 1/4$, $c = -1$, $\mu = 4$, $\varepsilon = 0.05$. There exist various periodic solutions for these values, we show the projection if $x(0) = 8$, $\dot{x}(0) = -8$, $y(0) = \dot{y}(0) = 2$

5.2 Numerical Example for $a \neq 0$

Starting at the parameter values $a = 2.4357,\ c = 1.61,\ \kappa = 0.25,\ \mu = 4$, two nontrivial equilibria are found using the equations above with amplitude-phase as follows: $r = 1.7738,\ \phi = -0.5332,\ R = 0.803154,\ \psi = -2.44185$. The other nontrivial equilibrium can easily be computed using the same equations above and is also found via continuation after hitting the fold bifurcation corresponding with the point labelled (LP) in Fig. 7. This figure is obtained via continuation, using the software package MATCONT [12]. The figure shows both equilibria collide and vanish at the critical parameter value $a_c = \pm 3.2969$. Note that the critical value a_c can be computed analytically. Indeed, the two nontrivial equilibria collide and disappear at a fold bifurcation. This happens if and only if $B^2 - 4AC = 0$. Working out this equation yields

$$a_c = \pm \frac{2c^2\mu}{3\sqrt{(4k^2-1)\left(-8c^2k\mu + 4c^4 + (4k^2-1)\mu^2\right)}}. \tag{25}$$

Substituting the parameter values as above in (25) shows that MATCONT as well as AUTO pinpoint the fold bifurcation with high accuracy. Using MATCONT we were able to establish that the stable nontrivial equilibrium undergoes a supercritical Hopf bifurcation at the parameter value $a_{h_1} = 3.2416714$, with negative normal form coefficient $l_1 = -0.142913099$ yielding a stable torus in the original system. The unstable equilibrium emerging from the fold undergoes as well a Hopf bifurcation at the critical value $a_{h_2} = 2.8325198$. This is a subcritical Hopf bifurcation with positive normal form coefficient $l_1 = 4.65799202$ yielding an unstable torus in the original system. The remaining bifurcations in the figure with negative values of the parameter a are symmetrical counterparts of the ones described above. Continuation of the locus of these bifurcations using AUTO, with the parameter c as second control parameters yields Fig. 7 and more detailed Fig. 8. Figure 9 shows the same detailed bifurcation diagram as in Fig. 8 but in the $ac-$parameter space. Note that when the parameter

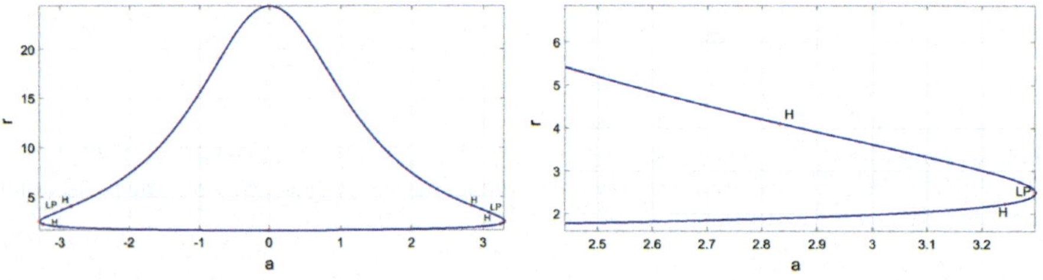

Fig. 7 Continuation of the nontrivial equilibrium with respect of the Van der Pol parameter a yielding a closed symmetrical curve (left). Right is a blow up of the region with interesting dynamics. LP corresponds with a fold bifurcation where two cycles collapse and disappear. H is a point where a Hopf bifurcation occurs yielding a torus

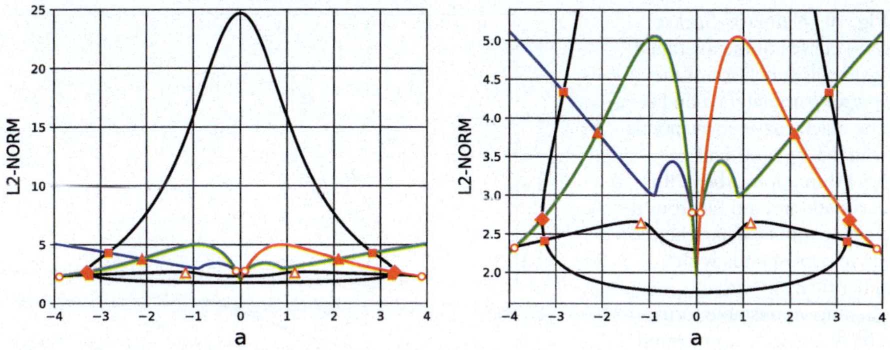

Fig. 8 Continuation using AUTO of the locus of Hopf- and limit point cycle bifurcations (H and LPC) obtained from Fig. 7 using the parameter c as second control parameter. The curves correspond with the co-dimension 2 bifurcations. Right is the same figure but zoomed in to the region with rich dynamics. Circle symbols correspond with Bogdanov–Takens bifurcations, filled squares are points where a Hopf bifurcation occurs, filled diamonds correspond with fold bifurcations (LP) as in Fig. 7, unfilled triangles represent a generalised Hopf bifurcation where the curve switches from sub- to supercritical Hopf and filled triangles correspond with Zero-Hopf bifurcations in the averaged system

Fig. 9 Detailed bifurcation diagram in the ac−parameter space

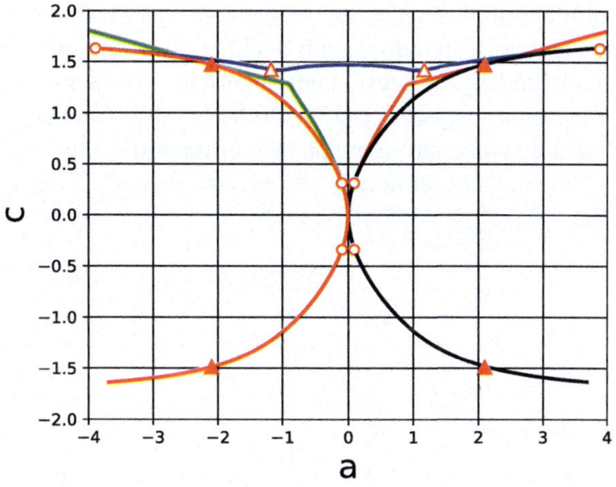

$a = 0$ the difference in amplitude between the two nontrivial solutions is maximal. This suggests the parameter a can be used in quenching the system's vibrations.

5.3 *A Stable 3-d Torus*

Starting at the parameter values $a = 1.6$, a bit away from the zero Hopf bifurcation, and continuing the nontrivial equilibrium of the averaged system using the

Fig. 10 Neimark–Sacker (NS) curves of the averaged system as functions of the control parameters a and c. The black curve corresponds with the supercritical NS-bifurcation. Above this curve and not too far from it a stable torus exists. The red curve corresponds with a subcritical NS-curve yielding an unstable torus. The filled circle corresponds to 1:3 resonance and filled diamonds refer to 1:4 resonance

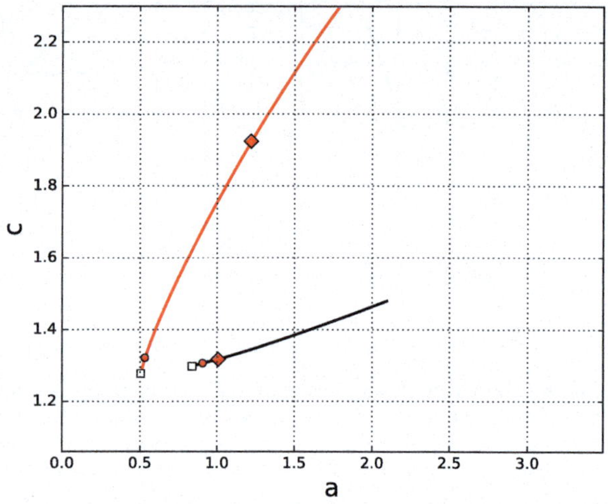

parameter c, the supercritical Hopf bifurcation found above is detected at the critical value $c_c = 1.39470$. The stable cycle emerging from the supercritical Hopf bifurcation is continued with respect to the parameter c. A supercritical Neimark–Sacker bifurcation is detected at the value $c_{NS1} = 1.40007$ yielding a torus in the averaged system corresponding with a stable $\mathbb{T}_3$ Torus in the original system. The unstable cycle undergoes subsequent subcritical Neimark–Sacker bifurcation at the parameter value $c_{NS2} = 2.18435$. Both Neimark–Sacker curves were continued in AUTO yielding some strong resonance bifurcations. See Fig. 10.

A Poincaré section ($x = 2$) shows a stable $\mathbb{T}_2$ torus in Fig. 11.

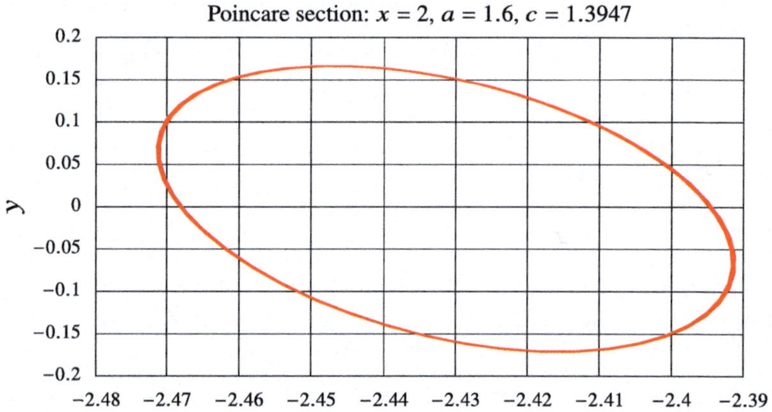

Fig. 11 Poincaré section ($x = 2$) showing a stable $\mathbb{T}_2$ torus in the original system (16) as predicted by the averaging method from the stable cycle emerging from the first supercritical Hopf bifurcation of the averaged system, with initial values $x(0) = 3$, $\dot{x}(0) = 0.4619$, $y(0) = -1.2352$, $\dot{y}(0) = 0.81037$. with parameters $\varepsilon = 0.01$, $a = 1.6$, $c = 1.3947$, $\kappa = 0.25$ and $\mu = 4$

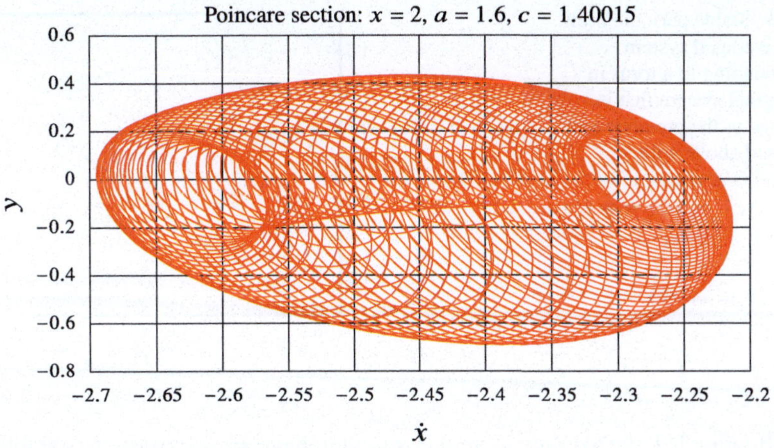

Fig. 12 $\mathbb{T}_3$ stable torus in the original system (16) as predicted by the supercritical Neimark–Sacker bifurcation in the averaged system. Initial values $x(0) = 3, \dot{x}(0) = 0.4619, y(0) = -1.2352, \dot{y}(0) = 0.81037$ with parameters $\varepsilon = 0.01, a = 1.6, c = 1.40015, \kappa = 0.25$ and $\mu = 4$

It is interesting to picture the $\mathbb{T}_3$ stable torus in the original system (16) that we found analytically as a supercritical Neimark–Sacker bifurcation in the averaged system. To visualise the torus in 4-space we use a three-dimensional picture obtained by the Poincaré section at $x = 2$, see Fig. 12. Note that finding these tori in a 4-d space can be quite difficult to locate even if the tori are stable. The averaging method is a powerful mathematical tool that helps to locate these invariant manifolds easily and accurately even if they are unstable.

5.4 *A Special Quasi-periodic Solution*

To be certain not to exclude solutions close to normal modes we will use other coordinates for the averaging analysis. We switch to the coordinates (x_1, x_2, y_1, y_2) instead of amplitude-phase coordinates by transforming:

$$\begin{aligned} x(t) &= x_1(t)\cos(t) + x_2(t)\sin(t), \quad \dot{x}(t) = -x_1(t)\sin(t) + x_2(t)\cos(t), \\ y(t) &= y_1(t)\cos(t) + y_2(t)\sin(t), \quad \dot{y}(t) = -y_1(t)\sin(t) + y_2(t)\cos(t). \end{aligned}$$

Averaging the variational equations over time yields the following system:

$$\begin{cases} \dot{x_1} = -\frac{1}{4}\varepsilon(-2cy_2 + 2kx_1 + x_2) \\ \dot{x_2} = -\frac{1}{4}\varepsilon(2cy_1 + 2kx_2 + x_1), \\ \dot{y_1} = -\frac{1}{8}\varepsilon\left(-4cx_2 + \mu y_1\left(y_1^2 + y_2^2 - 4\right) - 3ay_2\left(y_1^2 + y_2^2\right)\right) \\ \dot{y_2} = \frac{1}{8}\varepsilon\left(-4cx_1 - \mu y_2\left(y_1^2 + y_2^2 - 4\right) - 3ay_1\left(y_1^2 + y_2^2\right)\right) \end{cases} \tag{26}$$

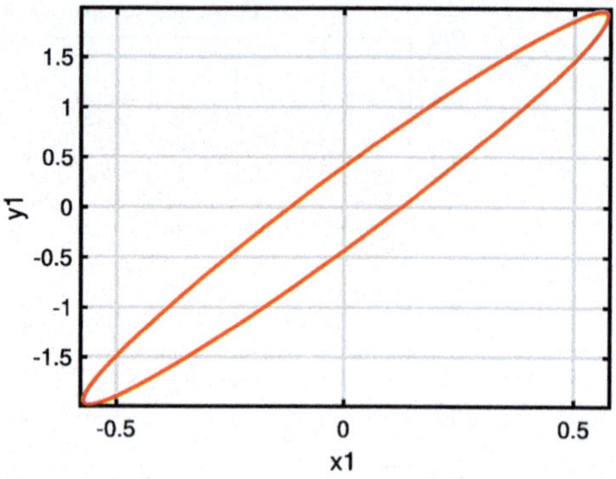

Fig. 13 Stable periodic orbit in the averaged system corresponding to a torus in the original system that is unrelated to the previous tori discussed above with parameters $a = 1.6,\ c = 1.732128,\ \kappa = 1.3,\ \mu = 4$

Following [6], the average system turns out to have an asymptotically stable periodic solution that is not related to the periodic solutions emanating from the Hopf bifurcations discussed above in the case $a = 0,\ c = 1.732128,\ \kappa = 1.3,\ \mu = 4$. This cycle in the averaged system was continued with respect to the parameter a until $a = 1.6$. The corresponding asymptotically stable periodic orbit in the averaged system is shown in Fig. 13. The corresponding stable $\mathbb{T}_2$ torus in the original system has been depicted numerically. See Fig. 14.

Remark
Looking for periodic solutions with $R = 2$ is equivalent to looking for solutions with $Z = 0$. A necessary and sufficient condition for this is to require $C = 0$ and ($B \neq 0$ or $A \neq 0$). This gives the following condition:

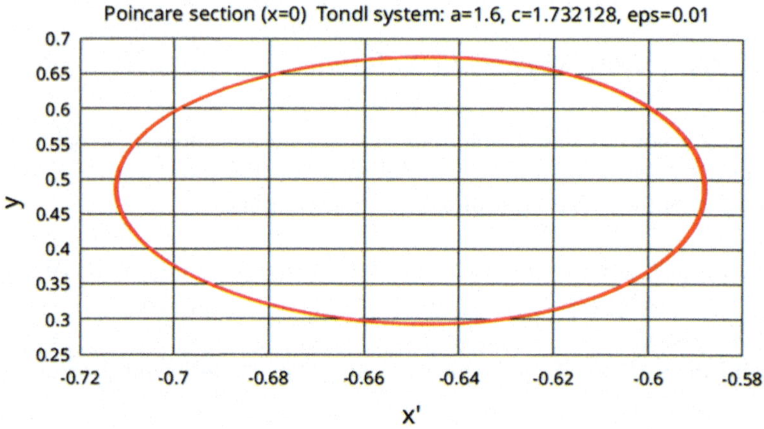

Fig. 14 Stable torus in the original system that is unrelated to the previous tori discussed above; parameters $a = 1.6,\ c = 1.732128,\ \kappa = 1.3,\ \mu = 4,\ \varepsilon = 0.01$

$$c = \pm\sqrt{\frac{3|a|}{2}}\left(1 - 4\kappa^2\right)^{1/4} \tag{27}$$

with

$$|\kappa| \leq 1/2. \tag{28}$$

The expression for the amplitude r becomes in this case:

$$r = 2\sqrt{6|a|}\left(1 - 4\kappa^2\right)^{-1/4}. \tag{29}$$

Note that in this special case the amplitudes become independent of the parameters μ and c. For the phase ϕ we find

$$\phi = \frac{1}{2}\arccos\left(\sqrt{1 - 4\kappa^2}\right). \tag{30}$$

The expression for the phase ψ becomes in this case:

$$\psi = \phi + \pi \mod 2\pi \tag{31}$$

In other words. Requiring $R = 2$ means $\chi = \pi \mod 2\pi$. With $\mu = 4$ we find two nontrivial equilbria.

$$(r_1, \phi_1, R_1, \psi_1) = \left(2.0616, \frac{\pi}{4}, 2.0616, \frac{3\pi}{4}\right); \tag{32}$$

$$(r_2, \phi_2, R_2, \psi_2) = \left(0.6597, -\frac{\pi}{4}, 1.9790, -\frac{3\pi}{4}\right). \tag{33}$$

5.5 *Iteration for Integral Equations*

In Sect. 5.1 we have obtained periodic solutions by looking for critical points (equilibria) of the averaged system. As described in Sect. 4.2, another method uses iteration of a corresponding integral equation while applying secularity conditions and expansion with respect to the small parameter. We will use the amplitude-phase representation for system (16) to formulate equivalent integral equations for the amplitudes r, R.

We start for $r, R > 0$ with the harmonic solutions $x^1 = r_0 \cos(t + \phi_0)$, $y^1 = R_0 \cos(t + \psi_0)$ with general initial values and substitute the expressions in the integrals of the form:

$$r(t) = r_0 + \varepsilon I_1(t),\ R(t) = R_0 + \varepsilon I_2(t).$$

We have from system (16) after transforming to variational equations:

$$\begin{cases} \dot{r} &= -\varepsilon \sin(t+\phi)[\kappa r \sin(t+\phi) - \frac{1}{2} r \cos(3t+\phi) - \frac{1}{2} r \cos(t-\phi) - cR\cos(t+\psi)], \\ \dot{R} &= \varepsilon \sin(t+\psi)[-\mu R \sin(t+\psi) + \frac{1}{2}\mu R^3 \sin(t+\psi) + \frac{1}{4} R^3 \sin(3t+3\psi) - \\ & \frac{1}{4}\mu R^3 \sin(t-\psi) - \frac{3}{4} a R^3 \cos(t+\psi) - \frac{1}{4} a R^3 \cos(3t+3\psi) - cr\cos(t+\phi)]. \end{cases} \tag{34}$$

$I_1(t), I_2(t)$ are obtained by integrating the righthand sides. Putting $\chi = \phi_0 - \psi_0$ we find when expanding to $O(\varepsilon)$ the secularity conditions:

$$\begin{cases} cR_0 \sin\chi &= r_0(\kappa - \frac{1}{2}\sin 2\phi_0), \\ cr_0 \sin\chi &= \mu R_0 (1 - \frac{1}{4} R_0^2). \end{cases} \tag{35}$$

The secularity conditions (35) lead to bounded solutions but not necessarily to periodic or quasi-periodic solutions; that would require conditions on the phases. We can solve for r_0, R_0:

$$r_0 = \frac{c}{\kappa - \frac{1}{2}\sin 2\phi_0} R_0 \sin\chi, \quad R_0 = 2\sqrt{1 - \frac{c^2}{\mu(\kappa - \frac{1}{2}\sin 2\phi_0)}\sin^2\chi}. \tag{36}$$

We have to exclude $\sin\chi = 0, t \geq 0$ as r_0 vanishes in this case. The set of solutions satisfying system (34) is very large. We have also to face the problem of persistence as the results of system (34) produce $O(\varepsilon)$ approximations, there may be higher order problems.

If we want $\chi(t)$ to be stationary we can add another condition that is appropriate for periodic and quasi-periodic solutions. Returning to the averaged system (17) we consider the equation for χ and will look for critical points r_0, R_0:

$$\dot{\chi} = \frac{\varepsilon}{2}\left(\frac{1}{2}\cos 2\phi_0 + \frac{cR_0}{r_0}\cos\chi - \frac{3a}{4}R_0^2 + \frac{cr_0}{R_0}\cos\chi\right). \tag{37}$$

The righthand side of (37) for χ has to vanish when substituting the values of R_0, ϕ_0, ψ_0 (or χ). With suitable parameter values c, a we find quasi- periodic solutions of system (16). With system (36) we have the condition:

$$\frac{1}{2}\cos 2\phi_0 + \frac{\cos\chi}{\sin\chi}(\kappa - \frac{1}{2}\sin 2\phi_0) + c^2 \frac{\sin\chi\cos\chi}{\kappa - \frac{1}{2}\sin 2\phi_0} - 3a + \frac{3ac}{\mu(\kappa - \frac{1}{2}\sin 2\phi_0)}\sin^2\chi = 0. \tag{38}$$

Parameter values satisfying (38) corresponding with real, positive r_0, R_0 of (36) produce bounded solutions of system (16). Choosing as an example $\phi_0 = 0$, $\kappa = 0.25$, $\mu = 4$ we have the condition:

$$\frac{1}{2} + \frac{\cos\chi}{4\sin\chi} + 4c^2 \sin\chi\cos\chi - 3a + 3ac\sin^2\chi = 0. \tag{39}$$

Comparison with the case $a = 0$.

In [6] it was shown for $a = 0$ that if $\cos\chi = 0$ and $\sin 2\phi_0 = -1$ we find a family of periodic solutions. In this case the solutions of system (16) synchronise if there is no interaction ($c = 0$). Equation (37) has righthand side zero so condition (38) is satisfied. The expressions (36) for r_0, R_0 become

$$r_0 = \frac{c}{\kappa + 0.5} R_0, \; R_0 = 2\sqrt{1 - \frac{c^2}{\mu(\kappa + 0.5)}}. \tag{40}$$

As in [6] these expressions for the periodic solutions can be used to look for Neimark–Sacker bifurcations. The analysis in the case $a \neq 0$ with $\cos\chi \neq 0$ is analytically more complicated.

6 Discussion and Conclusions

1. A classical topic in dynamics is the analysis of periodic solutions. Associated with this topic is the abundance of families of quasi-periodic solutions in models of mechanics. To illustrate this we presented a number of models and useful techniques, both analytic and numerical.
2. We started in Sect. 2 with a discussion of the Jafari–Sprott–Golpayegani project as these three-dimensional systems provide us with simple looking nonlinear systems that contain chaos and in general complex dynamics. Apart from the existing literature cited and the details we added for systems Sprott A (NE1), NE9 and NE8, the project has many exciting questions left.
3. In the formulation and analysis in Sect. 5 of interaction of a parametrically excited and a self-excited system we note that this example was inspired by studies of energy absorption (quenching) of undesirable oscillations caused by self-excitation, see for references [6]. Apart from the interesting Neimark–Sacker torus bifurcations in this case it is quite remarkable how a small detuning (by parameter a) of the self-excited Van der Pol-system produces many new phenomena.
4. The application of the iteration method for dimensions higher than two asks for more research.

References

1. C. Abdulwahed, F. Verhulst, Recurrent canards producing relaxation oscillations. Chaos **31** (2021). https://doi.org/10.1063/5.0040726
2. V.I. Arnold, V.V. Kozlov, A.I. Neihstadt, *Mathematical Aspects of Classical and Celestial Mechanics*. Encyclopaedia of Mathematical Sciences, ed. by V.I. Arnold. Dynamical Systems III (Springer, 1988)

3. E. Doedel, A.R. Champneys, T.F. Fairgrieve, Yu.A. Kuznetsov, B. Sandstede, X.J. Wang, AUTO97: *Continuation and Bifurcation Software for Ordinary Differential Equations (with Hom-Cont)*, silver edition (Concordia University, Montreal, Canada, 1997). Download versions at GifHub and SourceForge
4. T. Bakri, F. Verhulst, Bifurcations and quasi-periodic dynamics: torus breakdown. Z. Angew. Math. Phys. **65**, 1053–1076 (2014)
5. T. Bakri, Y.A. Kuznetsov, F. Verhulst, Torus bifurcations in a mechanical system. J. Dyn. Diff. Equ. **27**, 371–403 (2015)
6. T. Bakri, F. Verhulst, From A. Tondl's Dutch contacts to Neimark-Sacker-bifurcation. Appl. Comp. Mech. **16**, 87–100 (2022). https://doi.org/10.24132/acm.2022.770
7. T. Bakri, F. Verhulst, Time-reversal, tori families and canards in the Sprott A and NE9 systems. CHAOS (2022). https://doi.org/10.1063/s.0097508
8. M.-C. Ciocci, A. Litvak-Hinenzon, H.W. Broer, Survey on dissipative KAM theory including quasi-periodic bifurcation theory. Geometric Mechanics and Symmetry: The Peyresq Lectures **306**, 303–355 (2005)
9. S. Jafari, J.C. Sprott, S. Golpayegani, Elementary quadratic chaotic flows with no equilibria. Phys. Lett. A **377**, 699–702 (2013)
10. J.S.W. Lamb, J.A.G. Roberts, Time-reversal symmetry in dynamical systems: a survey. Physica D **112**, 61–39702 (1998)
11. J. Laskar, Frequency analysis of a dynamical system. Celest. Mech. Dyn. Astron. **56**, 191–196 (1993)
12. MATCONT, Numerical continuation and bifurcation program. http://www.matcont.ugent.be
13. M. Messias, A.C. Reinol, On the formation of hidden chaotic attractors and nested invariant tori in the Sprott A system. Nonlinear Dyn. **88**, 807–821 (2017)
14. M. Messias, A.C. Reinol, On the existence of periodic orbits and KAM tori in the Sprott A system: a special case of the Nosé-Hoover oscillator. Nonlinear Dyn. **92**, 1287–1297 (2018)
15. J.A.G. Roberts, G.R.W. Quispel, Chaos and time-reversal symmetry. Order and chaos in reversible dynamical systems. Phys. Rep. **216**, 63–177 (1992)
16. J.A. Sanders, F. Verhulst, J. Murdock, *Averaging Methods in Nonlinear Dynamical Systems*, rev. edn (Springer, 2007)
17. J.C. Sprott, Some simple chaotic flows. Phys. Rev. E **50**, R647–R650 (1994)
18. A. Tondl, On the interaction between self-excited and parametric vibrations, National Research Institute for Machine Design, Běchovice, Monographs and Memoranda 25, Prague (1978)
19. F. Verhulst, *Methods and Applications of Singular Perturbations* (Springer, New York etc., 2005)
20. F. Verhulst, *A Toolbox of Averaging Theorems, Ordinary and Partial Differential Equations*. Surveys and Tutorials in the Applied Mathematical Sciences (Springer, 2023)

Quasiperiodic Birhythmicity in a Multicycle van der Pol Oscillator

Mustapha Hamdi and Mohamed Belhaq

Abstract In this work, the dynamics of a time-delayed multicycle van der Pol oscillator is investigated in the case where the delay amplitude is harmonically modulated around a mean value with a certain amplitude and frequency. Using perturbation methods, approximate analytical solutions of quasiperiodic (QP) oscillations and their bifurcation equations are obtained and regions of existence of different solutions are determined. It is revealed that, while monorhythmic and birhythmic periodic oscillations are commonly present in the undelayed or delay-unmodulated multicycle van der Pol oscillator, the delay-modulated oscillator may exhibit *QP-monorhythmicity* or *QP-birhythmicity*. Based on the bifurcation diagrams, transitions between different rhythmicities are derived. The results show that for appropriate values of delay parameters, large-amplitude stable (QP-monorhythmic) oscillations can take place over a broadband of the modulation frequency near the *delay-induced parametric resonance*. Bistability generated by QP-birhythmicity can also occur for different values of delay parameters.

1 Introduction

In the classical van der Pol oscillator, one stable limit cycle may exist giving rise to a monorhythmic response. Birhythmicity, on the other hand, is characterized by the existence of two stable limit cycles with different amplitudes and frequencies separated by an unstable one. A simple oscillator that exhibits such a birhythmic behavior consists of a multicycle van der Pol oscillator in which the polynomial damping component is of sixth order. Over the past decades, attention has been devoted to investigate the birhythmic behavior in a multicycle van der Pol oscillator.

M. Hamdi (✉)
Faculty of Sciences and Technology, University Abdelmalek Essaâdi of Tétouan, Al Hoceima, Morocco
e-mail: hamustapha2000@yahoo.fr

M. Belhaq
Faculty of Sciences Ain Chock, University Hassan II of Casablanca, Casablanca, Morocco

M. Belhaq (ed.), *Advances in Nonlinear Dynamics and Control of Mechanical and Physical Systems*, Springer Proceedings in Physics 301,
https://doi.org/10.1007/978-981-99-7958-5_2

Such a birhythmic behavior is an usual phenomenon that occurs in biological [1, 2], mechanical [3, 4], chemical [5], biochemical [6, 7] and hydrodynamic systems [8].

It is known that in a multicycle van der Pol oscillator the parameter space can be divided into two different regions in which the system may exhibit birhythmicity (two stable limit cycles) or monorhythmicity; see for instance [9]. Depending on the nature of the problem, birhythmicity can be desirable or undesirable. For instance, it can be desirable in living systems as it enables the system to evolve in various modes of vibrations [5] according to the surrounding environment. Birhythmicity can also be regarded as unwanted regime in certain engineering applications [9] for which a small perturbation can switch the system response to unwanted vibrations. A typical example is the wind-induced birhythmic oscillator for which the system response may be attracted to a small or a large-amplitude limit cycle depending upon the initial state of the system. Therefore, it is important to figure out how to control birhythmicity by focusing on large-amplitude oscillations when desired, as in energy harvesting applications, or targeting small-amplitude motions usually required to quench vibrations.

The control of birhythmicity has received increasing attention and different techniques have been proposed. For instance, time-delayed feedback has been used [10] to suppress birhythmicity in the multicycle van der Pol oscillator. In [11], the conjugate self-feedback technique has been employed to suppress birhythmicity. The effect of filtered-feedback on the suppression of birhythmicity has been investigated in [12] and the combined influence of time delay and noise on birhythmicity has been considered in [13]. The case of a stochastic time-delayed birhythmic oscillator with fractional derivative has been tackled in [14], while the control of birhythmicity using two delays and colored noise was studied in [15]. Recently, the problem of suppressing birhythmicity has been addressed using damping modulation in a delayed multicycle van der Pol oscillator [16].

In the existing literature [10, 13–16], the delay introduced in the multicycle van der Pol equation was considered as time-invariant. The purpose of the present paper is to study the effect of a time-variant delay on the dynamics of a multicycle van der Pol oscillator. The concept of using time-variant delay has already been used to control stability in atomic force microscopy [17], in energy harvesting [18] and in other systems [19]. The time-variant delay is introduced by modulating the delay amplitude around a mean value with certain amplitude and frequency. The main objective is to investigate the effect of the modulation on the rhythmic response of the multicycle van der Pol oscillator.

The chapter is organized as follows: In the next section, approximation of periodic solutions of the time-delayed multicycle van der Pol oscillator is obtained using an averaging method. The obtained bifurcation equations are used to construct bifurcation diagrams for limit cycles. The effect of the modulation of the time delay on limit cycles is analyzed. In Sect. 3, a second-step perturbation method is applied on the slow flow equations to approximate QP solutions and their domain of existence. Section 4 provides bifurcation diagrams, shows different regions of periodic and QP rhythmicities and describes bifurcation transitions between limit cycle and tori branches. Section 4 concludes the work.

2 Periodic Rhythmicity

Consider a time-delayed birhythmic van der Pol oscillator for which the delay is introduced in the position and velocity such that the equation of motion is given by

$$\ddot{x} + x - \varepsilon(\alpha - \beta x^2 + \alpha_2 x^4 - \beta_2 x^6)\dot{x} = \varepsilon\lambda(t)x(t-\tau) + \varepsilon\lambda_1\dot{x}(t-\tau) \quad (1)$$

where α, β, α_2 and β_2 are the damping coefficients, $\lambda(t)$, λ_1 and τ denote, respectively, the delay amplitudes and time delay, and ε is a perturbation parameter introduced in the damping and delay parameters. An overdot denotes differentiation with respect to time t. We assume that the delay amplitude in the position $\lambda(t)$ is modulated around the nominal value λ_1, such that

$$\lambda(t) = \lambda_1 + \lambda_2 \cos \omega t \quad (2)$$

where λ_2 is the amplitude of the modulation and ω is the frequency. We consider the case of *a delay-induced parametric resonance* for which the frequency ω is near twice the natural frequency of the oscillator. Note that the case of a monocycle van der Pol oscillator (i.e., (1), (2) with $\alpha_2 = \beta_2 = 0$) has been analyzed in [19]. The resonance condition reads $1 = \left(\frac{\omega}{2}\right)^2 + \varepsilon\sigma$ where σ is a detuning parameter. To obtain the modulation equations of amplitude and phase of (1) near the delay-induced parametric resonance, we apply the Krylov–Bogoliubov method [20]. Equation (1) can be written as

$$\ddot{x} + \frac{\omega^2}{4}x = \varepsilon F(x, \dot{x}, t) \quad (3)$$

where

$$F(x, \dot{x}, t) = -\sigma x + (\alpha - \beta x^2 + \alpha_2 x^4 - \beta_2 x^6)\dot{x} + (\lambda_1 + \lambda_2 \cos \omega t)x(t-\tau) + \lambda_1\dot{x}(t-\tau) \quad (4)$$

The periodic solutions of (3) can be approximated by

$$x = R\cos\left(\frac{\omega}{2}t - \theta\right) \quad (5)$$

in which the amplitude R and the phase θ are time-dependent. Assume that over a period of oscillation $4\pi/\omega$ these variables do not change significantly and are governed by

$$\dot{R} = -\frac{\varepsilon}{2\pi}\int_0^{\frac{4\pi}{\omega}} \sin\left(\frac{\omega}{2}t - \theta\right)F(x, \dot{x}, t)dt \tag{6a}$$

$$R\dot{\theta} = -\frac{\varepsilon}{2\pi}\int_0^{\frac{4\pi}{\omega}} \cos\left(\frac{\omega}{2}t - \theta\right)F(x, \dot{x}, t)dt \tag{6b}$$

Integrating, truncating the Taylor series expansions for those terms containing time delay by assuming that the product $\varepsilon\tau$ is small [21] and fixing the parameters $\alpha = \beta = 1$, we obtain the following modulation equations

$$\dot{R} = \varepsilon(k_1 R + k_2 R^3 + \frac{\alpha_2}{16}R^5 - \frac{5\beta_2}{128}R^7 + k_3 R\cos 2\theta + k_4 R\sin 2\theta) \tag{7a}$$

$$R\dot{\theta} = \varepsilon(k_5 R + k_4 R\cos 2\theta - k_3 R\sin 2\theta) \tag{7b}$$

where $k_1 = \frac{1}{2} - \frac{\lambda_1}{\omega}\sin(\omega\tau/2) + \frac{\lambda_1}{2}\cos(\omega\tau/2)$, $k_2 = -\frac{1}{8}$, $k_3 = \frac{\lambda_2}{2\omega}\sin(\omega\tau/2)$, $k_4 = \frac{\lambda_2}{2\omega}\cos(\omega\tau/2)$ and $k_5 = \frac{-\sigma}{\omega} + \frac{\lambda_1}{\omega}\cos(\omega\tau/2) + \frac{\lambda_1}{2}\sin(\omega\tau/2)$. Due to the fact that the slow flow (7) is invariant under the transformation $\theta \to -\theta + \frac{\pi}{2}$, $k_3 \to -k_3$ and $k_5 \to -k_5$, system (7) can be rewritten in the form

$$\dot{R} = k_1 R + k_2 R^3 + \frac{\alpha_2}{16}R^5 - \frac{5\beta_2}{128}R^7 + sk_3 R\cos 2\theta + k_4 R\sin 2\theta \tag{8a}$$

$$R\dot{\theta} = sk_5 R + k_4 R\cos 2\theta - sk_3 R\sin 2\theta \tag{8b}$$

in which the new time scale $T = t/\varepsilon$ and $s = \pm 1$. This invariance property will be used to capture the modulation envelope of the QP response. Periodic motions of (3) corresponding to the equilibrium points $R =$ const, $\theta =$ const of the slow flow (8) are determined by setting $\dot{R} = \dot{\theta} = 0$. Eliminating θ and define $\rho = R^2$, the resulting algebraic equation in ρ reads

$$\rho^3 + A\rho^2 + B\rho + C_\pm = 0 \tag{9}$$

where $A = -\frac{8\alpha_2}{5\beta_2}$, $B = -\frac{128k_2}{5\beta_2}$ and $C_\pm = \frac{128}{5\beta_2}(\pm\sqrt{k_3^2 + k_4^2 - k_5^2} - k_1)$ with the condition $k_3^2 + k_4^2 - k_5^2 \geq 0$ to be satisfied. The real roots of (9) are determined using Cardano's method and considering the variable change $\rho' = \rho + \frac{A}{3}$. Consequently, (9) becomes

$$\rho'^3 + p\rho' + q_\pm = 0 \tag{10}$$

where $p = B + \frac{A^2}{3}$ and $q_\pm = \frac{2A^3}{27} - \frac{BA}{3} + C\pm$. System of equation (10) has three pairs of positive roots if its discriminant $\Delta_\pm = 4p^3 + 27q_\pm^2$ is negative. These solutions are given by

$$R_{1\pm} = \sqrt{\left(\frac{-q_\pm - \sqrt{\Delta_\pm/27}}{2}\right)^{1/3} + \left(\frac{-q_\pm + \sqrt{\Delta_\pm/27}}{2}\right)^{1/3} - A/3} \quad (11a)$$

$$R_{2\pm} = \sqrt{a^2\left(\frac{-q_\pm - \sqrt{\Delta_\pm/27}}{2}\right)^{1/3} + a\left(\frac{-q_\pm + \sqrt{\Delta_\pm/27}}{2}\right)^{1/3} - A/3} \quad (11b)$$

$$R_{3\pm} = \sqrt{a\left(\frac{-q_\pm - \sqrt{\Delta_\pm/27}}{2}\right)^{1/3} + a^2\left(\frac{-q_\pm + \sqrt{\Delta_\pm/27}}{2}\right)^{1/3} - A/3} \quad (11c)$$

with $a = e^{\frac{2\pi}{3}j}$ and $j = \sqrt{-1}$. The stability diagram of the limit cycle solutions is determined by the sign of the eigenvalues of the Jacobian matrix of the slow flow (8) evaluated in the $n_\pm th$ ($n = 1, 2, 3$) branch (11). These are the roots of the characteristic equation

$$S^2 - 2aS + b = 0 \quad (12)$$

where $a = k_1 + 2k_2R_n^2 + \frac{3\alpha_2}{16}R_n^4 - \frac{5\beta_2}{32}R_n^6$ and $b = (2k_2R_n^2 + \frac{\alpha_2}{4}R_n^4 - \frac{15\beta_2}{64}R_n^6)(2k_1 + 2k_2R_n^2 + \frac{\alpha_2}{8}R_n^4 - \frac{5\beta_2}{64}R_n^6)$.

The condition necessary for a Hopf bifurcation of periodic solutions is given by $a = 0$ when the condition $b > 0$ holds. On the other hand, the condition necessary for a saddle-node bifurcation of periodic solutions is given by $b = 0$.

Using the amplitude equation (9) and the saddle-node bifurcation condition, the well-known two-parameter bifurcation diagram in (α_2, β_2)-plane is shown in Fig. 1 for selected values of λ_1, λ_2, ω and τ. In this figure the white region corresponds to monorhythmicity and the magenta region to birhythmicity. A similar diagram has been reported in previous papers (see for instance [9]).

In what follows, the damping coefficients are fixed as $\alpha_2 = 0.15$, $\beta_2 = 0.004$. The two-parameter bifurcation diagram in the (λ_1, τ)-plane is shown in Fig. 2a for selected values of λ_2 and ω. The blue lines are given by Hopf condition and the

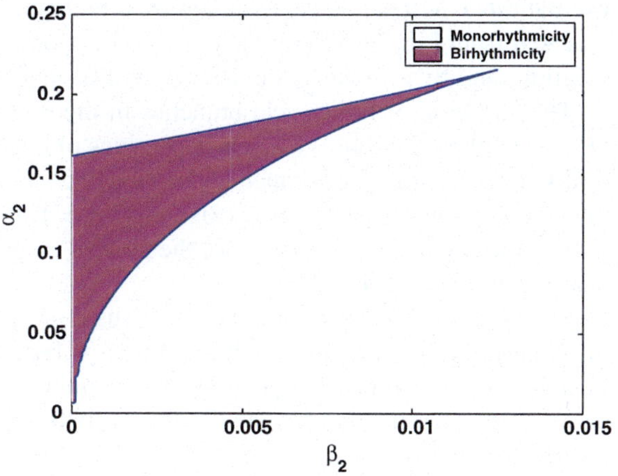

Fig. 1 Two-parameter bifurcation diagram in (α_2, β_2)-plane for $\lambda_1 = 0.8$, $\lambda_2 = 0.4$, $\tau = 1.2$ and $\omega = 1.9$. White zone for monorhythmicity, magenta zone for birhythmicity. (Color online)

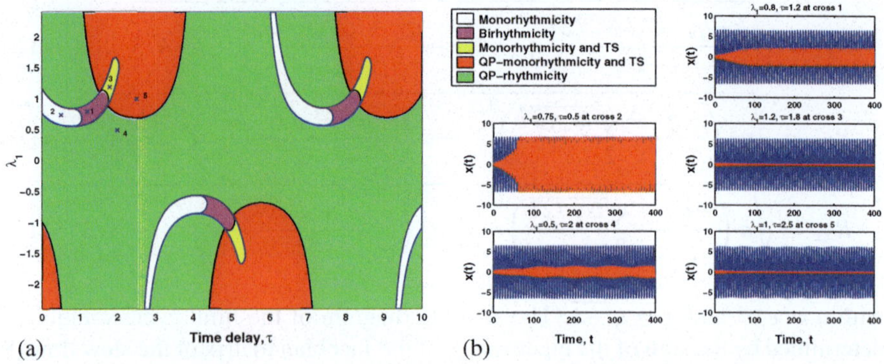

(a) (b)

Fig. 2 **a** Two-parameter bifurcation diagram for limit cycles in the plane (λ_1, τ) for $\omega = 2.1$ and $\lambda_2 = 0.4$. Black (blue) lines refer to Hopf (saddle-node) bifurcation. **b** Time histories corresponding to different crosses picked from (**a**); the legend indicates the nature of the response in each region. (Color online)

black lines are obtained using the saddle-node condition. Referring to this figure, different regions can be distinguished. In the magenta regions, there exist two stable limit cycles with small and large amplitudes separated by unstable one (birhythmic behavior). In the blank region, only the greater limit cycle persists (monorhythmicity), while in the yellow region, the greater limit cycle is present with a stable trivial solution; the response is monorhythmic or at rest depending on the initial state of the system. At the transition from the magenta to the white region, the small limit cycle disappears via a saddle-node bifurcation leaving only the greater one. As for the transition from the magenta to the yellow region, the small limit cycle undergoes a subcritical Hopf bifurcation giving rise to a stable trivial solution; the large-amplitude limit cycle persists. Time histories (obtained using `dde23` algorithm [22]) corresponding to crosses 1, 2, 3 picked from Fig. 2a and shown in Fig. 2b confirm the analytical predictions. Outside these regions (green and red regions), QP oscillations may take place as shown by time histories corresponding to crosses 4, 5 picked from Fig. 2a. To complete the bifurcation diagram in the (λ_1, τ)-plane, QP solutions and their regions of existence will be investigated in the next section.

The variation of limit cycle branches in the one-parameter bifurcation diagram (R, λ_1) is plotted in Fig. 3 for selected values of ω and λ_2 and different values of τ. Solid line refers to stable branch and dotted line to unstable one. For $\tau = 3$ (Fig. 3a) one stable (limit cycle) branch exists, for $\tau = 1.58$ (Fig. 3b) there are two stable limit cycles and for $\tau = 1$ (Fig. 3c) the system may exhibit one or two limit cycles depending on the value of λ_1.

The planar bifurcation diagram R versus λ_2 is plotted in Fig. 4 for selected set of parameters. It is observed that two limit cycles may exist for small values of λ_2 (birhythmicity), while only one limit cycle of great amplitude persists for large values of λ_2 (monorhythmicity); the small limit cycle being disappeared via a saddle-node bifurcation.

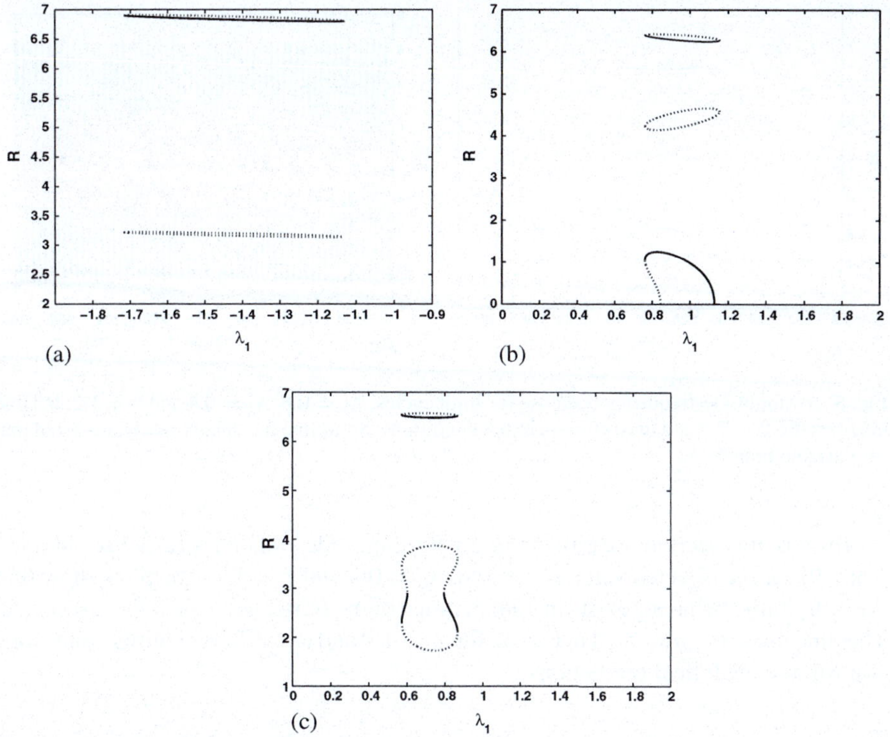

Fig. 3 The one-parameter bifurcation diagram of limit cycle branches in (R, λ_1)-plane for $\lambda_2 = 0.4$, $\omega = 1.9$ and **a** $\tau = 3$ (monorhythmiticiy), **b** $\tau = 1.58$ (birhythmiticiy), **c** $\tau = 1$ (monorhythmiticiy or birhythmiticiy). Solid line for stable and doted line for unstable

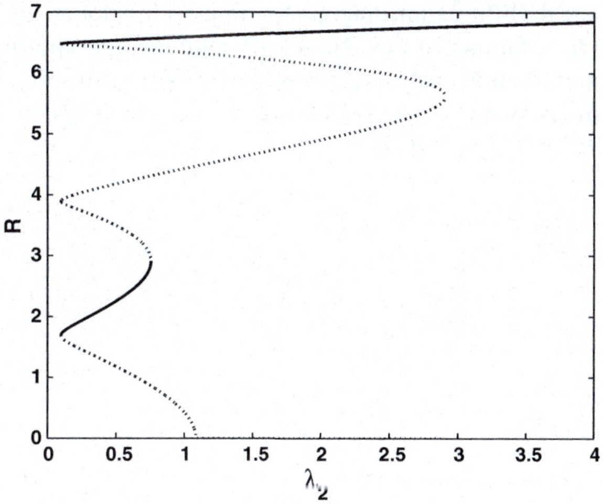

Fig. 4 The one-parameter bifurcation diagram of limit cycle branches in (R, λ_2)-plane for $\omega = 1.9$, $\lambda_1 = 0.8$, $\tau = 1.2$. Solid line for stable and doted line for unstable

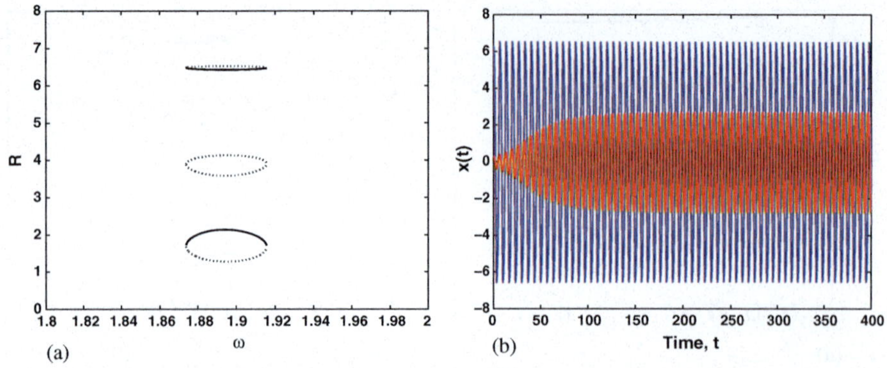

Fig. 5 **a** Amplitude-frequency response for limit cycles; $\lambda_1 = 0.8$, $\lambda_2 = 0.4$ and $\tau = 1.2$. **b** Time histories for $\omega = 1.9$ and two different initial conditions. Solid line for stable branch and doted line for unstable branch

To close this section, we plot in Fig. 5a the amplitude-frequency response obtained from (9) for the selected values $\lambda_1 = 0.8$, $\lambda_2 = 0.4$ and $\tau = 1.2$. The plots show that the birhythmic behavior exists in a small range of the frequency ω near the resonance. The time histories presented in Fig. 5b for $\omega = 1.9$ and two different initial conditions confirm the analytical predictions.

3 Quasiperiodic Rhythmicity

Next, QP-rhythmicity is investigated in the delayed birhythmic van der Pol oscillator and the bifurcation diagram for limit cycles plotted in Fig. 2a is completed by studying QP solutions of (1). These QP solutions are approximated using the second-step perturbation analysis [23]. We begin by transforming the modulation equations from the polar form (8) to the following Cartesian system using the variable change $u = R\cos\theta$, $v = -R\sin\theta$.

$$\begin{aligned}\frac{du}{dT} &= (sk_5 - k_4)v \\ &\quad + \eta\left\{(k_1 + sk_3)u + k_2u(u^2+v^2) + \frac{\alpha_2}{16}u(u^2+v^2)^2 - \frac{5\beta_2}{128}u(u^2+v^2)^3\right\}\end{aligned} \tag{13a}$$

$$\begin{aligned}\frac{dv}{dT} &= -(sk_5 + k_4)u \\ &\quad + \eta\left\{(k_1 - sk_3)v + k_2v(u^2+v^2) + \frac{\alpha_2}{16}v(u^2+v^2)^2 - \frac{5\beta_2}{128}v(u^2+v^2)^3\right\}\end{aligned} \tag{13b}$$

in which the bookkeeping parameter η is introduced in damping and nonlinearity. Periodic solutions of the slow flow system (13) correspond to QP solutions of the original system (1). Using the multiple scale technique [24], approximation of these periodic solutions can be sought in the form

$$u(T_1, T_2) = u_0(T_1, T_2) + \eta u_1(T_1, T_2) + O(\eta^2) \tag{14a}$$

$$v(T_1, T_2) = v_0(T_1, T_2) + \eta v_1(T_1, T_2) + O(\eta^2) \tag{14b}$$

where $T_1 = T$ and $T_2 = \eta T$. Introducing $D_i = \frac{\partial}{\partial T_i}$ yields $\frac{d}{dT} = D_1 + \eta D_2 + O(\eta^2)$, substituting (14) into (13) and collecting terms, one obtains up to the first order

$$D_1^2 u_0 + \nu^2 u_0 = 0, \tag{15a}$$

$$(sk_5 - k_4) v_0 = D_1 u_0 \tag{15b}$$

$$\begin{aligned} D_1^2 u_1 + \nu^2 u_1 = (sk_5 - k_4)\Big\{ &- D_2 v_0 + (k_1 - sk_3) v_0 + k_2 v_0 (u_0^2 + v_0^2) \\ &+ \frac{\alpha_2}{16} v_0 (u_0^2 + v_0^2)^2 - \frac{5\beta_2}{128} v_0 (u_0^2 + v_0^2)^3 \Big\} - D_1 D_2 u_0 \\ &+ D_2 \Big[(k_1 + sk_3) u_0 + k_2 u_0 (u_0^2 + v_0^2) \\ &+ \frac{\alpha_2}{16} u_0 (u_0^2 + v_0^2)^2 - \frac{5\beta_2}{128} u_0 (u_0^2 + v_0^2)^3 \Big] \end{aligned} \tag{16a}$$

$$\begin{aligned} (sk_5 + k_4) v_1 = &\ D_1 u_1 + D_2 u_0 - (k_1 + sk_3) u_0 - k_2 u_0 (u_0^2 + v_0^2) \\ &- \frac{\alpha_2}{16} u_0 (u_0^2 + v_0^2)^2 + \frac{5\beta_2}{128} u_0 (u_0^2 + v_0^2)^3 \end{aligned} \tag{16b}$$

where $\nu = \sqrt{k_5^2 - k_4^2}$ is the frequency of the slow flow limit cycle amplitude. Up to the first order the solution is given by

$$u_0(T_1, T_2) = r(T_2) \cos(\nu T_1 + \varphi(T_2)) \tag{17a}$$

$$v_0(T_1, T_2) = -\frac{\nu}{(sk_5 + k_4)} r_1(T_2) \sin(\nu T_1 + \varphi(T_2)) \tag{17b}$$

where r, φ are, respectively, the amplitude and the phase of the slow flow limit cycle. Substituting (17) into (16) and removing the generated secular terms yield the following slow-slow flow

$$\frac{dr}{dT_2} = k_1 r + \frac{k_2 D}{4} r^3 + \frac{\alpha_2 E}{256} r^5 - \frac{5\beta_2 F}{8192} r^7 \tag{18a}$$

$$\frac{d\varphi}{dT_2} = 0 \tag{18b}$$

in which $D = 2 + \frac{2\nu^2}{(sk_5-k_4)^2}$, $E = 6 + \frac{4\nu^2}{(sk_5-k_4)^2} + \frac{6\nu^4}{(sk_5-k_4)^4}$ and $F = 20 + \frac{12\nu^2}{(sk_5-k_4)^2} + \frac{12\nu^4}{(sk_5-k_4)^4} + \frac{20\nu^6}{(sk_5-k_4)^6}$. Equilibrium points of (18), corresponding to QP solutions of the original system (1), are obtained by setting $\frac{dr}{dT_2} = 0$. These are given by the steady-state solution $\rho = 0$ and real roots of the equation

$$\rho^3 + A'\rho^2 + B'\rho + C' = 0 \tag{19}$$

where $\rho = r^2$, $A' = -\frac{\alpha_2 E}{160\beta_2 F}$, $B' = -\frac{2048 k_2 D}{5\beta_2 F}$ and $C' = -\frac{8192 k_1}{5\beta_2 F}$. Equation (19) has three real roots r_1, r_2 and r_3 if its discriminant is negative. The expressions of these solutions read

$$r_1 = \sqrt{\left(\frac{-q' - \sqrt{\Delta/27}}{2}\right)^{1/3} + \left(\frac{-q' + \sqrt{\Delta/27}}{2}\right)^{1/3} - A'/3} \tag{20a}$$

$$r_2 = \sqrt{a^2\left(\frac{-q' - \sqrt{\Delta/27}}{2}\right)^{1/3} + a\left(\frac{-q' + \sqrt{\Delta/27}}{2}\right)^{1/3} - A'/3} \tag{20b}$$

$$r_3 = \sqrt{a\left(\frac{-q' - \sqrt{\Delta/27}}{2}\right)^{1/3} + a^2\left(\frac{-q' + \sqrt{\Delta/27}}{2}\right)^{1/3} - A'/3} \tag{20c}$$

where $\Delta = 4p'^3 + 27q'^2$, $p = B' + \frac{A'^2}{3}$, $q' = \frac{2A'^3}{27} - \frac{B'A'}{3} + C'$ and $a = e^{\frac{2\pi j}{3}}$.

Two of the solutions (stable and unstable) may coalesce at specific values of parameters, yielding a saddle-node bifurcation of QP solutions. This bifurcation is determined when (19) and the its derivative with respect to ρ

$$3\rho^2 + 2A'\rho + B' = 0 \tag{21}$$

are satisfied simultaneously [25]. Thus, the necessary conditions of saddle-node bifurcation of QP solutions are obtained by solving (21) for $\rho_{m\pm}$ $(m = 1, 2)$ and injecting it into (19). One obtains the following equation

$$\rho^3_{m\pm} + A'\rho^2_{m\pm} + B'\rho_{m\pm} + C' = 0 \tag{22}$$

Now, the bifurcation diagram for limit cycles presented in Fig. 2a is completed using the saddle-node bifurcation curves of QP solutions given by (22). These are presented in the bifurcation diagram (λ_1, τ) of Fig. 6a by blue lines. These blue lines separate the green region of Fig. 2a into a gray zone and a new green one and the red region of Fig. 2a into a cyan zone and a new red one. In the new green region, two stable tori are present producing *QP-birhythmicity* while in the gray region there exists one

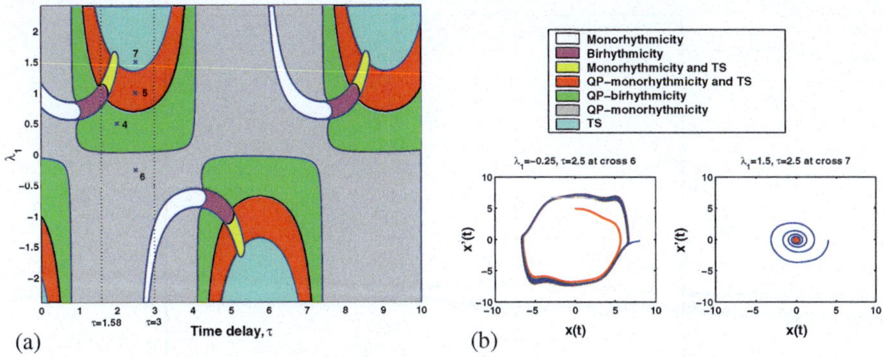

(a) (b)

Fig. 6 **a** Global two-parameter bifurcation diagram (for limit cycles and tori) in the (λ_1, τ)-plane for $\lambda_2 = 0.4$ and $\omega = 2.1$. **b** Legend indicating the nature of the response in different regions (top) and phase portraits (bottom) corresponding to crosses 6, 7 picked from (**a**). (Color online)

stable torus generating *QP-monorhythmicity*. On the other hand, in the red region there are a stable torus and a stable trivial solution, whereas in the cyan region only a stable trivial solution persists.

Figure 6b (bottom) presents phase portraits corresponding to crosses 6, 7 picked from Fig. 6a confirming the analytical prediction. Recall that the time histories related to crosses 4, 5 are shown in Fig. 2b.

The bifurcation diagram in Fig. 6a can be understood following the solution branches in the one-parameter bifurcation diagram (R, r) versus λ_1 drawn in Fig. 7 for a selected value $\tau = 1.58$ corresponding to the broken vertical line of Fig. 6a; R and r correspond to limit cycle and torus branches, respectively. Black lines refer to limit cycle branches (solid for stable and dotted for unstable), a green line corresponds to a stable torus (QP solution), while a red line represents an unstable torus. The boundaries of the modulation interval of QP solutions are obtained by setting $s = \pm 1$ in the QP-amplitude equation (19).

Increasing λ_1 from negative values in Fig. 7, the following transitions in system rhythmicity occur. For values of $\lambda_1 < 0.0388$ the trivial steady-state solution is unstable and the response stabilizes on a large-amplitude torus (QP-monorhythmicity, gray region in Fig. 6a). When λ_1 crosses the critical value $\lambda_1 = 0.0388$, two small-amplitude tori, one stable and one unstable are born via saddle-node bifurcation causing the system to generate QP-birhythmicity over the interval $[\lambda_1 = 0.0388, \lambda_1 = 0.7618]$ (green region in Fig. 6a). As λ_1 crosses the value $\lambda_1 = 0.7618$, the two stable limit cycles emerge (magenta region Fig. 6a) and the system behavior switches from QP-birhythmicity to birhythmicity with different amplitudes.

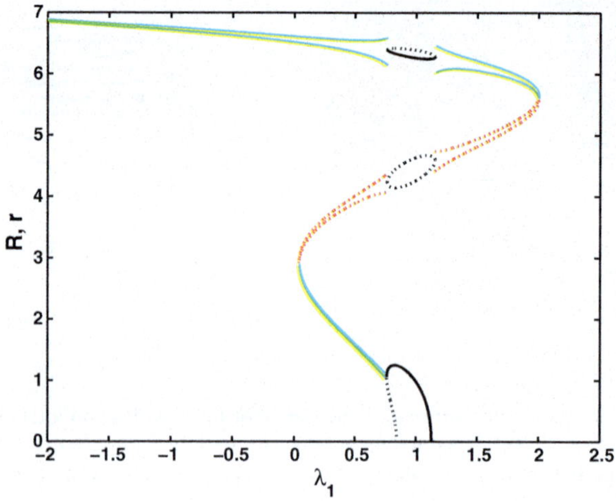

Fig. 7 One-parameter bifurcation diagram in $((R, r), \lambda_1)$-plane for $\lambda_2 = 0.4$, $\omega = 1.9$, $\tau = 1.58$ (refers to vertical dashed line in Fig. 6a). Solid black and green lines for stable, dashed black and red lines for unstable. Black for limit cycle branches, green and red for tori branches. (Color online)

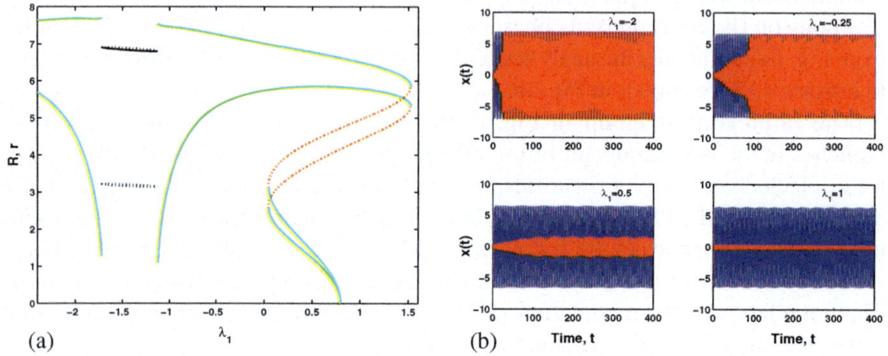

Fig. 8 **a** One-parameter bifurcation diagram in $((R, r), \lambda_1)$-plane for $\lambda_2 = 0.4$, $\omega = 1.9$, $\tau = 3$ (refers to vertical dashed line in Fig. 6). Solid black and green lines for stable, dashed black and red lines for unstable. Black for limit cycle branches, green and red for torus branches. **b** Time histories corresponding to $\lambda_1 = -2, -0.25, 0.5, 1$ plotted for two different initial conditions. (Color online)

We plot in Fig. 8a the solution branches in the bifurcation diagram $((R, r), \lambda_1)$ for another selected value $\tau = 3$, as indicated in Fig. 6a. Figure 8a shows that increasing λ_1 from larger negative values to $\lambda_1 = 0.0388$, the response stabilizes first on a large torus (gray region in Fig. 6a), then on a limit cycle (white region), and once again on a torus (gray region). When λ_1 crosses the value $\lambda_1 = 0.0388$, a small torus appears via a saddle-node bifurcation giving rise to QP-birhythmicity (green region). By further increasing λ_1, the amplitude of the small torus decreases and disappears at a certain critical value leaving only the large torus (red region in Fig. 6a). The large torus dies

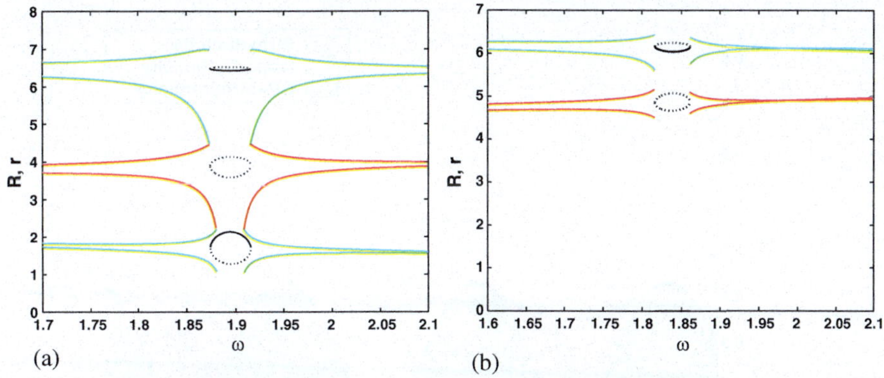

Fig. 9 Frequency-response curves for $\lambda_2 = 0.4$ and **a** $\lambda_1 = 0.8$, $\tau = 1.2$, **b** $\lambda_1 = 1.5$, $\tau = 1.58$. Solid and green for stable branches, dotted and red for unstable branches. (Color online)

through a saddle-node bifurcation and the system response is attracted by the trivial solution (cyan region in Fig. 6a). To validate the analytical prediction, time histories are presented in Fig. 8b for selected parameters λ_2, ω, τ and for two different initial conditions. It can be seen from the plots that for $\lambda_1 = -2$ or -0.25, the system response is QP-monorhythmic (one large torus), for $\lambda_1 = 0.5$ the system has a QP-birhythmic behavior (two tori) and for $\lambda_1 = 1$, large-amplitude QP-monorhythmicity is present with a stable trivial steady state.

Figure 9a shows the amplitude-frequency response curves for selected values of $\lambda_1 = 0.8$, $\lambda_2 = 0.4$ and $\tau = 1.2$ picked within the QP-birhythmic region in Fig. 7a. The plots in this figure reveal that there is a birhythmic behavior in a small range of the frequency near the resonance and a QP-birhythmic behavior over a broad range of the frequency away from the resonance.

In Fig. 9b is presented the amplitude-frequency response for $\lambda_1 = 1.5$, $\lambda_2 = 0.4$ and $\tau = 1.58$ picked in the region of QP-monorhythmicity of Fig. 7a. In this case, there exists a monorhythmic behavior in a small interval near the resonance and a QP-monorhythmicity elsewhere.

Finally, we draw in Fig. 10 the two-parameter bifurcation diagrams λ_2 versus τ. Three regions of monorhythmic, birhythmic and QP-birhythmic behaviors can be distinguished in this figure. It is observed that for small values of the modulated amplitude λ_2 and different values of time delay τ, monorhythmic, birhythmic or QP-birhythmic behavior may take place, whereas for large values of λ_2, only monorhythmic behavior prevails.

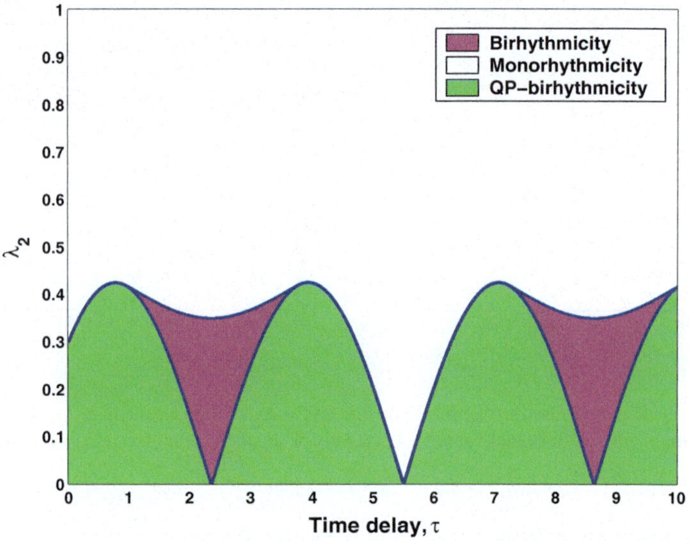

Fig. 10 Bifurcation diagram in (λ_2, τ)-plane for $\lambda_1 = 0.15$, $\omega = 2$. (Color online)

4 Conclusions

We have studied the effect of the modulated time delay on the rhythmic behavior of a multicycle van der Pol oscillator. The study has been performed near the delay-induced parametric resonance for which the frequency of the modulation is nearly twice the natural frequency of the oscillator. Using perturbation methods and numerical simulations, approximations of periodic and QP solutions as well as their bifurcation equations have been obtained. The utilization of the bifurcation equations allowed the determination of bifurcation diagrams and the different regions in which different rhythmicities can occur. The transition between the different rhythmicities occurs through Hopf, Neimark–Sacker and saddle-node bifurcations. The study has revealed that a small modulation of time delay in the multicycle van der Pol oscillator may generate QP-monorhythmicity and QP-birhythmicity. The results have also shown that a suitable choice of the modulation is able to direct the system to small or large-amplitude periodic or QP rhythmicity. The large-amplitude QP-monorhythmic response, occurring in a broad range of the frequency, may have potential applications in systems ruled by multicycle van der Pol oscillators under delay-modulated control. This large-amplitude can be exploited, for instance, to scavenge energy with good performance over a broad range of the frequency. On the other hand, targeting small-amplitude QP-rhythmic response can also be of importance in some applications for which quenched vibrations are desired.

References

1. A. Goldbeter, Computational approaches to cellular rhythms. Nature **420**, 238 (2002)
2. S. Kar, D.S. Ray, Large fluctuations and nonlinear dynamics of birhythmicity. Europhys. Lett. **67**, 137 (2004)
3. J.M.T. Thompson, H.B. Stewart, *Nonlinear Dynamics and Chaos*, 2nd edn. (Wiley, Chichester, 1986)
4. C.A.K. Kwuimy, C. Nataraj, *In Structural Nonlinear Dynamics and Diagnosis*, ed. by M. Belhaq, vol. 168 (Springer, Switzerland, 2015), p. 97
5. M. Alamgir, I.R. Epstein, Systematic design of chemical oscillators. 17. Birhythmicity and compound oscillation in coupled chemical oscillators: chlorite-bromate-iodide system. J. Amer. Chem. Soc. **105**, 2500 (1983)
6. F. Kaiser, S. Eichwald, Bifurcation structure of a driven, multi-limit-cycle van der pol oscillator (i): the superharmonic resonance structure. Int. J. Bifurc. Chaos **1**, 485 (1991)
7. C. Eichwald, F. Kaiser, Bifurcation structure of a driven multi-limit-cycle van der Pol oscillator (ii): symmetry-breaking crisis and intermittency. Int. J. Bifurc. Chaos **1**, 711 (1991)
8. Y.H. Shiau, Y.F. Peng, R.R. Hwang, C.K. Hu, Multistability and symmetry breaking in the two-dimensional flow around a square cylinder. Phys. Rev. E **60**, 6188 (1999)
9. H.G. Enjieu Kadji, J.B. Chabi Orou, R. Yamapi, P. Woafo, Nonlinear dynamics and strange attractors in the biological system. Chaos Solitons Fractals **32**, 862 (2007)
10. P. Ghosh, S. Sen, S.S. Riaz, D.S. Ray, Controlling birhythmicity in a self-sustained oscillator by time-delayed feedback. Phys. Rev. E **83**(3), 036205 (2011)
11. D. Biswas, T. Banerjee, J. Kurths, Control of birhythmicity through conjugate self-feedback: theory and experiment. Phys. Rev. E **94**, 042226 (2016)
12. D. Biswas, T. Banerjee, J. Kurths, Effect of filtered feedback on birhythmicity: suppression of birhythmic oscillation. Phys. Rev. E **99**, 062210 (2019)
13. Z. Ma, L. Ning, Bifurcation regulations governed by delay self-control feedback in a stochastic birhythmic system. Int. J. Bifurc. Chaos **27**, 1750202 (2017)
14. Q. Guo, Z. Sun, W. Xu, Bifurcations in a fractional birhythmic biological system with time delay. Commun. Nonlinear Sci. Numer. Simul. **72**, 318 (2019)
15. Y. Sun, L. Ning, Bifurcation analysis of a self-sustained birhythmic oscillator under two delays and colored noises. Int. J. Bifurc. Chaos **30**, 2050013 (2020)
16. S. Saha, S. Chakraborty, G. Gangopadhyay, Suppressing birhythmicity by parametrically modulating nonlinearity in limit cycle oscillators. Physica D **416**, 132793 (2021)
17. I. Kirrou, M. Belhaq, Control of bistability in non-contact mode atomic force microscopy using modulated time delay. Nonlinear Dyn. **81**, 607 (2015)
18. M. Belhaq, M. Hamdi, Energy harvesting from quasi-periodic vibrations. Nonlinear Dyn. **86**, 2193 (2016)
19. M. Hamdi, M. Belhaq, Quasi-periodic vibrations in a delayed van der Pol oscillator with time-periodic delay amplitude. J. Vib. Control **24**, 57265734 (2015)
20. N. Krylov, N.N. Bogoliubov, *Introduction to Non-linear Mechanics* (Princeton University, 1943)
21. S. Virkus, R. Rand, The dynamics of two coupled van der Pol oscillators with delay coupling. Nonlinear Dyn. **30**, 205 (2002)
22. L.F. Shampine, S. Thompson, Appl. Numer. Math. **37**, 441 (2000)
23. M. Belhaq, M. Houssni, Quasi-periodic oscillations, chaos and suppression of chaos in a nonlinear oscillator driven by parametric and external excitations. Nonlinear Dyn. **18**, 1 (1999)
24. H. Nayfeh, D.T. Mook, *Nonlinear Oscillations* (Wiley, New York, 1979)
25. A.H. Nayfeh, B. Balachandran, *Applied Nonlinear Dynamics: Analytical, Computational and Experimental Methods* (Wiley, London, 2008)

Incremental Harmonic Balance with Two Time Scales for a Nonlinear Quasi-Periodic Mathieu Equation

B. X. Zhang, J. L. Huang, and W. D. Zhu

Abstract Quasi-periodic solution of a damped quasi-periodic (QP) Mathieu's equation with cubic nonlinearity is obtained by using the incremental harmonic balance (IHB) with two time scales. It is found that the QP equation consists of stable and unstable QP solutions and every solution consists of carrier frequencies and their spaced uniformly sidebands. The IHB with two time scales is applied to trace the response curve of the amplitude of the parameter excitation. Besides, three types of QP solutions are found, which all agree very well with the results from numerical integration. While the perturbation method using the double-step method of multiple scales (MMS) obtains only one type of QP solutions, since the small excitation frequency is nearly the natural frequency of the second perturbation equation. Finally the stability of QP solutions is analyzed by the Floquet theory.

1 Introduction

Mathieu's [1] equation is governed by the equation with periodically varying coefficients

$$\ddot{x} + (\delta + \epsilon \cos t)\, x = 0, \tag{1}$$

Esmailzadeh and Nakhaie jazar [2] found necessary and sufficient conditions to exist periodic response of Mathieu–Duffing equation and used Schauder's fixed-point theorem to very its validity.

$$\ddot{x} + (\delta + \epsilon \cos t)\, x + \xi x^3 = 0, \tag{2}$$

B. X. Zhang · J. L. Huang (✉) · W. D. Zhu
Department of Applied Mechanics and Engineering, Sun Yat-sen University, Guangzhou 510275, P.R. China
e-mail: huangjl@mail.sysu.edu.cn

B. X. Zhang
e-mail: zhangbx25@mail2.sysu.edu.cn

M. Belhaq (ed.), *Advances in Nonlinear Dynamics and Control of Mechanical and Physical Systems*, Springer Proceedings in Physics 301,
https://doi.org/10.1007/978-981-99-7958-5_3

where ξ is a coefficient of the cubic nonlinear term. Ng and Rand [3] used a perturbation method, which is second-order averaging, to analyze bifurcations of Mathieu–Duffing equation. Zounes and Rand [4] investigated transition curves of the QP Mathieu's equation (3) by harmonic balance utilizing Hill's determinants and the perturbation method.

$$\ddot{x} + (\delta + \epsilon (\cos t + \cos \omega t))\, x = 0, \tag{3}$$

where ω is an irrational excitation frequency. Belhaq et al. [5, 6] developed a new perturbation method by using the double-step method of multiple scales (MMS) to analyze the damped nonlinear QP Mathieu's equation

$$\ddot{x} + c\dot{x} + \left(\tilde{\omega}_0^2 + \alpha \cos(2\omega t) + \beta \cos(2\omega_d t)\right) x + \xi x^3 = 0, \tag{4}$$

where two excitation frequencies ω and ω_d are incommensurable. Zounes and Rand [7] used Lie transform with elliptic functions to investigate subharmonic resonances of an undamped equation of (4). All these problems require a theoretical framework to understand their responses, but unfortunately there isn't an effective analytical method for analyzing nonlinear QP Mathieu's equations; this motivates one to provide a semi-analytical method to obtain accurate solutions of a damped nonlinear QP Mathieu's equation.

Recently, some authors of this work presented an incremental harmonic balance (IHB) method with two time scales [8] that was used to analyze QP motions of multiple-degree-of-freedom (DOF) nonlinear systems with internal resonances and later extended to calculate QP motions of a van der Pol-Mathieu equation in [9], where the IHB method with two time scales can be used to find all frequencies of responses and their corresponding amplitudes. The advantage of the double-step MMS [6] is that it can provide analytical results. However, the double-step MMS is only applicable to QP solutions whose Fourier spectra include a few number of frequency components and a weakly nonlinear equation such as (4) with small parameters. When any excitation frequency is near any natural frequency, the double-step MMS is invalid. So the double-step MMS obtains only one type of QP solution of (4) that actually has three types of QP solutions as discussed in Sect. 4. The aim of this work is to obtain all QP solutions of the damped nonlinear QP Mathieu's equation by using the IHB method with two time scales, where some solutions cannot be obtained from the double-step MMS and to analyze the stability of solutions. The corresponding Fourier spectra are calculated, which greatly facilitate the understanding of the behavior of the damped nonlinear QP Mathieu's equation.

2 Formulation of the IHB Method with Two Time Scales for QP Solutions

Fourier spectra of quasi-periodic solutions consist of uniformly spaced sidebands surrounding carrier frequencies which was previously observed by some authors of this work from numerical integration (NI) for other systems and by Kreider and Nayfeh [10] from experimental investigation. Therefore, spectra of QP solutions contain the two basic frequencies ω and ω_d, where ω_d is the equal distance or its half between every two adjacent frequencies in sidebands as discussed in Sect. 4. Following previous development by some authors of this work [8], one introduces two new time variables defined as

$$\tau_1 = \omega t, \quad \tau_2 = \omega_d t, \tag{5}$$

where $0 \leq \tau_1, \tau_2 \leq 2\pi$. Then $x(t)$ can be rewritten as a function of τ_1 and τ_2:

$$x(t) = x(\omega t, \omega_d t) = x(\tau_1, \tau_2) \tag{6}$$

Assume the following operator relations:

$$\begin{aligned} \Theta &= \frac{d}{dt} = \omega \frac{\partial}{\partial \tau_1} + \omega_d \frac{\partial}{\tau_2}, \\ \Theta^2 &= \frac{d^2}{d^2 t} = \omega^2 \frac{\partial^2}{\partial \tau_1^2} + 2\omega\omega_d \frac{\partial^2}{\partial \tau_1 \partial \tau_2} + \omega_d^2 \frac{\partial^2}{\partial \tau_2^2}. \end{aligned} \tag{7}$$

Substituting (5) and (7) into (4) yields

$$\Theta^2 x + c\Theta x + \left(\tilde{\omega}_0^2 + \alpha \cos(2\tau_1) + \beta \cos(2\tau_2)\right) x + \gamma x^3 = 0, \tag{8}$$

The first step of the IHB method with two time scales is the Newton–Raphson iterative procedure. Let $(x_0, \omega_0, \alpha_0, \omega_{d0}, \beta_0)$ denote a guessed solution firstly; small increments $(\Delta x, \Delta\omega, \Delta\alpha, \Delta\omega_d, \Delta\beta)$ are added to the guessed solution $(x_0, \omega_0, \alpha_0, \omega_{d0}, \beta_0)$. Neglecting small increments of higher orders, one can obtain the linearized incremental equation

$$\begin{aligned} &\Theta_0^2 \Delta x + c\Theta_0 \Delta x + \left(\tilde{\omega}_0^2 + \alpha_0 \cos(2\tau_1) + \beta_0 \cos(2\tau_2)\right) \Delta x + 3\xi x_0^2 \Delta x \\ &= R - \left[2\left(\omega_0 \frac{\partial^2 x_0}{\partial \tau_1^2} + \omega_{d0} \frac{\partial^2 x_0}{\partial \tau_1 \partial \tau_2}\right) + c\frac{\partial x_0}{\partial \tau_1}\right] \Delta\omega \\ &\quad - \left[2\left(\omega_{d0} \frac{\partial^2 x_0}{\partial \tau_2^2} + \omega_0 \frac{\partial^2 x_0}{\partial \tau_1 \partial \tau_2}\right) + c\frac{\partial x_0}{\partial \tau_2}\right] \Delta\omega_d \\ &\quad - x_0 \cos(2\tau_1) \Delta\alpha - x_0 \cos(2\tau_2) \Delta\beta, \end{aligned} \tag{9}$$

where

$$R = -\left(\Theta_0^2 x_0 + c\Theta_0 x_0 + \left(\tilde{\omega}_0^2 + \alpha_0 \cos(2\tau_1) \ + \beta_0 \cos(2\tau_2)\right) x_0 + \xi x_0^3\right) \quad (10)$$

is a correction term that would vanish if the current solution were exact.

The second step of the IHB method with two time scales is the Galerkin procedure. QP solutions of the damped nonlinear QP Mathieu's equation can be obtained by expanding the displacement $x_0(\tau_1, \tau_2)$ in multiple Fourier series with finite terms in the hyper-time domain $0 < \tau_1, \tau_2 < 2\pi$ with $\tau_1 = \omega t$ and $\tau_2 = \omega_d t$. One assumes

$$\begin{aligned} x_0 = & \sum_{j_2=-m_1}^{m_1} a_{1,j_2} \cos(\tau_1 + j_2\tau_2) + \sum_{j_2=-m_2}^{m_2} a_{2,j_2} \cos(3\tau_1 + j_2\tau_2) \\ & + \sum_{j_2=-m_1}^{m_1} b_{1,j_2} \sin(\tau_1 + j_2\tau_2) + \sum_{j_2=-m_2}^{m_2} b_{2,j_2} \sin(3\tau_1 + j_2\tau_2) = \mathbf{C}_s\mathbf{A}, \end{aligned} \quad (11)$$

where

$$\begin{aligned} \mathbf{C}_s &= \left[\mathbf{C}_1\ \mathbf{C}_2\ \mathbf{S}_1\ \mathbf{S}_2\right], \\ \mathbf{A} &= \left[\mathbf{A}_1^{\mathrm{T}}\ \mathbf{A}_2^{\mathrm{T}}\ \mathbf{B}_1^{\mathrm{T}}\ \mathbf{B}_2^{\mathrm{T}}\right]^{\mathrm{T}}, \end{aligned} \quad (12)$$

in which

$$\begin{aligned} \mathbf{C}_i &= \left[\cos((2i-1)\tau_1 - m_i\tau_2) \cdots \cos((2i-1)\tau_1) \cdots \cos((2i-1)\tau_1 + m_i\tau_2)\right], \\ \mathbf{S}_i &= \left[\sin((2i-1)\tau_1 - m_i\tau_2) \cdots \sin((2i-1)\tau_1) \cdots \sin((2i-1)\tau_1 + m_i\tau_2)\right], \\ \mathbf{A}_i &= \left[a_{i,-m_i} \cdots a_{(2i-1),0} \cdots a_{(2i-1),m_i}\right]^{\mathrm{T}} \\ \mathbf{B}_i &= \left[b_{i,-m_i} \cdots b_{(2i-1),0} \cdots b_{(2i-1),m_i}\right]^{\mathrm{T}}, \quad i = 1, 2. \end{aligned} \quad (13)$$

The increment Δx is expressed in the form

$$\Delta x = \mathbf{C}_s \Delta\mathbf{A}, \quad (14)$$

where $\Delta\mathbf{A} = \left[\Delta\mathbf{A}_1^{\mathrm{T}}\ \Delta\mathbf{A}_2^{\mathrm{T}}\ \Delta\mathbf{B}_1^{\mathrm{T}}\ \Delta\mathbf{B}_2^{\mathrm{T}}\right]^{\mathrm{T}}$. Substituting (11) and (14) into (9) and applying the Galerkin procedure to balance the harmonic terms yield

$$\begin{aligned} & \int_0^{2\pi}\int_0^{2\pi} \delta(\Delta x)\left\{\Theta_0^2\Delta x + c\Theta_0\Delta x + \left[\tilde{\omega}_0^2 + \alpha_0\cos(2\tau_1) \ + \beta_0\cos(2\tau_2)\right]\Delta x + 3\xi x_0^2\Delta x\right\} d\tau_1 d\tau_2 \\ & = \int_0^{2\pi}\int_0^{2\pi} \delta(\Delta x)\left\{R - \left[2\left(\omega_0\frac{\partial^2 x_0}{\partial\tau_1^2} + \omega_{d0}\frac{\partial^2 x_0}{\partial\tau_1\partial\tau_2}\right) + c\frac{\partial x_0}{\partial\tau_1}\right]\Delta\omega\right. \\ & \left. - \left[2\left(\omega_{d0}\frac{\partial^2 x_0}{\partial\tau_2^2} + \omega_0\frac{\partial^2 x_0}{\partial\tau_1\partial\tau_2}\right) + c\frac{\partial x_0}{\partial\tau_2}\right]\Delta\omega_d \ - x_0\cos(2\tau_1)\Delta\alpha - x_0\cos(2\tau_2)\Delta\beta\right\} d\tau_1 d\tau_2. \end{aligned} \quad (15)$$

One can easily obtain a linear matrix in the form

$$\mathbf{K}_A \Delta \mathbf{A} = \mathbf{R} - \mathbf{R}_\omega \Delta\omega - \mathbf{R}_{\omega_d} \Delta\omega_d - \mathbf{R}_\alpha \Delta\alpha - \mathbf{R}_\beta \Delta\beta, \tag{16}$$

$\mathbf{K}_A$ is the tangent matrix, $\mathbf{R}$ is the corrective vector, and $\mathbf{R}_\omega$, $\mathbf{R}_{\omega_d}$, $\mathbf{R}_\alpha$, and $\mathbf{R}_\beta$ are the vectors due to unit changes of $\Delta\omega$, $\Delta\omega_d$, $\Delta\alpha$, and $\Delta\beta$, respectively. Three ways of incrementation process, i.e., the amplitude increment, the frequency increment, and arc-length increment can be employed in order to trace the frequency response curve in this work, as presented in [11, 12].

3 Stability of QP Solutions

The Floquet theory extended to stability analysis of QP solutions in [13]. Based on the approximation of QP solutions by periodic solutions, Floquet theory can be applied to study the stability of QP solutions in this work. One introduces a fundamental frequency defined as

$$\tilde{\omega} = \min\{|m\omega + n\omega_d| \mid m, n \in \mathcal{Z}\} \tag{17}$$

where $\tilde{\omega}$ is infinitely small, i.e., it is essentially 0, since the ratio of the two parametric excitation frequencies is incommensurable. However, when one approximates the ratio as a rational value in computer, the fundamental frequency $\tilde{\omega}$ is a very small value and nonzero. One can assume

$$p_1 = \frac{\omega}{\tilde{\omega}},\ p_2 = \frac{\omega_d}{\tilde{\omega}},\ p_1, p_2 \in \mathcal{N}. \tag{18}$$

With approximation of the frequencies in (18), the displacement $x_0(\tau_1, \tau_2)$ in the multiple Fourier series in the form of (11) can be transferred to a Fourier series in the form

$$\begin{aligned} x_0 = &\sum_{j_2=-m_1}^{m_1} a_{1,j_2} \cos(p_1 + j_2 p_2)\tau + \sum_{j_2=-m_2}^{m_2} a_{2,j_2} \cos(3p_1 + j_2 p_2)\tau \\ &+ \sum_{j_2=-m_1}^{m_1} b_{1,j_2} \sin(p_1 + j_2 p_2)\tau + \sum_{j_2=-m_2}^{m_2} b_{2,j_2} \sin(3p_1 + j_2 p_2)\tau, \end{aligned} \tag{19}$$

where $\tau = \tilde{\omega}t$. One introduce a small perturbation $\Delta x = s(t)e^{\lambda t}$ on x_0 to get the stability of x_0:

$$x = x_0 + \Delta x = x_0 + s(t)e^{\lambda t}, \tag{20}$$

where $s(t) = \tilde{\mathbf{C}}_s \tilde{\mathbf{U}}$, in which

$$\begin{aligned}\tilde{\mathbf{C}}_s = \big[&\cos\left(p_1 - m_1 p_2\right)\tilde{\omega}t \cdots \cos p_1\tilde{\omega}t \cdots \cos\left(p_1 + m_1 p_2\right)\tilde{\omega}t \\ &\cos\left(3p_1 - m_1 p_2\right)\tilde{\omega}t \cdots \cos 3p_1\tilde{\omega}t \cdots \cos\left(3p_1 + m_1 p_2\right)\tilde{\omega}t \\ &\sin\left(p_1 - m_1 p_2\right)\tilde{\omega}t \cdots \sin p_1\tilde{\omega}t \cdots \sin\left(p_1 + m_1 p_2\right)\tilde{\omega}t \\ &\sin\left(3p_1 - m_1 p_2\right)\tilde{\omega}t \cdots \sin 3p_1\tilde{\omega}t \cdots \sin\left(3p_1 + m_1 p_2\right)\tilde{\omega}t\big]\end{aligned} \tag{21}$$

and $\tilde{\mathbf{U}}$ is a nonzero vector, is a periodic term and $e^{\lambda t}$ is a decay term. Substituting (20) into (4), neglecting all the terms containing increment products, and noting that x_0 satisfies (4) yield

$$\Delta\ddot{x} + c\Delta\dot{x} + \left[\tilde{\omega}_0^2 + \alpha\cos\left(2p_1\tilde{\omega}t\right) + \beta\cos\left(2p_2\tilde{\omega}t\right) + 3\xi x_0^2\right]\Delta x = 0. \tag{22}$$

Differentiating Δx with respect to time yields

$$\begin{aligned}\Delta\dot{x} &= \tilde{\omega}\tilde{\mathbf{C}}_s\tilde{\mathbf{U}}e^{\lambda t} + \lambda\tilde{\mathbf{C}}_s\tilde{\mathbf{U}}e^{\lambda t}, \\ \Delta\ddot{x} &= \tilde{\omega}^2\tilde{\mathbf{C}}_s\tilde{\mathbf{U}}e^{\lambda t} + 2\lambda\tilde{\omega}\tilde{\mathbf{C}}_s\tilde{\mathbf{U}}e^{\lambda t} + \lambda^2\tilde{\mathbf{C}}_s\tilde{\mathbf{U}}e^{\lambda t}.\end{aligned} \tag{23}$$

Substituting (23) and (11) into (22) yields

$$\begin{aligned}&\Big(\lambda^2\tilde{\mathbf{C}}_s + 2\lambda\tilde{\omega}\tilde{\mathbf{C}}_s + \tilde{\omega}^2\tilde{\mathbf{C}}_s'' + c\lambda\tilde{\mathbf{C}}_s' + c\tilde{\omega}\tilde{\mathbf{C}}_s' + \Big(\tilde{\omega}_0^2 + \alpha\cos\left(2p_1\tilde{\omega}t\right) \\ &\quad + \beta\cos\left(2p_2\tilde{\omega}t\right) + 3\xi\tilde{\mathbf{C}}_s\tilde{\mathbf{A}}\tilde{\mathbf{C}}_s\tilde{\mathbf{A}}\Big)\tilde{\mathbf{C}}_s\Big)\tilde{\mathbf{U}} = 0.\end{aligned} \tag{24}$$

Using the Galerkin procedure as shown in Sect. 2 to balance harmonic terms yields

$$\left(\lambda^2\tilde{\mathbf{M}} + \lambda\tilde{\mathbf{C}} + \tilde{\mathbf{K}}\right)\tilde{\mathbf{U}} = \mathbf{0}, \tag{25}$$

where

$$\begin{aligned}\tilde{\mathbf{M}} &= \int_0^{2\pi} \tilde{\mathbf{C}}_s^{\mathrm{T}}\tilde{\mathbf{C}}_s d\tau, \\ \tilde{\mathbf{C}} &= \int_0^{2\pi} \tilde{\mathbf{C}}_s^{\mathrm{T}}\left(2\tilde{\omega}\tilde{\mathbf{C}}_s + c\tilde{\mathbf{C}}_s\right)d\tau, \\ \tilde{\mathbf{K}} &= \int_0^{2\pi} \tilde{\mathbf{C}}_s^{\mathrm{T}}\Big(\tilde{\mathbf{C}}_s'' + c\tilde{\omega}\tilde{\mathbf{C}}_s' + \big(\tilde{\omega}_0^2 + \alpha\cos\left(2p_1\tilde{\omega}t\right) \\ &\qquad + \beta\cos\left(2p_2\tilde{\omega}t\right) + 3\xi\tilde{\mathbf{C}}_s\mathbf{A}\tilde{\mathbf{C}}_s\tilde{\mathbf{A}}\big)\tilde{\mathbf{C}}_s\Big)d\tau.\end{aligned} \tag{26}$$

The eigenvalue equation of (25) can be reformulated as

$$\mathbf{B}_1\psi_k = \lambda_k\mathbf{B}_2\psi_k, \tag{27}$$

where ψ_k is the kth eigenvector associated with the eigenvalue λ_k that represents the Floquet exponent, and

$$\mathbf{B}_1 = \begin{bmatrix} \mathbf{0} & \mathbf{I} \\ -\tilde{\mathbf{K}} & \mathbf{0} \end{bmatrix}, \ \mathbf{B}_2 = \begin{bmatrix} \mathbf{I} & \mathbf{0} \\ \tilde{\mathbf{C}} & \tilde{\mathbf{M}} \end{bmatrix}. \tag{28}$$

According to Hill method, Floquet exponents λ_k of the stability criteria is that the two eigenvalues with the smallest imaginary part in modulus [13, 14]. If all real parts are negative in two eigenvalues, the QP solution is stable, otherwise the solution is unstable.

4 Numerical Results and Discussion

QP solutions of the damped nonlinear QP Mathieu equation are investigated in this section. Parameters of the equation are $c = 0.001$, $\tilde{\omega}_0^2 = 1.1$, $\alpha = 0.08$, $\omega = 1.2075$, $\omega_d = \sqrt{3}/20$, and $\gamma = 0.5$. In the QP solution process of the IHB method with two time scales, the integer multiple j_1 of the first time scale τ_1 is taken as $j_1 = 1, 3$. It is found that there exist three types of QP solutions related to integer multiples j_2 of the second time scale $\tau_2 = \omega_d t$, i.e., type I: $j_2 = \cdots - 3, -2, -1, 0, 1, 2, 3, \ldots$, type II: $j_2 = \cdots - 6, -4, -2, 0, 2, 4, 6, \ldots$, and type III: $j_2 = \ldots - 5, -3, -1, 1, 3, 5, \ldots$; (11) can then be written as

$$x_0 = \sum_{j_2=-m_1}^{m_1} A_{1,j_2} \cos\left(\tau_1 + j_2\tau_2 + \varphi_{1,j_2}\right) + \sum_{j_2=-m_2}^{m_2} A_{2,j_2} \cos\left(3\tau_1 + j_2\tau_2 + \varphi_{3,j_2}\right), \tag{29}$$

where

$$A_{i,j_2} = \sqrt{a_{i,j_2}^2 + b_{i,j_2}^2}, \quad \varphi_{i,j_2} = atan2\left(-b_{i,j_2}, a_{i,j_2}\right), \quad i = 1, 3, \tag{30}$$

in which (m_1, m_2) are natural numbers, even positive integers, and odd positive integers for type I, type II, and type III, respectively. A noticeable and interesting feature of the damped nonlinear QP Mathieu equation, namely the presence of co-existing three types of QP solutions, is obtained from the IHB method with two time scales. But only type I QP solutions are obtained from the double-step MMS as discussed in [6], while spaced distance $2\omega_d$ in sidebands of type II and III QP solutions is about the natural frequency of the second MMS method, they are hardly obtained.

4.1 Type I: Spectra of QP Solutions that Contain the Carrier Frequencies $(\omega, 3\omega)$ with Uniformly Spaced Distance ω_d Apart in Sidebands

The multiple Fourier series of (29) is used to continue the QP solution with uniformly spaced distance ω_d apart in sidebands and $m_1 = m_2 = 6$ so that with total harmonic terms retained $\tilde{n}_c = \tilde{n}_s = 2m_1 + 2m_2 + 2 = 26$. Figure 1 shows solution diagrams of the damped nonlinear QP Mathieu's equation with $\omega = 1.2075$, $\omega_d = \sqrt{3}/20$, $c = 0.001$, $\tilde{\omega}_0 = 1.1$, $\alpha = 0.08$, and $\xi = 0.5$ from the three methods, i.e., the IHB method with two time scales, NI, and the double-step MMS, where amplitudes $A_{1,0}$, $A_{1,1}$, $A_{1,-1}$, $A_{1,2}$, and $A_{1,-2}$ defined in (29) correspond to harmonic terms $\cos\left(\tau_1 + \varphi_{1,0}\right)$, $\cos\left(\tau_1 + \tau_2 + \varphi_{1,1}\right)$, $\cos\left(\tau_1 - \tau_2 + \varphi_{1,-1}\right)$, $\cos\left(\tau_1 + 2\tau_2 + \varphi_{1,2}\right)$, and $\cos\left(\tau_1 - 2\tau_2 + \varphi_{1,-2}\right)$, respectively. It can be seen from Fig. 1 that the results from the IHB method with two time scales are nearly identical to those from NI for the whole range of the parametric excitation magnitude β, while the results from the double-step MMS shows noticeable difference from the results from NI. Presented in Fig. 2a are time histories with the parametric excitation magnitude $\beta = 0.04$ calculated by the three methods, i.e., the IHB method with two time scales, NI, and the double-step MMS. In Fig. 2d, the corresponding spectra is shown. It can be seen from Fig. 2 that the result from the IHB method with two time scales agrees very well with that from NI, whereas the result from the double-step MMS shows noticeable deviations.

4.2 Type II: Spectra of the QP Solutions that Contain the Carrier Frequencies $(\omega, 3\omega)$ with Uniformly Spaced Distance $2\omega_d$ Apart in Sidebands

Spectra of the QP solutions in this case contain the carrier frequency $(\omega, 3\omega)$ with uniformly spaced distance $2\omega_d$ apart in sidebands; (29) can then be written as

$$
\begin{aligned}
x_0 = {} & A_{1,0}\cos\left(\tau_1 + \varphi_{1,0}\right) + A_{1,\pm 2}\cos\left(\tau_1 \pm 2\tau_2 + \varphi_{1,\pm 2}\right) \\
& + A_{1,\pm 4}\cos\left(\tau_1 \pm 4\tau_2 + \varphi_{1,\pm 4}\right) + \cdots \\
& + A_{2,0}\cos\left(3\tau_1 + \varphi_{3,0}\right) + A_{2,\pm 2}\cos\left(3\tau_1 \pm 2\tau_2 + \varphi_{3,\pm 2}\right) \\
& + A_{2,\pm 4}\cos\left(3\tau_1 \pm 4\tau_2 + \varphi_{3,\pm 4}\right) + \cdots ,
\end{aligned}
\tag{31}
$$

Figure 3 shows the QP solution curves of amplitudes $A_{1,0}$, $A_{1,2}$, and $A_{1,-2}$ corresponding to harmonic terms $\cos\left(\tau_1 + \varphi_{1,0}\right)$, $\cos\left(\tau_1 + 2\tau_2 + \varphi_{1,2}\right)$, and $\cos\left(\tau_1 - 2\tau_2 + \varphi_{1,-2}\right)$, respectively, versus the parametric excitation magnitude β of the damped nonlinear QP Mathieu's equation with $\omega = 1.2075$, $\omega_d = \sqrt{3}/20$, $c = 0.001$, $\tilde{\omega}_0 = 1.1$, $\alpha = 0.08$, and $\xi = 0.5$. The curves from the IHB method with

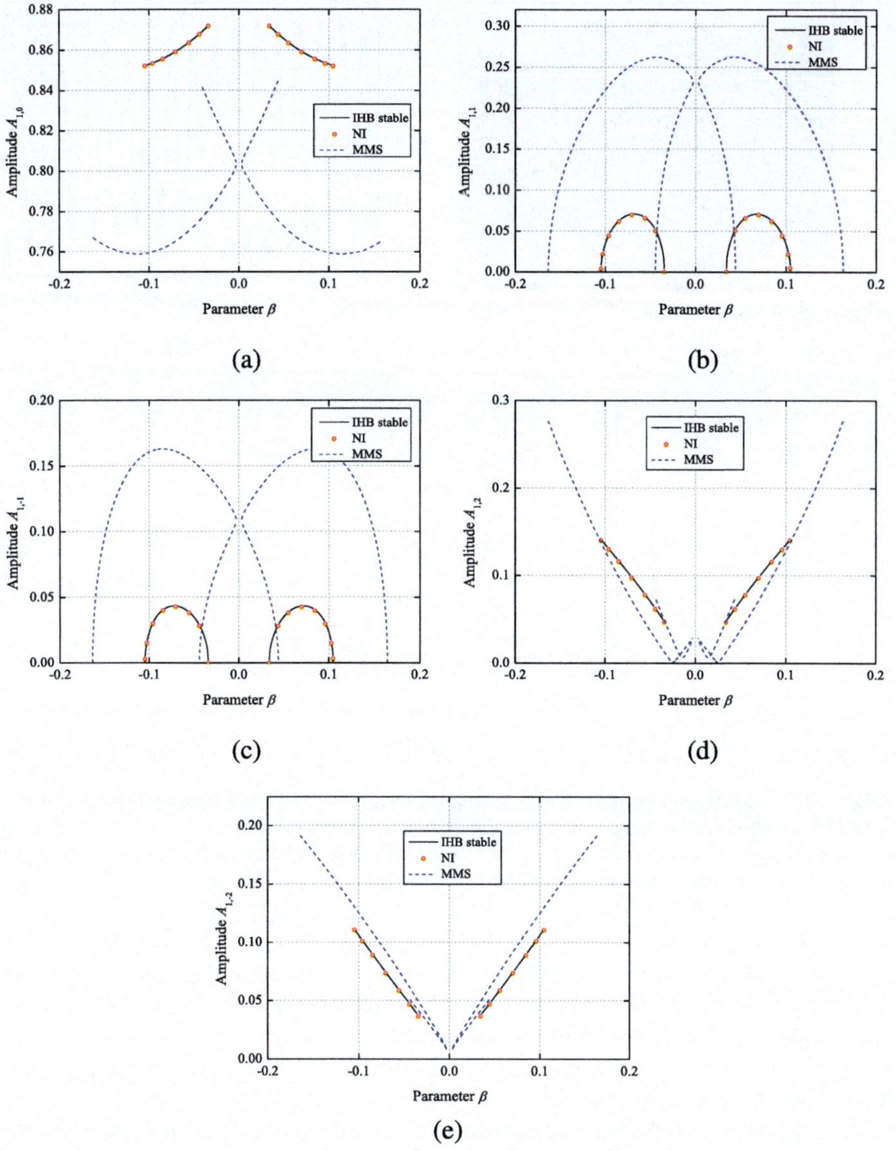

Fig. 1 Amplitudes of type I QP solutions of the damped nonlinear QP Mathieu's equation versus the parametric excitation magnitude β with $\omega = 1.2075$, $\omega_d = \sqrt{3}/20$, $c = 0.001$, $\tilde{\omega}_0 = 1.1$, $\alpha = 0.08$, and $\xi = 0.5$ from the three methods, i.e., the IHB method with two time scales, numerical integration (NI), and the double-step MMS: **a** $\beta \sim A_{1,0}$, **b** $\beta \sim A_{1,1}$, **c** $\beta \sim A_{1,-1}$, **d** $\beta \sim A_{1,2}$, and **e** $\beta \sim A_{1,-2}$

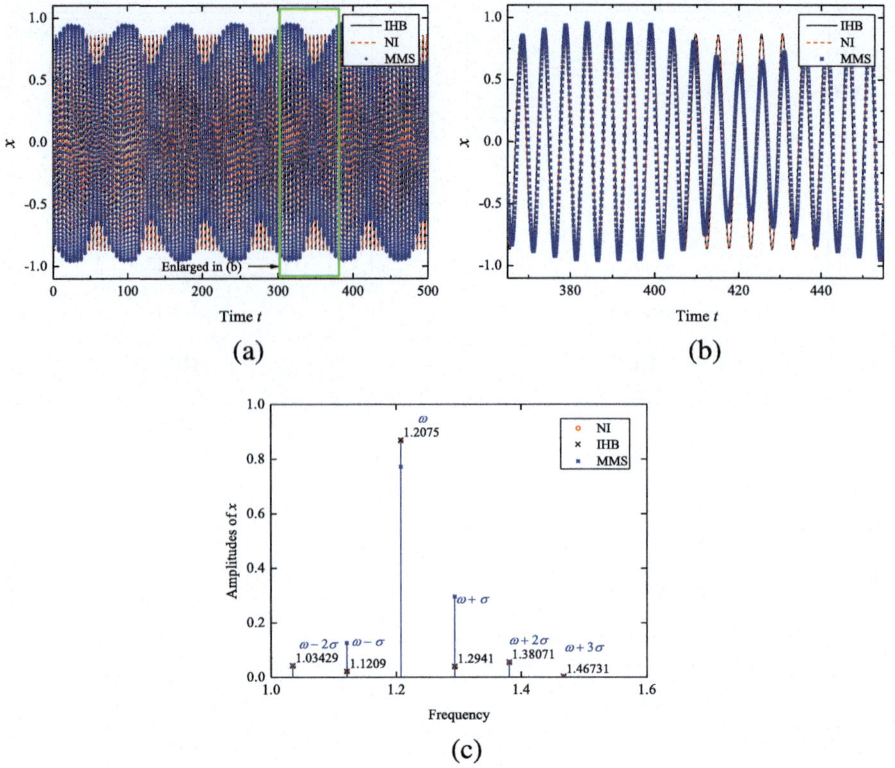

Fig. 2 IHB method with two time scales, numerical integration (NI), and the double-step MMS for type I QP solutions of the damped nonlinear QP Mathieu's equation with $\omega = 1.2075, \omega_d = \sqrt{3}/20$, $c = 0.001$, $\tilde{\omega}_0 = 1.1$, $\alpha = 0.08$, and $\xi = 0.5$ at the parametric excitation magnitude $\beta = 0.04$: **a** time histories, **b** enlargement of a zone highlighted in (**a**) and **c** Fourier spectra

two time scales have three closed parts. The green short dot lines represent the type I QP response curves. It can be seen that the results from the IHB method with two time scales agree very well with those from NI.

Figure 4 shows time histories and spectra of the QP solution of the damped nonlinear QP Mathieu's equation with $\omega = 1.2075$, $\omega_d = \sqrt{3}/20$, $c = 0.001$, $\tilde{\omega}_0 = 1.1$, $\alpha = 0.08$, and $\xi = 0.5$ at the parametric excitation magnitude $\beta = 0.6$ calculated by the IHB method with two time scales and NI.

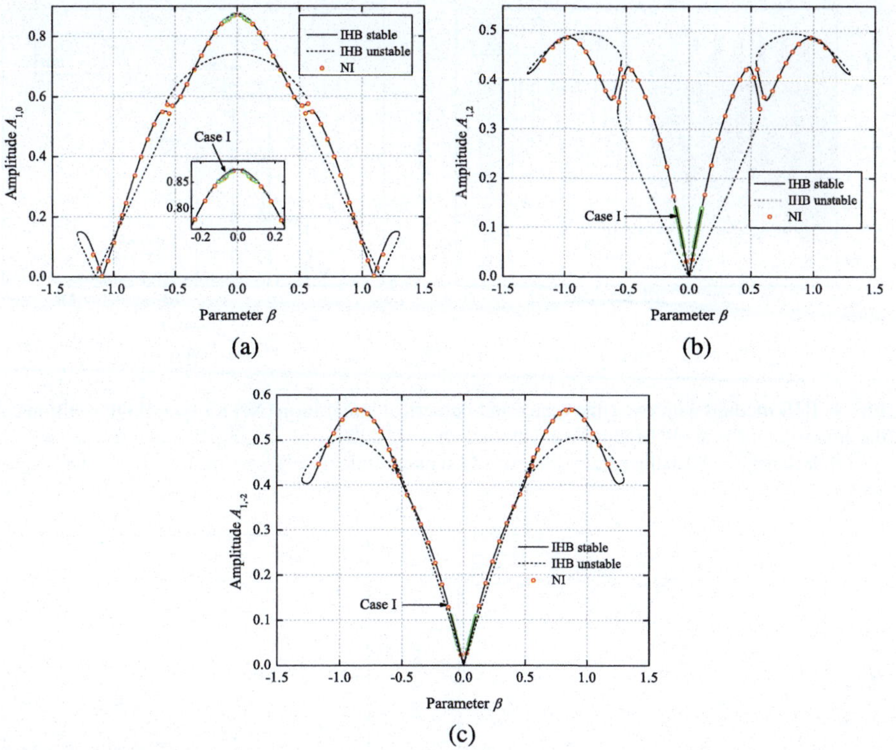

Fig. 3 Amplitudes of type II QP solutions versus the parametric excitation magnitude β of the damped nonlinear QP Mathieu's equation with $\omega = 1.2075$, $\omega_d = \sqrt{3}/20$, $c = 0.001$, $\tilde{\omega}_0 = 1.1$, $\alpha = 0.08$, and $\xi = 0.5$ from the IHB method with two time scales and numerical integration (NI); the QP solutions shown in green short dot lines are type I solutions: **a** $\beta \sim A_{1,0}$, **b** $\beta \sim A_{1,2}$, and **c** $\beta \sim A_{1,-2}$

4.3 Type III: Spectra of the QP Solutions that Do Not Contain the Frequencies $(\omega, 3\omega)$ with Uniformly Spaced Distance $2\omega_d$ Apart in Sidebands

Spectra of the QP solutions in this case do not have the frequencies $(\omega, 3\omega)$ with uniformly spaced distance $2\omega_d$ apart in sidebands; (29) can then be written as

$$\begin{aligned} x_0 = {} & A_{1,\pm 1}\cos\left(\tau_1 \pm \tau_2 + \varphi_{1,\pm 1}\right) + A_{1,\pm 3}\cos\left(\tau_1 \pm 3\tau_2 + \varphi_{1,\pm 3}\right) \\ & + A_{1,\pm 5}\cos\left(\tau_1 \pm 5\tau_2 + \varphi_{1,\pm 5}\right) + \cdots + A_{2,\pm 1}\cos\left(3\tau_1 \pm \tau_2 + \varphi_{3,\pm 1}\right) \\ & + A_{2,\pm 3}\cos\left(3\tau_1 \pm 3\tau_2 + \varphi_{3,\pm 3}\right) + A_{2,\pm 5}\cos\left(3\tau_1 \pm 5\tau_2 + \varphi_{3,\pm 5}\right) + \cdots, \end{aligned} \tag{32}$$

Figure 5 shows the QP response curves of amplitudes $A_{1,-1}$, $A_{1,1}$, and $A_{1,3}$ corresponding to harmonic terms $\cos\left(\tau_1 - \tau_2 + \varphi_{1,-1}\right)$, $\cos\left(\tau_1 + \tau_2 + \varphi_{1,1}\right)$, and

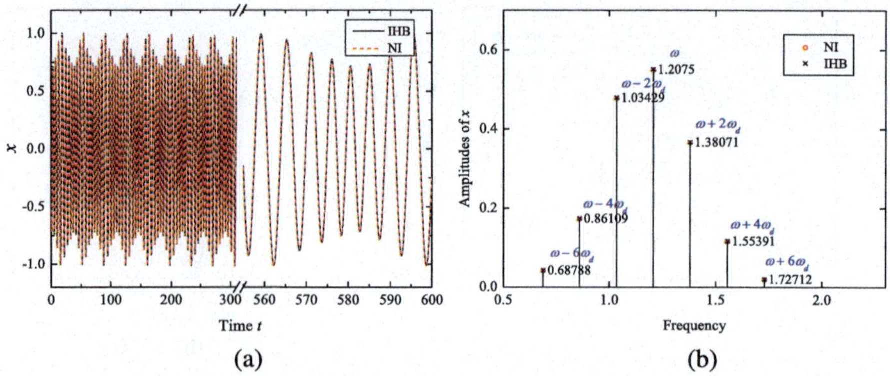

Fig. 4 IHB method with two time scales and numerical integration (NI) for type II QP solutions of the damped nonlinear QP Mathieu's equation with $\omega = 1.2075, \omega_d = \sqrt{3}/20, c = 0.001, \tilde{\omega}_0 = 1.1, \alpha = 0.08,$ and $\xi = 0.5$ at the parametric excitation magnitude $\beta = 0.6$: **a** time histories and **b** Fourier spectra

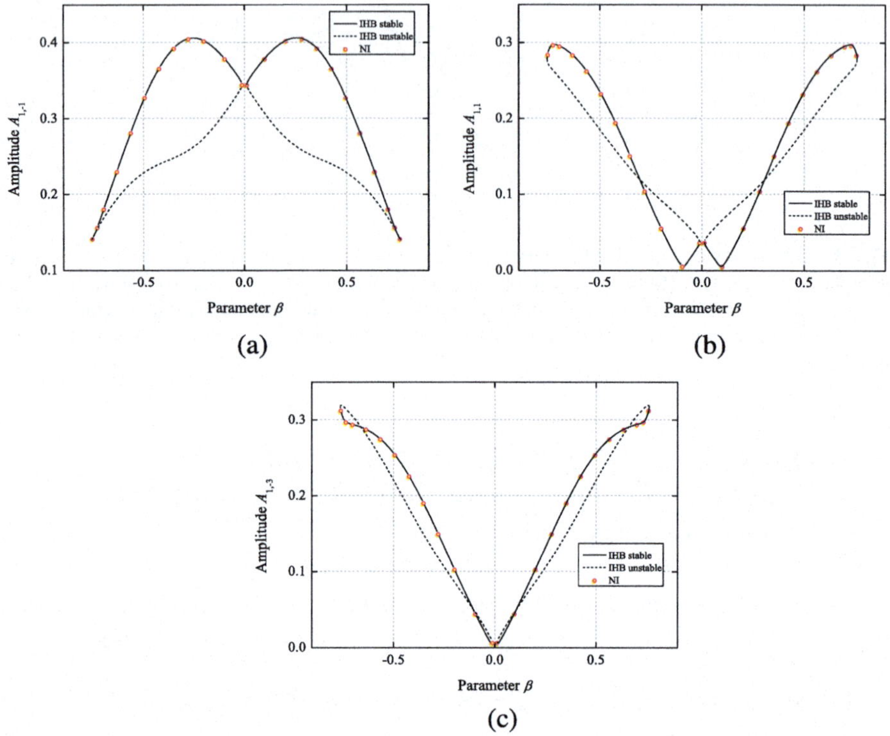

Fig. 5 Amplitudes of type III QP solutions versus the parametric excitation magnitude β of the damped nonlinear QP Mathieu's equation with $\omega = 1.2075$, $\omega_d = \sqrt{3}/20$, $c = 0.001$, $\tilde{\omega}_0 = 1.1$, $\alpha = 0.08$, and $\xi = 0.5$ from the IHB method with two time scales and numerical integration (NI): **a** $\beta \sim A_{1,-1}$, **b** $\beta \sim A_{1,1}$, and **c** $\beta \sim A_{1,-3}$

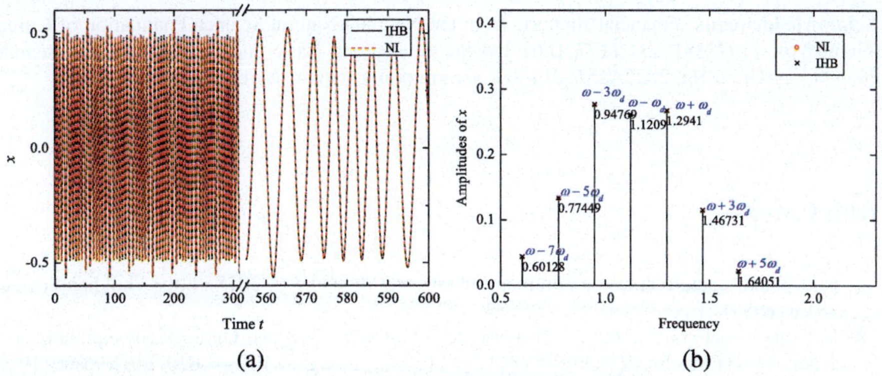

Fig. 6 The IHB method with two time scales and numerical integration (NI) for type III QP solutions of the damped nonlinear QP Mathieu's equation with $\omega = 1.2075$, $\omega_d = \sqrt{3}/20$, $c = 0.001$, $\tilde{\omega}_0 = 1.1$, $\alpha = 0.08$, and $\xi = 0.5$ at the parametric excitation magnitude $\beta = 0.6$: **a** time histories and **b** Fourier spectra

$\cos\left(\tau_1 + 3\tau_2 + \varphi_{1,3}\right)$, respectively, versus the parametric excitation magnitude β of the damped nonlinear QP Mathieu's equation with $\omega = 1.2075$, $\omega_d = \sqrt{3}/20$, $c = 0.001$, $\tilde{\omega}_0 = 1.1$, $\alpha = 0.08$, and $\xi = 0.5$. The curves from the IHB method with two time scales have two closed parts. It can be seen that the results from the IHB method with two time scales agree very well with those from NI. Figure 6 shows time histories, and spectra of the QP response of the damped nonlinear QP Mathieu's equation with $\omega = 1.2075$, $\omega_d = \sqrt{3}/20$, $c = 0.001$, $\tilde{\omega}_0 = 1.1$, $\alpha = 0.08$, and $\xi = 0.5$ at the parametric excitation magnitude $\beta = 0.6$ calculated by the IHB method with two time scales and NI.

5 Conclusion

The IHB method with two time scales is an efficient, straightforward, and reliable method for obtaining the QP solutions of the damped nonlinear QP Mathieu's equation whose spectra contain sidebands, uniformly spaced with a distance ω_d or $2\omega_d$ apart, around the carrier frequencies ω and 3ω. Three types of QP solutions are obtained by using the present method. Based on approximation of QP solutions by periodic solutions with a large period, Floquet theory is applied to conduct stability analysis of the damped nonlinear QP Mathieu's equation and to determine the bifurcation points. The accuracy of the present method is higher than that of the double-step MMS for the first type QP solutions. The IHB method with two time scales can also be used to obtain the other two types of QP solutions, while the double-step MMS cannot be used to obtain these solutions.

Acknowledgements Financial supports from the National Natural Science Foundation of China (Grant Nos. 11972381 and 11772100) and the Guangdong Basic and Applied Basic Research Foundation (Grant No. 2022A1515011809) are gratefully acknowledged.

References

1. R.H. Rand, Lecture notes on nonlinear vibrations (Version 53) (2015). https://hdl.handle.net/1813/28989
2. E. Esmailzadeh, G. Nakhaie-jazar, Periodic solution of a Mathieu-Duffing type equation. Int. J. Non-Linear Mech. **32**(5), 905–912 (1997). https://doi.org/10.1016/s0020-7462(96)00119-9
3. L. Ng, R. Rand, Bifurcations in a Mathieu equation with cubic nonlinearities. Chaos Solitons Fractals **14**(2), 173–181 (2002). https://doi.org/10.1016/s0960-0779(01)00226-0
4. R.S. Zounes, R.H. Rand, Transition curves for the quasi-periodic Mathieu equation. SIAM J. Appl. Math. **58**(4), 1094–1115 (1998). https://doi.org/10.1137/s0036139996303877
5. M. Belhaq, K. Guennoum, M. Houssni, Asymptotic solutions for a damped non-linear quasi-periodic Mathieu equation. Int. J. Non-Linear Mech. **37**(3), 445–460 (2002). https://doi.org/10.1016/s0020-7462(01)00020-8
6. K. Guennoun, M. Houssni, M. Belhaq, Quasi-periodic solutions and stability for a weakly damped nonlinear quasi-periodic Mathieu equation. Nonlinear Dyn.**27**(3), 211–236 (2002). https://doi.org/10.1023/a:1014496917703
7. R.S. Zounes, R.H. Rand, Global behavior of a nonlinear quasiperiodic Mathieu equation. Nonlinear Dyn. **27**(1), 87–105 (2002). https://doi.org/10.1023/a:1017931712099
8. J.L. Huang, W.D. Zhu, An incremental harmonic balance method with two timescales for quasiperiodic motion of nonlinear systems whose spectrum contains uniformly spaced sideband frequencies. Nonlinear Dyn. **90**(2), 1015–1033 (2017). https://doi.org/10.1007/s11071-017-3708-6
9. J.L. Huang, T. Wang, W.D. Zhu, An incremental harmonic balance method with two time-scales for quasi-periodic responses of a Van der Pol-Mathieu equation. Int. J. Non-Linear Mech. **135**, 103767 (2017). https://doi.org/10.1016/j.ijnonlinmec.2021.103767
10. W. Kreider, A.H. Nayfeh, Experimental investigation of single-mode responses in a fixed-fixed buckled beam. Nonlinear Dyn. **15**(2), 155–177 (1998). https://doi.org/10.1023/a:1008231012968
11. J. Ren, W. Fan, W.D. Zhu, Comparison between the incremental harmonic balance method and alternating frequency/time-domain method. J. Vibr. Acoust. **143** (2021)
12. J. Ren, W.D. Zhu, An optimized efficient Galerkin averaging-incremental harmonic balance method for high-dimensional spatially discretized models of continuous systems based on parallel computing. J. Comput. Nonlinear Dyn. **16** (2021)
13. W. Kreider, A.H. Nayfeh, The continuation and stability analysis methods for quasi-periodic solutions of nonlinear systems. Nonlinear Dyn. **100**(2), 1469–1496 (2020). https://doi.org/10.1007/s11071-020-05497-7
14. G. Moore, Floquet theory as a computational tool. SIAM J. Numer. Anal. **42**(6), 2522–2568 (2005). https://doi.org/10.1137/s0036142903434175

Effect of a Periodic Gravitational Excitation with Frequency Modulation on Convective Instability in Porous Media

Karam Allali and Mohamed Belhaq

Abstract This work deals with the influence of a periodic gravitational excitation on convective instability in porous media. It is assumed that the frequency of the periodic excitation is modulated around a mean value with a certain amplitude and frequency. In the suggested model, we consider the coupling between the heat equation, concentration equation, and the hydrodynamics equations. Two approximations are assumed, the first represents the incompressibility and is formulated by the Boussinesq approach, while the second represents the effect of the porous matrix on the flow and is formulated by Darcy law. Linear stability analysis of the problem is fulfilled and the convective instability boundary is determined using numerical simulations. Results show that the convective instability boundary can be greatly influenced by modulating the frequency of the periodic gravitational modulation. Specifically, it is demonstrated that the modulation of the frequency of the periodic gravitational excitation has a stabilizing or a destabilizing effect depending on its amplitude or frequency.

1 Introduction

The reaction front is known to be the region between the reactant and the product during a chemical process. The propagation of reaction fronts can be influenced by many factors. One important factor among them is the force caused by some external excitations such as gravitational modulation. The vibrations can influence the dynamical and thermal prosperities of the fluid flow. Many works [1–5] have

K. Allali (✉)
Laboratory of Mathematics Computer Science and Applications, Faculty of Sciences and Technologies, Hassan II University of Casablanca, PO Box 146, 20650 Mohammedia, Morocco
e-mail: karam.allali@fstm.ac.ma

M. Belhaq
Department of Physics, Faculty of Sciences Ain Chock, Hassan II University of Casablanca, PO Box 5366, 20100 Casablanca, Morocco
e-mail: mbelhaq@yahoo.fr

M. Belhaq (ed.), *Advances in Nonlinear Dynamics and Control of Mechanical and Physical Systems*, Springer Proceedings in Physics 301,
https://doi.org/10.1007/978-981-99-7958-5_4

highlighted the effect of vibrations on fluids convective instability. The influence of vibrations on reaction front convective instability, when both reactant and product are in liquid phases, was studied in [6], while when the reactant is liquid and the product is solid, the same problem was tackled in [7]. The authors of these two works have used Navier-Stokes equations to model the propagating polymerization front and demonstrated numerically that the vibrations play an essential role in enhancing the convective instability.

Reaction front may also propagate in porous matrix [8–10]. The modelization of this kind of problem often uses Darcy's law to illustrate mathematically the hydrodynamics in porous media. The vibrations under the form of gravitational modulation have been utilized in order to check the impact of the amplitude and the frequency of vibration on the convective instability of the reaction front in porous media [11]. The linear stability analysis of the problem has been fulfilled and the onset of thermal convection was examined. Later, the same problem was tackled in [12] by considering the effect of the quasi-periodic gravitational modulation instead of the periodic one. It was shown that the ratio of the two quasi-periodic incommensurate frequencies plays an essential role in controlling the reaction front convective instability.

The amplitude and the frequency remain two important keys in controlling the stability of the studied system. These two parameters can themselves be modulated and different studies have highlighted this issue with the different consequences on the system stability or instability. Indeed, the effect of a periodic gravitational vibration with amplitude modulation was studied in [13]. The authors have demonstrated that the modulated amplitude gives rise to certain stability regions of the propagating reaction front. Amplitude modulation control has shown its effectiveness in many other scientific domains. For example, amplitude modulation can suppress or enhance bursting oscillations in a certain mechanical oscillator [14]. Amplitude modulation is studied in many other domains such as acoustics and optics [15–17]. Like amplitude modulation, the modulation of frequency has a great impact on science and industry. For example in radio broadcasting the frequency modulation is utilized since it offers tangible results in radio transmission and in avoiding undesirable interferences [18]. Moreover, several other applications are also observed in radar, in EEG, and in seismic studies [19–21].

Motivated by the previous works, the present paper is concerned with the influence of a periodic gravitational modulation on the convective instability in porous media. Here, we consider that the frequency of the gravitation modulation can itself be modulated around a mean value with a certain amplitude and frequency. To this end, we assume that the upward propagating front is influenced by an external excitation b, vertical to the interface between the reactant and the product. Hence, the upward propagating reaction front will be subject to the modulated gravitation $g + bg$, where g is the classical earth gravity. The time dependence of the external excitation is given by $b(t) = \lambda_1 sin(\nu_m(t)t)$; where λ_1 and $\nu_m(t)$ are the amplitude and the time-dependent frequency of the gravitation modulation, respectively. Moreover, we assume that the frequency $\nu_m(t)$ is modulated around a certain mean value ν_1, such that $\nu_m(t) = \nu_1 + \lambda_2 cos(\nu_2 t)$, where λ_2 and ν_2 are the amplitude and the frequency of the modulated frequency $\nu_m(t)$, respectively. In our case, we consider that the hydrodynamics of the propagating reaction front is described by Darcy law, while the fluid incompressibility is represented by Boussinesq approximation. The boundary of the convective stability will be found using the linear stability analysis of the original model. The interface problem will be given and the reduced problem will be solved numerically in order to investigate the impact of the gravitational frequency modulation on the convective instability.

The organization of this paper is as follows. First, we introduce the propagating reaction front model in the next Section. Next, Sect. 3 presents the linear stability analysis of the problem using the infinite narrow reaction zone method. The formulation of the interface problem is given in the same section. The numerical simulations along with discussions are provided in Sect. 4. Finally, a conclusion is given in Sect. 5.

2 Problem Formulation

We assume that the reaction front propagates oppositely to the gravity direction, in an upward manner. Moreover, we consider that the porous matrix is filled by an incompressible fluid and the overall physical system is subjected to gravitational modulation with a modulated frequency. The upward reaction front under gravitational modulation with modulated frequency is illustrated in Fig. 1. The physical model consists of the heat equation coupled with the equation of hydrodynamics under Darcy and Boussineq approximations and can be written as

$$\frac{\partial \theta}{\partial t} + \mathbf{u}.\nabla\theta = \kappa\Delta\theta + qK(\theta)\phi(C), \tag{1}$$

$$\frac{\partial C}{\partial t} + \mathbf{u}.\nabla C = d\Delta C + K(\theta)\phi(C), \tag{2}$$

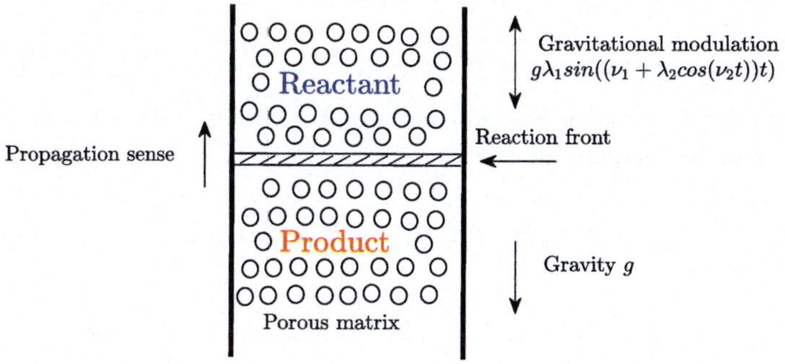

Fig. 1 The propagating reaction front under gravitational modulation with modulated frequency

$$\mathbf{u}+\frac{K}{\mu}\nabla p=\frac{g\beta K}{\mu}\rho(\theta-\theta_0)(1+\lambda_1\sin((\nu_1+\lambda_2\cos(\nu_2 t))t), \tag{3}$$

$$\nabla.\mathbf{u}=0, \tag{4}$$

with the following boundary conditions:

$$\theta=\theta_i\ ,C=1 \text{ and } \mathbf{u}=0 \text{ when } y\to+\infty, \tag{5}$$
$$\theta=\theta_b\ ,C=0 \text{ and } \mathbf{u}=0 \text{ when } y\to-\infty. \tag{6}$$

where θ represents the temperature, C stands for the depth of conversion, $\mathbf{u}$ is the fluid velocity, p is the pressure, κ is the coefficient of thermal diffusivity, d is the diffusion coefficient, q is the adiabatic heat release, g is the gravity acceleration, ρ is the density, β is the coefficient of thermal expansion, μ is the viscosity and γ is the unit vector in the upward direction. In addition, θ_0 is the mean value of temperature, θ_i is an initial temperature while θ_b is the temperature of the burned mixture given by $\theta_b=\theta_i+q$. The function $K(\theta)\phi(C)$ is the reaction rate such that the temperature dependence is given by the classical Arrhenius law [22]:

$$K(\theta)=k_0\exp\left(-\frac{E}{R_0\theta}\right). \tag{7}$$

Here, E represents the activation energy, R_0 stands for the universal gas constant and k_0 represents the pre-exponential factor. We recall that the asymptotic analysis of this problem will be carried out for activation energy sufficiently large and we consider the following zero-order reaction:

$$\phi(C)=\begin{cases}1 \text{ if } C<1\\ 0 \text{ if } C=1\end{cases}. \tag{8}$$

In order to give the model in its dimensionless form, we will introduce the following spatial variables $\tilde{x}=\frac{xc_0}{\kappa}$, $\tilde{y}=\frac{yc_0}{\kappa}$, time $\tilde{t}=\frac{tc_0^2}{\kappa d}$, velocity $\frac{\mathbf{u}}{c_0}$, pressure $\frac{p\kappa\mu}{K}$ with $c_0=c/\sqrt{2}$ and frequencies $\sigma_1=\frac{\kappa}{c_0^2}\nu_1$, $\sigma_2=\frac{\kappa}{c_0^2}\nu_2$. By denoting $T=\frac{\theta-\theta_b}{q}$ and for simplicity, we keep the same notation for variables. Hence, one obtains the following system of equations

$$\frac{\partial T}{\partial t}+\mathbf{u}\nabla T=\Delta T+W_Z(T)\,\phi(C), \tag{9}$$

$$\frac{\partial C}{\partial t}+\mathbf{u}\nabla C=\Lambda\Delta C+W_Z(T)\,\phi(C), \tag{10}$$

$$\mathbf{u} + \nabla p = R_p\,(T + T_0)\begin{pmatrix}0\\1\end{pmatrix}(1 + \lambda_1 \sin((\sigma_1 + \lambda_2 \cos(\sigma_2 t))t), \tag{11}$$

$$div\,(v) = 0, \tag{12}$$

subjected to the following conditions at infinity:

$$T = -1, C = 0 \text{ and } \mathbf{u} = 0 \text{ when } y \to +\infty, \tag{13}$$
$$T = 0, C = 1 \text{ and } \mathbf{u} = 0 \text{ when } y \to -\infty. \tag{14}$$

Here $\Lambda = d/\kappa$ stands for the inverse of the Lewis number, $\mathcal{R}_p = \dfrac{K c_0^2 P^2 \mathcal{R}}{\mu^2}$, where $\mathcal{R}$ is the Rayleigh number and P is the Prandtl number defined, respectively, by $R = \dfrac{g\beta q\kappa^2}{\mu c_1^3}$ and $P = \dfrac{\mu}{\kappa}$. In addition, we will use the parameters $\delta = \dfrac{R_0\theta_b}{E}$ and $T_0 = \dfrac{\theta_b - \theta_0}{q}$. The reaction rate is formulated as

$$W_Z(T) = Z\exp\Big(\frac{T}{Z^{-1} + \delta T}\Big), \tag{15}$$

where $Z = \dfrac{qE}{R_0\theta_b^2}$ represents the Zeldovich number.

The linear stability analysis will be fulfilled in the case of zero Lewis number ($\Lambda = 0$), which corresponds to a liquid mixture.

3 The Linear Stability Analysis

3.1 The Infinitely Narrow Reaction Zone Approximation

In order to obtain the interface problem, the linear stability analysis of the original system (9)–(14) is performed, by considering the physical property of the reaction zone which is supposed to be sufficiently infinitely narrow such that the reaction force is neglected outside this reaction zone. We recall that this approach is called Zeldovich-Frank-Kamenetskii approach, which is a well-known approach used for examining many propagating reaction fronts as well as their scientific physical applications [22, 23]. The interface problem will be obtained by applying an asymptotic analysis assuming that $\epsilon = \dfrac{1}{Z}$ is a very small parameter. We first denote by $\zeta(t, x)$ the reaction zone location. Hence, the new independent variable in the direction of the propagating reaction front is given by

$$y_1 = y - \zeta(t, x). \tag{16}$$

Then, the new functions $T_1, C_1, \mathbf{u}_1, p_1$ are introduced, such that

$$\begin{aligned} T(t, x, y) = T_1(t, x, y_1), \quad C(t, x, y) = C_1(t, x, y_1), \\ \mathbf{u}(t, x, y) = \mathbf{u}_1(t, x, y_1), \quad p(t, x, y) = p_1(t, x, y_1). \end{aligned} \tag{17}$$

Subsequently, the (9)–(14) can be re-written in the following form (where the index 1 for the independent variables will be omitted):

$$\frac{\partial T}{\partial t} - \frac{\partial T}{\partial y_1}\frac{\partial \zeta}{\partial t} + \mathbf{u}.\tilde{\nabla} T = \tilde{\Delta} T + W_Z(T)\phi(C), \tag{18}$$

$$\frac{\partial C}{\partial t} - \frac{\partial C}{\partial y_1}\frac{\partial \zeta}{\partial t} + \mathbf{u}.\tilde{\nabla} C = W_Z(T)\phi(C), \tag{19}$$

$$\mathbf{u} + \tilde{\nabla} p = \mathcal{R}_p(T + T_0)(1 + \lambda_1 sin((\sigma_1 + \lambda_2 cos(\sigma_2 t))t), \tag{20}$$

$$\frac{\partial v_x}{\partial x} - \frac{\partial v_x}{\partial y_1}\frac{\partial \zeta}{\partial x} + \frac{\partial v_y}{\partial y_1} = 0, \tag{21}$$

where we have

$$\tilde{\Delta} = \frac{\partial^2}{\partial x^2} + \frac{\partial^2}{\partial y_1^2} - 2\frac{\partial \zeta}{\partial x}\frac{\partial^2}{\partial x \partial y_1} + \left(\frac{\partial \zeta}{\partial x}\right)^2 \frac{\partial^2}{\partial y_1^2} - \frac{\partial^2 \zeta}{\partial x^2}\frac{\partial}{\partial y_1}, \tag{22}$$

$$\tilde{\nabla} = \left(\frac{\partial}{\partial x} - \frac{\partial \zeta}{\partial x}\frac{\partial}{\partial y_1}, \frac{\partial}{\partial y_1}\right). \tag{23}$$

The matched asymptotic expansions will be used to approximate the jump conditions and therefore to resolve the interface problem. To this purpose, we will seek the outer solution to the problem in the form

$$\begin{aligned} T = T^0 + \epsilon T^1 + \cdots, \quad C = C^0 + \epsilon C^1 + \cdots, \\ \mathbf{u} = \mathbf{u}^0 + \epsilon \mathbf{u}^1 + \cdots, \quad p = p^0 + \epsilon p^1 + \cdots \end{aligned} \tag{24}$$

Here $(T^0, C^0, \mathbf{u}^0)$ is the dimensionless form to the basic solution.

To obtain the inner solution, the stretching coordinate $\eta = y_1/\epsilon$ is introduced, we will seek the inner solution under the following form

$$\begin{aligned} &T = \epsilon \tilde{T}^1 + \cdots, \quad C = \tilde{C}^0 + \epsilon \tilde{C}^1 + \cdots, \\ &\mathbf{u} = \tilde{\mathbf{u}}^0 + \epsilon \tilde{\mathbf{u}}^1 + \cdots, \quad p = \tilde{p}^0 + \epsilon \tilde{p}^1 + \cdots, \quad \zeta = \tilde{\zeta}^0 + \epsilon \tilde{\zeta}^1 + \cdots \end{aligned} \tag{25}$$

By replacing the above expansions, into (18)–(21), the following first-order inner problem is obtained:

$$\left(1+\left(\frac{\partial\tilde{\zeta}^0}{\partial x}\right)^2\right)\frac{\partial^2\tilde{T}^1}{\partial\eta^2}+\exp\left(\frac{\tilde{T}^1}{1+\delta\tilde{T}^1}\right)\phi(\tilde{C}^0)=0, \tag{26}$$

$$-\frac{\partial\tilde{C}^0}{\partial\eta}\frac{\partial\tilde{\zeta}^0}{\partial\eta}-\frac{\partial\tilde{C}^0}{\partial\eta}\left(\tilde{v}_x^0\frac{\partial\tilde{\zeta}^0}{\partial x}-\tilde{v}_y^0\right)=\exp\left(\frac{\tilde{T}^1}{1+\delta\tilde{T}^1}\right)\phi(\tilde{C}^0), \tag{27}$$

$$\frac{\partial\tilde{p}^0}{\partial\eta}=0, \tag{28}$$

$$\tilde{v}_x^0+\frac{\partial\tilde{p}^0}{\partial x}-\frac{\partial\tilde{\zeta}^0}{\partial t}\frac{\partial\tilde{p}^1}{\partial\eta}=0, \tag{29}$$

$$\tilde{v}_y^0+\frac{\partial\tilde{p}^1}{\partial\eta}=-\mathcal{R}_pT_0(1+\lambda_1\sin((\sigma_1+\lambda_2\cos(\sigma_2t))t), \tag{30}$$

$$-\frac{\partial\tilde{v}_x^0}{\partial\eta}\frac{\partial\tilde{\zeta}^0}{\partial x}+\frac{\partial\tilde{v}_y^0}{\partial\eta}=0. \tag{31}$$

However, the matching conditions take the following form

$$\eta\to+\infty:\ \tilde{T}^1\sim T^1|_{y_1=0+}+\eta\frac{\partial T^0}{\partial y_1}|_{y_1=0+},\ \tilde{C}^0\to0,\ \tilde{\mathbf{u}}^0\to\mathbf{u}^0|_{y_1=0+}, \tag{32}$$

$$\eta\to-\infty:\ \tilde{T}^1\to T^1|_{y_1=0-},\ \tilde{C}^0\to1,\ \tilde{\mathbf{u}}^0\to\mathbf{u}^0|_{y_1=0-}. \tag{33}$$

Noticing that from (28), we obtain the fact that $\tilde{p}^0$ does not depend on η, which signifies that the pressure is obviously continuous through the reaction interface. Next, we denote by s the following quantity

$$s=\tilde{v}_x^0\frac{\partial\tilde{\zeta}^0}{\partial x}-\tilde{v}_y^0, \tag{34}$$

From (31), we see that s does not depend on η. Also, from (29), (30) and (34) we can easily remark that $\tilde{v}_x^0$ and $\tilde{v}_y^0$ do not depend on η, which gives the continuity of the velocity through the front interface.

The jump conditions for the temperature are derived from (26). Likewise, from (27) it follows that $\tilde{C}^0$ is a monotone function and $0<\tilde{C}^0<1$. Since we consider zero-order reaction, we have $\phi(\tilde{C}^0)\equiv1$ and we conclude from (26) that $\tilde{T}^1$ is also a monotone function. Multiplying (26) by $\dfrac{\partial\tilde{T}^1}{\partial\eta}$ and integrating yields

$$\left(\frac{\partial \tilde{T}^1}{\partial \eta}\right)^2\Big|_{\eta=+\infty} - \left(\frac{\partial \tilde{T}^1}{\partial \eta}\right)^2\Big|_{\eta=-\infty} = -\frac{2}{A}\int_{-\infty}^{T^1} \exp\left(\frac{\tau}{1+\delta\tau}\right) d\tau, \tag{35}$$

where

$$A = 1 + \left(\frac{\partial \tilde{\zeta}^0}{\partial x}\right)^2. \tag{36}$$

Subtracting (26) from (27) and integrating gives

$$\frac{\partial \tilde{T}^1}{\partial \eta}\Big|_{\eta=+\infty} - \frac{\partial \tilde{T}^1}{\partial \eta}\Big|_{\eta=-\infty} = -\frac{1}{A}\left(\frac{\partial \tilde{\zeta}^0}{\partial t} + s\right). \tag{37}$$

Using the matching conditions and truncating the expansion:

$$T^0 \approx T, \;\; T^1|_{y_1=0-} \approx ZT|_{y_1=0} \;\; \zeta^0 \approx \zeta, \;\; \mathbf{u} \approx \mathbf{v}^0, \tag{38}$$

the following jump conditions are obtained

$$\left(\frac{\partial T}{\partial y_1}\right)^2\Big|_{y_1=0+} - \left(\frac{\partial T}{\partial y_1}\right)^2\Big|_{y_1=0-} = 2Z\left(1 + \left(\frac{\partial \zeta}{\partial x}\right)^2\right)^{-1} \int_{-\infty}^{T|_{y_1=0}} \exp\left(\frac{\tau}{Z^{-1}+\delta\tau}\right) d\tau, \tag{39}$$

$$\frac{\partial T}{\partial y_1}\Big|_{y_1=0+} - \frac{\partial T}{\partial y_1}\Big|_{y_1=0-} = -\left(1 + \left(\frac{\partial \zeta}{\partial x}\right)^2\right)^{-1} \left(\frac{\partial \zeta}{\partial t} + (v_x \frac{\partial \zeta}{\partial x} - v_y)\Big|_{y_1=0}\right). \tag{40}$$

3.2 The Interface Problem and Traveling Wave Solution

The interest of the interface problem is to give two systems of equations, the first concerns the reactant while the second system of equations is related to the product along with the jump conditions.

For the reactant (the unburnt medium, $y > \zeta$), we have

$$\frac{\partial T}{\partial t} + \mathbf{u}.\nabla T = \Delta T, \tag{41}$$

$$C \equiv 0, \tag{42}$$

$$\mathbf{u} + \nabla p = \mathcal{R}_p(T + T_0)(1 + \lambda_1 \sin((\sigma_1 + \lambda_2 \cos(\sigma_2 t))t), \tag{43}$$

$$\nabla.\mathbf{u} = 0. \tag{44}$$

In the product (burnt medium, $y < \zeta$), we obtain the following

$$\frac{\partial T}{\partial t} + \mathbf{u}.\nabla T = \Delta T, \tag{45}$$

$$C \equiv 1, \tag{46}$$

$$\mathbf{u} + \nabla p = \mathcal{R}_p(T + T_0)(1 + \lambda_1 \sin((\sigma_1 + \lambda_2 \cos(\sigma_2 t))t), \tag{47}$$

$$\nabla.\mathbf{u} = 0. \tag{48}$$

In addition, we complete these systems by the following jump conditions at the located interface ($y = \zeta$)

$$[T] = 0, \quad \left[\frac{\partial T}{\partial y}\right] = \frac{\frac{\partial \zeta}{\partial t}}{1 + \left(\frac{\partial \zeta}{\partial x}\right)^2}, \tag{49}$$

$$\left[\left(\frac{\partial T}{\partial y}\right)^2\right] = -\frac{2Z}{1 + (\frac{\partial \zeta}{\partial x})^2} \int_{-\infty}^{T(\zeta)} \exp\Big(\frac{s}{1/Z + \delta s}\Big) ds, \tag{50}$$

$$[\mathbf{u}] = 0, \tag{51}$$

where [ ] denotes the quantity $[g] = g|_{\zeta - 0} - g|_{\zeta + 0}$. The resulting free boundary problem is accompanied by the conditions at infinity:

$$y \to +\infty, \quad T = -1 \text{ and } \mathbf{u} = 0, \tag{52}$$

$$y \to -\infty, \quad T = 0 \text{ and } \mathbf{u} = 0. \tag{53}$$

The linear stability analysis is carried out for the steady-state solutions for the founded interface problem. Indeed, the resulting interface problem admits a traveling wave solution as follows:

$$T(t, x, y) = T_s(y - ut), \quad C(t, x, y) = C_s(y - ut) \text{ and } \mathbf{u} = 0, \tag{54}$$

for which the steady-state solution T_s is formulated by

$$T_s(t, y) = \begin{cases} 0 & \text{if } y < 0 \\ e^{-uy} - 1 & \text{if } y > 0 \end{cases}, \tag{55}$$

while the other steady-state solution C_s is given by

$$C_s(t, y) = \begin{cases} 1 & \text{if } y < 0 \\ 0 & \text{if } y > 0 \end{cases}. \tag{56}$$

Here u represents the velocity of the stationary front.

Introducing the moving frame new coordinates such as $y_1 = y - ut$, the above traveling wave problem will be transformed as

$$\frac{\partial T}{\partial t} + u\frac{\partial T}{\partial y} + \mathbf{u}.\nabla T = \Delta T, \tag{57}$$

$$\mathbf{u} + \nabla p = \mathcal{R}_p(T + T_0)(1 + \lambda_1 \sin((\sigma_1 + \lambda_2 \cos(\sigma_2 t))t), \tag{58}$$

$$\nabla.\mathbf{u} = 0, \tag{59}$$

along with the jump conditions (49)–(51).

Next, we introduce a small perturbation for the linear stability analysis. To this purpose, we consider a perturbation of the reaction front under the following form

$$\zeta(t, x) = ut + \xi(t, x), \text{ with } \xi(t, x) = \epsilon_1(t)e^{ikx}. \tag{60}$$

The stability of the solution is fulfilled by assuming the solution of the problem in the perturbed form:

$$T = T_s + \tilde{T}, \quad v = v_s + \tilde{v}, \tag{61}$$

where

$$\begin{aligned} \tilde{T}(t, x, y) &= T_j(y, t)e^{ikx}, \text{ for } j = 1, 2, \\ \tilde{v}(t, x, y) &= v_j(y, t)e^{ikx}, \text{ for } j = 1, 2. \end{aligned} \tag{62}$$

Here the index $j = 1$ is for solutions when $z < 0$ and $j = 2$ is for those ones when $z > 0$. For convenience, the pressure p is eliminated and the component v_x of the velocity from the interface problem applying two times the operator $curl$. Hence, we arrive to the following problem

For the product ($y < 0$):

$$v_1'' - k^2 v_1 = -\mathcal{R}_p k^2(1 + \lambda_1 \sin((\sigma_1 + \lambda_2 \cos(\sigma_2 t))t)T_1, \tag{63}$$

$$\frac{\partial T_1}{\partial t} - T_1'' - uT_1' + k^2 T_1 = 0. \tag{64}$$

For the reactant ($y > 0$):

$$v_2'' - k^2 v_2 = -\mathcal{R}_p k^2(1 + \lambda_1 \sin((\sigma_1 + \lambda_2 \cos(\sigma_2 t))t)T_2, \tag{65}$$

$$\frac{\partial T_2}{\partial t} - T_2'' - uT_2' + k^2 T_2 = u \exp(-uy)v_2, \tag{66}$$

Taking into account the fact that

$$T\,|_{\xi=\pm 0} = T_s(\pm 0) + \xi T_s'(\pm 0) + \tilde{T}(\pm 0), \tag{67}$$

and

$$\frac{\partial T}{\partial y}\,|_{\xi=\pm 0} = T_s'(\pm 0) + \xi T_s''(\pm 0) + \frac{\partial \tilde{T}}{\partial y}(\pm 0), \tag{68}$$

we arrive to the following jump conditions:

$$T_2(0,t) - T_1(0,t) = u\epsilon_1(t), \tag{69}$$

$$T_2'(0,t) - T_1'(0,t) = -\epsilon_1(t)u^2 - \epsilon_1'(t) + v_1(0,t), \tag{70}$$

$$\epsilon_1(t)u^2 + T_2'(0,t) = -\frac{Z}{u}T_1(0,t), \tag{71}$$

$$v_2^{(i)}(0,t) = v_1^{(i)}(0,t) \quad i = 0, 1. \tag{72}$$

4 Numerical Results and Discussion

In order to determine the boundaries to the convective instability, we solve numerically the resulting problem (63)–(66) along with the jump conditions (69)–(72) for given values of the parameters. Specifically, for different amplitudes and frequencies and also for different values of Rayleigh number $\mathcal{R}_p$. The numerical accuracy is obtained by decreasing successively the time and space steps.

Figure 2 depicts the time evolution of the maximum of temperature for two different values of Rayleigh number $\mathcal{R}_p$. First, it is observed that both curves are oscillating during the progress of time. This is because the propagating reaction front is subjected to the gravitational modulation. The overall observation is that when the Rayleigh number is $\mathcal{R}_p = 25.5$, the maximum of temperature is decreasing. Meanwhile, a slight increase of Rayleigh number ($\mathcal{R}_p = 26$) leads to an increase of the maximum of temperature. The first value leads to the stability of the reaction front, while the second stands for the instability of the reaction front. We have found that the critical value of Rayleigh number is approximately $\mathcal{R}_c \approx 25.75$. This is the limit value of Rayleigh number to consider in order to avoid instability.

Next, we change only one parameter value which is the amplitude λ_1 and keeping the others fixed. Indeed, Fig. 3 shows the time evolution of the maximum of temperature for three different values of Rayleigh number $\mathcal{R}_p$. Again, since the propagating reaction front is subjected to the gravitational modulation with modulated frequency, the three curves are oscillating during the progress of time. When $\mathcal{R}_p = 25.5$, the maximum of temperature is decreasing during the evolution of time. However,

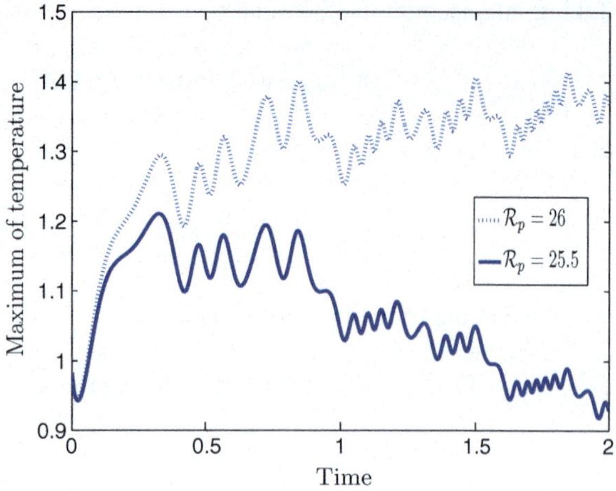

Fig. 2 Maximum of temperature as function of time for $\lambda_1 = 0.2$, $\sigma_1 = 20$, $\lambda_2 = 10$, $\sigma_2 = 10$ and for two different values of Rayleigh number $\mathcal{R}_p$

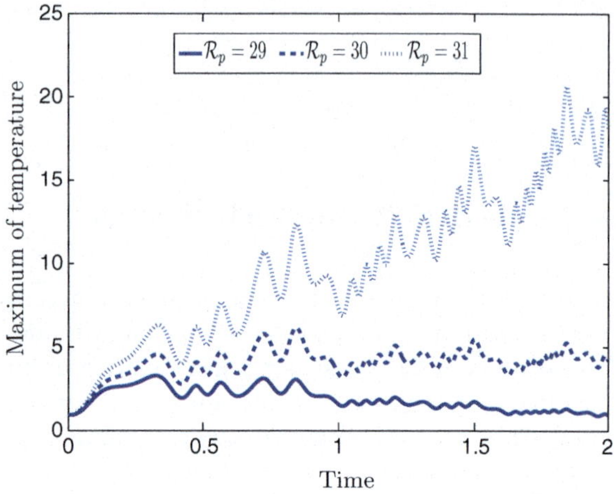

Fig. 3 Maximum of temperature as function of time for $\lambda_1 = 1$, $\sigma_1 = 20$, $\lambda_2 = 10$, $\sigma_2 = 10$ and for three different values of Rayleigh number $\mathcal{R}_p$

when $\mathcal{R}_p = 31$, the maximum of temperature is decreasing during the evolution of time. One can conclude that the critical value of Rayleigh number is approximately $\mathcal{R}_c \approx 30.2$. Moreover, an increase of the amplitude λ_1 has a destabilizing effect on the reaction front.

On the other hand, the frequency σ_1 and the modulated frequency σ_2 have an interesting effect in stabilizing the reaction front. Clearly, an increase in these frequencies leads to a reaction front stabilization. This is clearly depicted in Fig. 4 concerning the effect of σ_1 and Fig. 5 concerning the effect of σ_2. By increasing one of those two frequencies, the overall behavior of the temperature maximum decreases as function of time. Another remark that can be exhibited from the plots is that by increasing the frequency σ_2, oscillations inside one single period increase too.

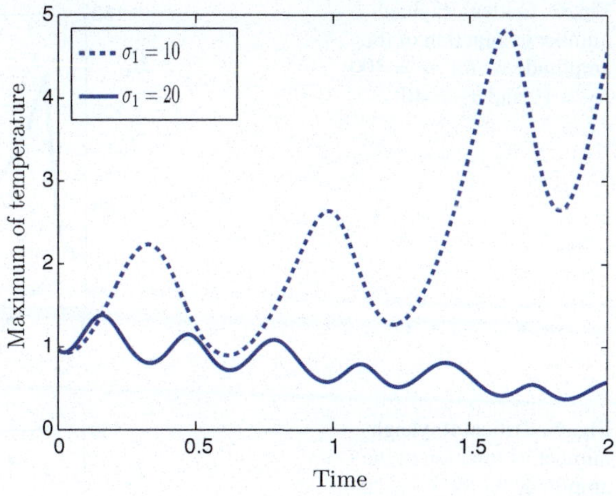

Fig. 4 Maximum of temperature as function of time for $\lambda_1 = 0.5$, $\lambda_2 = 0.5$, $\sigma_2 = 10$, $\mathcal{R}_p = 24$ and for two different values of Rayleigh number σ_1.

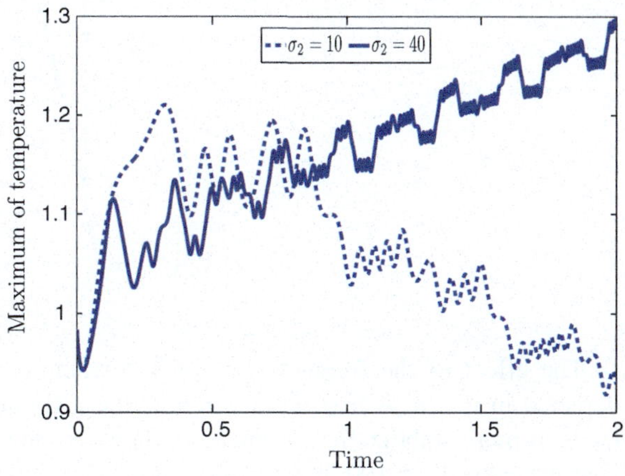

Fig. 5 Maximum of temperature as function of time for $\lambda_1 = 0.2$, $\sigma_1 = 20$, $\lambda_2 = 10$, $\mathcal{R}_p = 25.5$ and for two different values of the frequency σ_2

The effect of the amplitude λ_1 on the reaction front convective instability is shown in Fig. 6. First, we observe that for zero gravitational modulation amplitude ($\lambda_1 = 0$) we find the same critical value of Rayleigh number as in the previous works [11, 12], i.e., $\mathcal{R}_c \approx 26$. From this figure, we see that small amplitudes stabilize more the reaction front, while high amplitudes have a destabilizing effect on the reaction front. Moreover, very high amplitude values destabilize the propagating reaction front. The pick of stability here is reached for a relatively small amplitude. Beyond this, the reaction front losses gradually its stability. The effect of the amplitude of the modulated frequency λ_2 on the reaction front convective instability is depicted in Fig. 7. The oscillating behavior of the critical Rayleigh number is observed in the plot. This figure shows that an increase of the amplitude λ_2 stabilizes the reaction front.

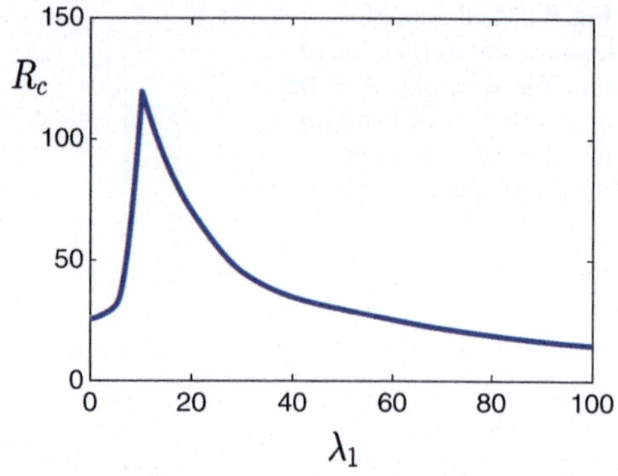

Fig. 6 Critical Rayleigh number as function of the amplitude λ_1 for $\sigma_1 = 300$, $\lambda_2 = 10$ and $\sigma_2 = 40$

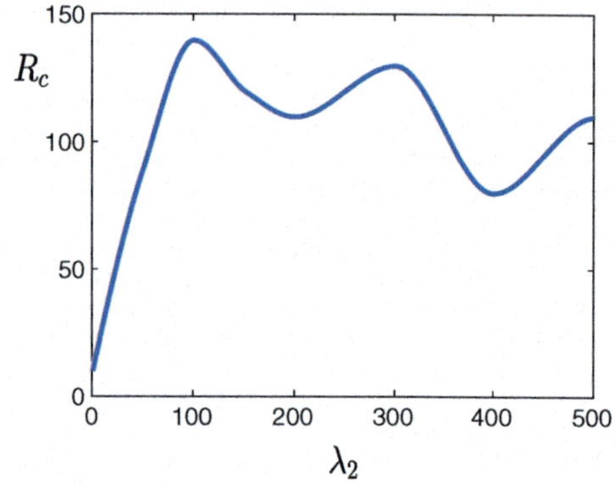

Fig. 7 Critical Rayleigh number as function of the amplitude λ_2 for $\lambda_1 = 20$, $\sigma_1 = 10$ and $\sigma_2 = 100$

The effect of the frequency σ_2 on the reaction front convective instability is depicted in Fig. 8. A first observation is that a slight increase of the frequency σ_2 leads to more stabilization of the reaction front. A zone of stabilization is situated near the value $\sigma_2 \approx 100$. Beyond this value, the reaction front losses considerably its stability. One can conclude that increasing slightly σ_2 in a certain interval stabilizes the reaction front. Figure 9 depicts the same phenomena, i.e., the critical Rayleigh number as a function of the frequency σ_2. It is worthy to notice that in this case, alternate regions of stability appear.

5 Conclusion

In this paper, we have examined the effect of a periodic gravitational modulation on the convection instability of an upwarding reaction front. The considered model consists of the equation of heat and the equations of hydrodynamics under Boussinesq and Darcy approximations. It is supposed that the frequency of the periodic gravitational excitation is itself modulated around a mean value with a certain amplitude and frequency. Attention is focused on the effect of the modulated frequency

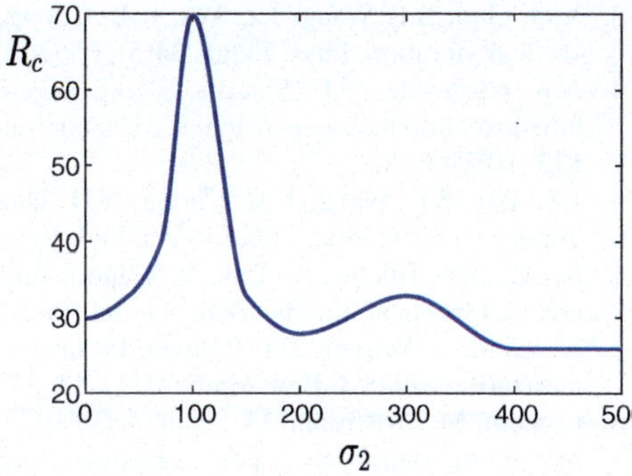

Fig. 8 Critical Rayleigh number as function of the frequency σ_2 for $\lambda_1 = 5$, $\sigma_1 = 300$ and $\lambda_2 = 10$

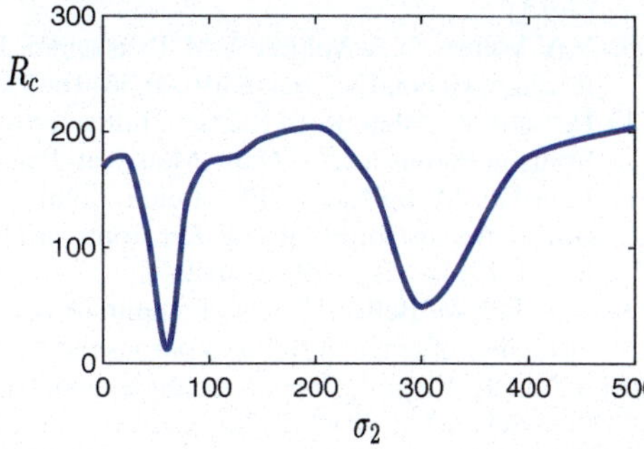

Fig. 9 Critical Rayleigh number as function of the frequency σ_2 for $\lambda_1 = 10$, $\sigma_1 = 300$ and $\lambda_2 = 10$

of the gravitational modulation on the reaction front. The linear stability analysis of the problem is carried out. In order to obtain the interface problem, the reduced differential equations along with the jump equations are solved numerically via finite difference scheme approximations. It was shown that the modulated frequency of the periodic gravitational excitation plays an essential role in stabilization the reaction front. Indeed, the results demonstrated that increasing slightly the amplitude λ_2 of the modulated frequency ν_m form zero, stabilizes the reaction front. On the other hand, it was demonstrated that for a fixed value of λ_2, the frequency σ_2 may have a stabilizing or a destabilizing effect. For a small value of the amplitude λ_1 of the gravitational modulation, σ_2 may stabilize the reaction front in a certain interval of σ_2. For large values of λ_1, the frequency σ_2 produces a destabilizing effect on the reaction front in certain alternate regions of σ_2.

References

1. N. Hosseini, M.D. Griffith, J.S. Leontini, Flow-induced vibrations in long rows of cylinders and their links to convective instabilities. Int. J. Heat Fluid Flow **94**, 108922 (2022)
2. I. Simanovskii, A. Nepomnyashchy, A. Viviani, P. Queeckers, A. Parente, Instabilities of a droplet on a liquid substrate heated from below under the action of vibration. Int. J. Non-Linear Mech. **152**, 104387 (2023)

3. X.Q. Guo, B.F. Wang, J.Z. Wu, K.L. Chong, Q. Zhou, Turbulent vertical convection under vertical vibration. Phys. Fluids **34**(5), 055106 (2022)
4. A.K. Kushwaha, Y.D. Sharma, Significance of vertical vibration on the stability of thermo-bioconvection in a suspension of oxytactic microorganisms. Int. Commun. Heat Mass Transf. **133**, 105943 (2022)
5. J.Z. Wu, B.F. Wang, K.L. Chong, Y.H. Dong, C. Sun, Q. Zhou, Vibration-induced 'anti-gravity'tames thermal turbulence at high Rayleigh numbers. J. Fluid Mech. **951**, A13 (2022)
6. K. Allali, F. Bikany, A. Taik, V. Volpert, Influence of vibrations on convective instability of reaction fronts in liquids. Math. Model Nat. Phenom. **5**(7), 35–41 (2010)
7. K. Allali, V. Volpert, J.A. Pojman, Influence of vibrations on convective instability of polymerization fronts. J. Eng. Math. **41**(1), 13–31 (2001)
8. L. Salhi, M. El-Amrani, M. Seaid, A Galerkin-characteristic unified finite element method for moving thermal fronts in porous media. J. Comput. Appl. Math. **404**, 113159 (2022)
9. M. Wangen, Stability of reaction-fronts in porous media. Appl. Math. Model. **37**(7), 4860–4873 (2013)
10. V.A. Volpert, V.A. Volpert, V.M. Ilyashenko, J.A. Pojman, Frontal polymerization in a porous medium. Chem. Eng. Sci. **53**(9), 1655–1665 (1998)
11. H. Aatif, K. Allali, K. El Karouni, Influence of Vibrations on convective instability of reaction fronts in Porous media. Math. Mod. Nat. Phen. **5**(5), 123–37 (2010)
12. K. Allali, M. Belhaq, K. El Karouni, Influence of quasi-periodic gravitational modulation on convective instability of reaction fronts in porous media. Commun. Nonlinear Sci. Numer. Simul. **17**(4), 1588–1596 (2012)
13. K. Allali, M. Belhaq, Effect of amplitude modulation of gravitational vibration on convective instability of reaction fronts in porous media. Appl. Math. Sci. **7**(21), 1007–1022 (2013)
14. Y. Yu, N. Wang, Z. Chen, Amplitude modulation control method for bursting dynamics under time-delayed feedback. J. Vib. Control. **28**(23–24), 3550–3558 (2022)
15. K. Conrady, K. Bolin, A. Sjöblom, A. Rutgersson, Amplitude modulation of wind turbine sound in cold climates. Appl. Acoust. **158**, 107024 (2020)
16. X. Guo, L. Zhang, J. Liu, L. Chen, L. Fan, T. Liu, Residual amplitude modulation suppression to 6x10-7 in frequency modulation, in *Twelfth International Conference on Information Optics and Photonics*, vol. 12057. (SPIE, 2021), pp. 9–14
17. B. Ghosh, S. Chandra, S.N. Paul, Amplitude modulation of electron plasma waves in a quantum plasma. Phys. Plasmas **18**(1), 012106 (2011)
18. L. Yu, R. Wang, L. Lu, Y. Zhu, C. Wu, B. Zhang, P. Wang, Stable radio frequency dissemination by simple hybrid frequency modulation scheme. Opt. Lett. **39**(18), 5255–5258 (2014)
19. J. Vovnoboy, R. Levinger, N. Mazor, D. Elad, A dual-loop synthesizer with fast frequency modulation ability for 77/79 GHz FMCW automotive radar applications. IEEE J. Solid-State Circuits **53**(5), 1328–1337 (2018)
20. X.Q. Liu, H.L. Zhou, W.H. Cao, H. Li, Y.H. Li, A.D. Ji, Gauss linear frequency modulation wavelet transforms and its application to seismic phases identification. Acta Seism. Sin. **15**, 636–645 (2002)
21. M.J. Henry, J. Obleser, Frequency modulation entrains slow neural oscillations and optimizes human listening behavior. Proc. Natl. Acad. Sci. **109**(49), 20095–20100 (2012)
22. Y.B. Zeldovich, G.I. Barenblatt, V.B. Librovich, G.M. Makhviladze, *The Mathematical Theory of Combustion and Explosions* (Consultants Bureau Plenum, New York, 1985)
23. Y.B. Zeldovich, D.A. Frank-Kamenetsky, The theory of thermal propagation of flames. Zh Fiz Khim **12**, 100–5 (1938)

Widening of Anti-resonances by Increasing the Amplitude of Parametric Excitation

Zacharias Kraus and Peter Hagedorn

Abstract Anti-resonance is a feature of parametrically excited systems, which introduces a stabilizing effect to the system dynamics. To be applicable in real systems, it must be robust against parameter variation. One way to increase robustness is to widen the frequency range in which the anti-resonance can be utilized for vibration mitigation. In this work, a two-dimensional system is investigated. Lyapunov characteristic exponents (LCEs) serve as an indicator for the damping behavior of the system. An analytical approximation of the LCEs in the vicinity of the anti-resonance is derived using the method of normal forms. It is found that an increase in the amplitude of the parametric excitation leads to a widening of the anti-resonance with respect to the excitation frequency.

1 Introduction

Anti-resonance is a stabilizing phenomenon, observable in multidimensional linear systems. It can only occur in systems that may be represented by equations of motion with time-dependent coefficients, also called *parametrically excited* systems. Additionally, a coupling between degrees of freedom (DoFs) is needed. In contrast to fundamental resonances, which appear at specific rational fractions of the eigenfrequencies, anti-resonance may emerge at the sum or difference of the coupled DoFs' eigenfrequencies [9].

Parametric excitation has been known since the formulation of Mathieu and Hills equations in the nineteenth century. Early studies on parametric resonance involving two eigenfrequencies, conducted in the 1940s by Cesari [2] and Mettler [13], focused on the unstabilizing aspect only. Research on stabilization through parametric excitation can be traced back to the works of Tondl [18] and Schmieg [17] in the late 1970s, although their results were not yet interpreted as anti-resonance. It was as late

Z. Kraus (✉) · P. Hagedorn
Technical University of Darmstadt, Dynamics and Vibrations Group, FNB, Dolivostraße 15, 64293 Darmstadt, Germany
e-mail: kraus@dyn.tu-darmstadt.de

M. Belhaq (ed.), *Advances in Nonlinear Dynamics and Control of Mechanical and Physical Systems*, Springer Proceedings in Physics 301,
https://doi.org/10.1007/978-981-99-7958-5_5

as 1998 that anti-resonance was formulated explicitly in [19]. In the following years, anti-resonance was studied systematically, with notable work by Tondl, Dohnal, and Karev. Especially stability boundary curves and, for the description of stabilization, Lyapunov characteristic exponents (LCEs) are extensively studied with numerical and semi-analytical methods for different systems (see, e.g., [1, 8, 14, 21]). An overview of the literature is given in [20].

The stabilizing properties of anti-resonance were successfully shown in experiments as described in [5, 7, 10, 16]. One field of application that is especially interesting is rotordynamics, where a periodic variation of parameters is commonly seen due to rotary motion. In recent works, parametric excitation is deliberately introduced to rotary systems to help to mitigate vibrations via anti-resonance [7, 12, 15]. Since the excitation frequency where anti-resonance occurs depends on several system parameters, which may be hard to identify in real systems, robustness against parameter variation is a key aspect when realizing stabilization. For systems with damping coefficients of different signs, stability borders have been derived in the vicinity of an anti-resonance in [4]. However, in the context of vibration mitigation for already stable solutions, no studies are known to the authors.

In this contribution, there will be a particular focus on robustness regarding variations in the excitation frequency. The LCEs will be used as an indicator for stability, a common approach when assessing combination resonances [4, 8, 9, 12, 21]. If the largest LCE is greater than $0\,\mathrm{s}^{-1}$, amplitudes become unbounded and the solution unstable. Solutions are stable if all LCEs are less than $0\,\mathrm{s}^{-1}$. As the LCEs are linked to the decay of amplitudes over time, a reduction of LCEs implies increased stability, which is desirable in the context of vibration mitigation. The LCEs of a two-dimensional, parametrically excited system will be analyzed. For a thorough insight into each parameter's influence, a (semi-)analytical approach is indispensable, as numerical methods only provide information on one system configuration at a time. One possibility is the method of normal forms, which application on parametrically excited systems is described in detail in [3, 9]. The method of normal forms will be employed in Sect. 2 to derive a semi-equivalent, autonomous system. From the eigenvalues of this system, an approximate solution of the LCEs near the anti-resonance will be derived in Sect. 3. For verification of the results, the approximate solution will be simplified for two limit cases in Sect. 4 and compared to numerical results in Sect. 5.

2 Normal Form

The equations of motion of a two-dimensional system with coupling through parametric excitation can be written as

$$\ddot{\mathbf{u}} + \begin{bmatrix} \delta_1 & 0 \\ 0 & \delta_2 \end{bmatrix} \dot{\mathbf{u}} + \left(\begin{bmatrix} \omega_1^2 & 0 \\ 0 & \omega_2^2 \end{bmatrix} + \begin{bmatrix} 0 & \varepsilon_{12} \\ \varepsilon_{21} & 0 \end{bmatrix} \cos(\Omega t) \right) \mathbf{u} = \mathbf{0}. \tag{1}$$

The vector $\mathbf{u} = (u_1, u_2)^T$ contains the two degrees of freedom that are damped with the damping factors δ_i and have the eigenfrequencies ω_i. It should be noted that the system is normalized with the mass matrix, leaving the coefficients of the second derivative to be 1. Parametric excitation is achieved by adding periodic terms with frequency Ω and amplitudes $\varepsilon_{12}, \varepsilon_{21}$ to the stiffness. For an illustration of the LCEs' behavior, Floquet analysis is applied to the system in (1) using the values

$$\omega_1 = 1.3\,\text{Hz}, \quad \omega_2 = 3.0\,\text{Hz}, \quad \delta_1 = 0.15\,\frac{1}{\text{s}}, \quad \delta_2 = 0.05\,\frac{1}{\text{s}}. \tag{2}$$

Figure 1 shows the LCEs for different values of $\varepsilon^2 = \varepsilon_{12}\varepsilon_{21}$. Far away from any resonances, the LCEs approach constant values determined by the damping of the DoF: $\lambda_i = -\frac{\delta_i}{2}$. Parametric excitation may lead to three kinds of resonances, all visible in Fig. 1. A fundamental resonance is visible in the upper LCE at $\Omega = \omega_2$. Due to the parametric excitation matrix being populated only on the off-diagonal, fundamental resonances are only evoked by higher-order terms. Therefore, a relatively large amplitude ε^2 of parametric excitation terms is required for the fundamental resonance to become visible. It should also be noted that a large amplitude ε^2 causes the resonance frequency to shift. The second type of resonance, a combination resonance, can be found at the difference frequency $\Delta_{21} = \omega_2 - \omega_2$. It is caused by coupling the two DoF in (1) through the off-diagonal parametric excitation terms. The combination resonance leads to a destabilization of the solutions, which can even cause instability, as can be seen for $\varepsilon^2 = 4\,\text{s}^{-4}$ in Fig. 1, where the largest LCE surpasses $0\,\text{s}^{-1}$. The third type of resonance, which is subject of this contribution, is anti-resonance, located at the sum frequency $\Omega = \Sigma_{21}$ in Fig. 1. At the anti-resonance, the largest LCE decreases, corresponding with a stabilization of the solutions. Maximal stabilization is obtained at $\Omega = \Sigma_{21}$ if the amplitude ε is greater than a critical amplitude $\varepsilon_{\text{crit.}}$, as described in [9]. In this case, both LCEs meet at $-\frac{\delta_1+\delta_2}{4}$. In the vicinity of the sum frequency, parametric excitation still exerts influence on the LCEs, as their values gradually approach $-\frac{\delta_1+\delta_2}{4}$. Figure 1 suggests that the frequency range where anti-resonance affects the LCEs broadens with increasing amplitude. This would imply an increase of robustness against variations in excitation frequency and is to be investigated analytically in this contribution.

To derive analytical expressions for the LCEs, (1) must first be transformed into an autonomous first-order system. According to the method of normal forms, as described in [9], the identity $\cos(\Omega t) = \frac{1}{2}\left(e^{i\Omega} + e^{-i\Omega}\right)$ is used and the state space is extended by two variables $q_z, q_{\bar{z}} = q_5, q_6$ and one dummy variable $q_d = q_7$. Thereby it is transformed into an equivalent, nonlinear first-order system with $n_1 = 7$ degrees of freedom:

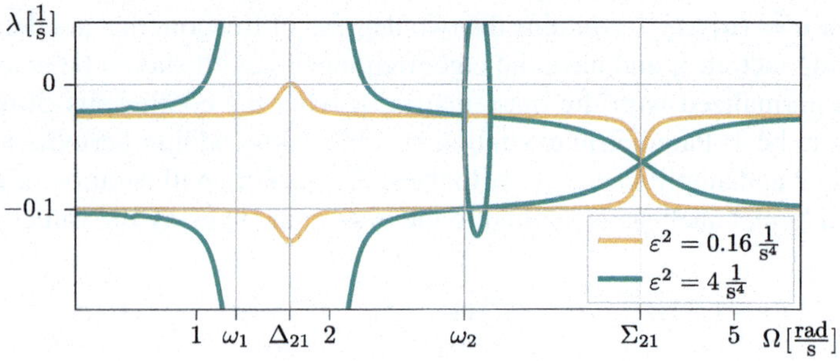

Fig. 1 LCEs calculated numerically with Floquet theory for $\varepsilon^2 = 0.16\,\mathrm{s}^{-4}$ (beige) and $4\,\mathrm{s}^{-4}$ (green)

$$\dot{\mathbf{q}} = \begin{bmatrix} \dot{q}_1 \\ \dot{q}_2 \\ \dot{q}_3 \\ \dot{q}_4 \\ \dot{q}_5 \\ \dot{q}_6 \\ \dot{q}_7 \end{bmatrix} = \begin{bmatrix} q_3 \\ q_4 \\ -\omega_1^2 q_1 - \frac{1}{2}\varepsilon_{12}(q_5 + q_6)q_2 - \delta_1 q_3 q_7 \\ -\omega_2^2 q_2 - \frac{1}{2}\varepsilon_{21}(q_5 + q_6)q_1 - \delta_2 q_4 q_7 \\ i\Omega q_5 \\ -i\Omega q_6 \\ 0 \end{bmatrix} = \mathbf{A}\mathbf{q} + \mathbf{f}(\mathbf{q}) . \tag{3}$$

Considering just the linear part $\mathbf{Aq}$ of (3), eigenvectors are derived, which are then used to transform the system into modal coordinates:

$$\dot{\mathbf{x}} = \mathbf{\Lambda x} + \mathbf{F}_2\mathbf{x}^2 + \mathbf{F}_3\mathbf{x}^3 + \mathcal{O}(\mathbf{x}^4) , \tag{4}$$

where the linear part is described by a diagonal matrix, consisting of eigenvalues λ_i, $i \in \{1, 2, \dots, n_1\}$:

$$\mathbf{\Lambda} = \mathrm{diag}(\lambda_i) = \mathrm{diag}(i\omega_1, -i\omega_1, i\omega_2, -i\omega_2, i\Omega, -i\Omega, 0) . \tag{5}$$

Modal transformation of the nonlinear terms $\mathbf{f}(\mathbf{q})$ leads to an equivalent series of nonlinear terms $\mathbf{F}_p\mathbf{x}^p$ in modal coordinates. The vectors $\mathbf{x}^p = \{x_j^p\}_{j\in\{1,2,\dots,n_p\}}$ contain all $n_p = \binom{n_1+p-1}{p}$ monomes with exponents m_i of p-th order

$$\mathbf{x}^p = \{\prod_{i=1}^{n_1} x^{m_i} \mid \sum_{i=1}^{n_1} m_i = p,\ m_i \in \mathbb{Z}\}. \tag{6}$$

The matrices $\mathbf{F}_p$ have the dimensions $n_1 \times n_p$ and contain all coefficients of $\mathbf{x}^p$.

The normal form is developed around a specific excitation frequency Ω. To obtain an approximation near the anti-resonance, $\Omega = \omega_1 + \omega_2$ is chosen. The normal form will only contain resonant terms, so all terms that are non-resonant for the chosen excitation frequency are eliminated in a near-identity transformation. Matrices $\mathbf{F}_p$ are split into coefficients $\mathbf{H}_p$ of resonant monomes and coefficients $\mathbf{G}_p$ of non-resonant monomes. Resonant monomes are those which lead to time-dependent terms in the normal form if the corresponding entry in $\mathbf{F}_p$ is not zero. Therefore, they can be found by applying the resonance condition

$$\{x_j^p \mid \sum_{i=1}^{n_1} m_i \lambda_i - \lambda_j \overset{!}{=} 0,\ m_z - m_{\overline{z}} \neq 0\} \tag{7}$$

to all monomes in $\mathbf{x}^p$. The condition $m_z - m_{\overline{z}} \neq 0$ is to guarantee a resonant monome contains at least one of the extended state variables representing time dependency, but not both to the same degree, as this would cause them to cancel in the normal form. Finally, the normal form is constructed according to

$$\dot{\mathbf{y}} = \mathbf{\Lambda y} + \mathbf{H}_2 \mathbf{y}^2 + \mathcal{O}(\mathbf{y}^3)\,. \tag{8}$$

For the system in (3) with resubstituting $y_{5,6} = e^{\pm i\Omega}$, $y_7 = \delta$, this yields

$$\begin{bmatrix} \dot{y}_1(t) \\ \dot{y}_2(t) \\ \dot{y}_3(t) \\ \dot{y}_4(t) \end{bmatrix} = \begin{bmatrix} i\omega_1 - \frac{\delta_1}{2} & 0 & 0 & -\frac{ie^{it\Omega}\varepsilon_{12}}{4\omega_2} \\ 0 & -\frac{\delta_1}{2} - i\omega_1 & \frac{ie^{-it\Omega}\varepsilon_{12}}{4\omega_2} & 0 \\ 0 & -\frac{ie^{it\Omega}\varepsilon_{21}}{4\omega_1} & i\omega_2 - \frac{\delta_2}{2} & 0 \\ \frac{ie^{-it\Omega}\varepsilon_{21}}{4\omega_1} & 0 & 0 & -\frac{\delta_2}{2} - i\omega_2 \end{bmatrix} \begin{bmatrix} y_1(t) \\ y_2(t) \\ y_3(t) \\ y_4(t) \end{bmatrix}, \tag{9}$$

which is separable into two uncoupled systems in $y_1,\ y_4$ and $y_2,\ y_3$. The eigenvalues of these two systems have complex conjugated imaginary parts and equal real parts. Since only the LCEs, identical to the real parts of the eigenvalues [11], are of interest, analysis of one of the two systems is sufficient. For the following calculations the system $y_1,\ y_4$ will be chosen. Using the transformation

$$y_1 = e^{\frac{1}{2}i\Omega t}u_1, \quad y_4 = e^{-\frac{1}{2}i\Omega t}u_2 \tag{10}$$

an autonomous system is derived:

$$\begin{bmatrix} \dot{u}_1 \\ \dot{u}_2 \end{bmatrix} = \begin{bmatrix} \frac{1}{2}(-i(\Omega - 2\omega_1) - \delta_1) & -\frac{i\varepsilon_{12}}{4\omega_2} \\ \frac{i\varepsilon_{21}}{4\omega_1} & \frac{1}{2}(i(\Omega - 2\omega_2) - \delta_2) \end{bmatrix} \begin{bmatrix} u_1 \\ u_2 \end{bmatrix} \tag{11}$$

The eigenvalues of this system, expressed with the detuning parameter $\sigma = \Omega - \omega_1 - \omega_2$, are

$$\frac{-\delta_1 - \delta_2 + 2i(\omega_1 - \omega_2)}{4} \pm \frac{1}{4}\sqrt{\frac{\varepsilon_{12}\varepsilon_{21}}{\omega_1\omega_2} + [\delta_1 - \delta_2 + 2i\sigma]^2}. \tag{12}$$

3 Approximate Lyapunov Characteristic Exponents Near Anti-resonance

Lyapunov characteristic exponents can be derived from (12) by considering only the real part of the eigenvalues. This yields

$$\lambda_{1,2} = \frac{-\delta_1 - \delta_2}{4} \pm \frac{1}{4} \operatorname{Re}\left(\sqrt{a + ib}\right) \tag{13}$$

$$a = \frac{\varepsilon_{12}\varepsilon_{21}}{\omega_1\omega_2} + (\delta_1 - \delta_2)^2 - 4\sigma^2 \tag{14}$$

$$b = 4\,(\delta_1 - \delta_2)\,\sigma \tag{15}$$

The real part of a complex square root can be written as

$$\operatorname{Re}\left(\sqrt{a + ib}\right) = \sqrt[4]{a^2 + b^2}\,\frac{a + \sqrt{a^2 + b^2}}{\sqrt{2\left(a^2 + b^2\right) + 2a\sqrt{a^2 + b^2}}}. \tag{16}$$

An expression for the LCEs may be derived by inserting (16) into (13). However, this would lead to a very complex expression, which does not allow for an intuitive understanding of the phenomena. Therefore, (16) will be simplified using the following three assumptions. First, analysis is limited to the vicinity of the sum resonance frequency:

$$\Omega \approx \omega_1 + \omega_2 \leftrightarrow \sigma \approx 0 \tag{17}$$

Second, it is assumed that

$$\begin{pmatrix} 0 & \varepsilon_{12} \\ \varepsilon_{21} & 0 \end{pmatrix} \text{ neg. definite,} \tag{18}$$

which implies $\varepsilon_{12}\varepsilon_{21} < 0\,\mathrm{s}^{-4}$. This assures that the anti-resonance lies at the sum resonance frequency. A violation of this assumption, i.e., $\varepsilon_{12}\varepsilon_{21} > 0\,\mathrm{s}^{-4}$, would cause anti-resonance and combination resonance to swap frequencies, with the anti-resonance appearing at the difference frequency and combination resonance at the sum frequency [6]. By assuming the parametric excitation matrix to be positive definite, the results of this contribution can be applied to anti-resonance at the difference frequency. The third assumption,

$$-\frac{\varepsilon_{21}\varepsilon_{12}}{\omega_1\omega_2} > (\delta_1 - \delta_2)^2\,, \tag{19}$$

states that the parametric excitation is large compared to the difference of damping terms. This ensures that $\varepsilon > \varepsilon_{\text{crit}}$ (see [4, 9]), meaning the LCEs at the sum frequency have already reached the point of the strongest stabilization. If these assumptions are applied to a and b in (13), it can be easily seen that

$$a^2 \gg b^2, \tag{20a}$$

$$a < 0\,\mathrm{s}^{-2}. \tag{20b}$$

Now the assumptions in (20) will be used to simplify Equation (16). With (20a) and $\sqrt{a^2} = -a$ due to (20b), the fourth root can be simplified to

$$\sqrt[4]{a^2 + b^2} \approx \sqrt[4]{a^2} = \sqrt{-a}. \tag{21}$$

Because $\frac{b^2}{a^2}$ is small as per (20a), the Taylor series $\sqrt{1 + \frac{b^2}{a^2}} = 1 + \frac{b^2}{2a^2} + \mathcal{O}\left(\left(\frac{b^2}{a^2}\right)^3\right)$ can be used to approximate

$$a + \sqrt{a^2 + b^2} = a\left[1 - \sqrt{1 + \frac{b^2}{a^2}}\right] \approx a\left[1 - \left(1 + \frac{b^2}{2a^2}\right)\right] = -\frac{b^2}{2a}. \tag{22}$$

Finally, (20a) is made use of again to allow an extension by the very small term $\frac{b^4}{4}$, which is used to derive a binomial that cancels with the square root

$$2a\sqrt{a^2 + b^2} = -2\sqrt{a^4 + b^2a^2} \approx -2\sqrt{a^4 + b^2a^2 + \left(\frac{b^2}{2}\right)^2} = -2\left(a^2 + b^2\right). \tag{23}$$

By applying (21)–(23), (16) can be simplified to

$$\mathrm{Re}\left(\sqrt{a + ib}\right) \approx \sqrt{-a}\,\frac{-\dfrac{b^2}{2a}}{\sqrt{2a^2 + 2b^2 - 2a^2 - b^2}} = \sqrt{-a}\,\frac{-b^2}{2a\sqrt{b^2}} = \sqrt{-\frac{b^2}{4a}}. \tag{24}$$

Inserting into (13), an approximate solution for the LCEs around an anti-resonance is obtained:

$$\lambda_{1,2} \approx -\frac{\delta_1 + \delta_2}{4} \pm \frac{1}{4}\sqrt{\frac{4\left(\delta_1 - \delta_2\right)^2 \sigma^2}{4\sigma^2 - \frac{\varepsilon_{12}\varepsilon_{21}}{\omega_1\omega_2} - \left(\delta_1 - \delta_2\right)^2}} \tag{25}$$

4 Analysis of Limit Cases and Stability Borders

To illustrate the change of $\lambda_{1,2}$ for varying Ω, (25) will be analyzed for two limit cases. The first limit case is Ω far away from the resonance frequency $\Omega \approx \omega_1 + \omega_2$, the second Ω in close vicinity to it. Additionally, (25) is applied to unstable systems to find the stability borders regarding σ, between which the anti-resonance causes the solutions to become stable.

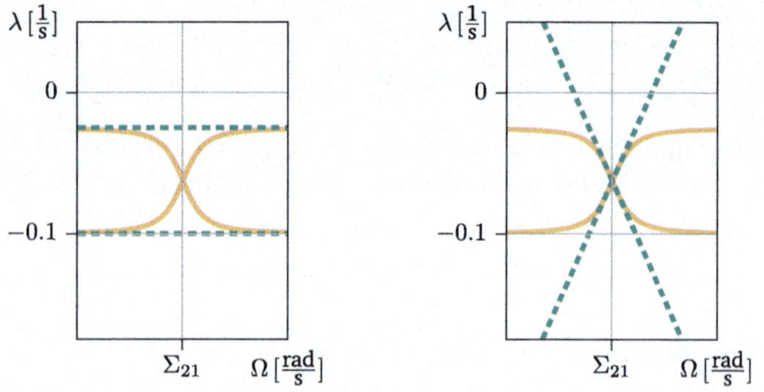

(a) Limit case Ω far away from Σ_{21} (b) Limit case $\Omega \approx \Sigma_{21}$

Fig. 2 LCEs near anti-resonance calculated with Floquet theory (beige) and approximate solution in limit cases (green, dashed)

For Ω far away from $\omega_1 + \omega_2$, the expressions $\frac{\varepsilon_{12}\varepsilon_{21}}{\omega_1\omega_2}$ and $(\delta_1 - \delta_2)^2$ are negligible in the denominator. Then (25) simplifies to

$$\lambda_{1,2} \approx -\frac{\delta_1 + \delta_2}{4} \pm \frac{1}{4}\sqrt{\frac{(\delta_1 - \delta_2)^2}{1}} \tag{26}$$

$$\lambda_1 \approx -\frac{\delta_1}{2}, \quad \lambda_2 \approx -\frac{\delta_2}{2}. \tag{27}$$

The LCEs far away from resonances are constant and equal to half the negative damping of the corresponding degree of freedom. This is in agreement with the results found in [9, 10]. Figure 2a compares the LCEs far away from the resonance with the LCEs computed with numerically Floquet theory. It can be seen that the numerical solution approaches the limits given in (27) asymptotically, showing that (27) gives an excellent approximation of the LCEs for Ω far away from $\omega_1 + \omega_2$.

In the case of $\Omega \approx \omega_1 + \omega_2$, the parameter σ^2 becomes negligible and the tangents of (25) at $\Omega = \omega_1 + \omega_2$ approach

$$\lambda_{1,2} \approx -\frac{\delta_1 + \delta_2}{4} \pm \frac{1}{4}\sqrt{\frac{4\,(\delta_1 - \delta_2)^2}{-\frac{\varepsilon_{12}\varepsilon_{21}}{\omega_1\omega_2}}}\,|\sigma|\,. \tag{28}$$

Equation (28) are lines in respect to Ω with a slope proportional to $1/\sqrt{\varepsilon_{12}\varepsilon_{21}}$, as visible in Fig. 2b. For increasing ε^2, the tangent's slopes decrease, causing a "widening" of the anti-resonance. This effect can be used to increase the robustness of the anti-resonance against variations in the excitation frequency Ω as discussed earlier.

The results can be applied to unstable solutions as well, i.e., if one of the damping constants $\delta_i < 0\,\mathrm{s}^{-1}$. It is known that anti-resonance may stabilize the solutions in this case. To find the interval in Ω where the solution is stable, Max $(\lambda_{1,2})$ is set to $0\,\mathrm{s}^{-1}$ and (25) is solved for σ. This leads to

$$\sigma_{1,2} = \pm\frac{\delta_1+\delta_2}{2}\sqrt{\frac{(\delta_1-\delta_2)^2+\frac{1}{4}\frac{\varepsilon_{12}\varepsilon_{21}}{\omega_1\omega_2}}{\delta_1\delta_2}}. \tag{29}$$

The stability borders were also derived in [4] by applying averaging methods. The expressions differ slightly regarding the squared damping term. However, the resulting difference in $\sigma_{1,2}$ is very small for the assumed parameter ranges. The discrepancy between the results can be explained by different simplifications being made in both derivations.

5 Numerical Verification

To verify the approximation in (25), it is compared against LCEs computed by applying Floquet theory to (1). The values in (2) will be used for the calculations. Figure 3 shows the LCEs calculated using the approximate solution. The approximation is very close to the solution found with Floquet theory in close vicinity of the resonance frequency as well as far away from it. A slight deviation can be noticed only in the transition zone between the two near-linear sections. This deviation diminishes as ε is increased. Furthermore, as only the sum resonance frequency Σ_{21} was considered while deriving the normal form, the resonance at the difference frequency Δ_{21} and fundamental resonances are not included in the normal form nor in the approximate solution. The results shown in Fig. 3 support the conclusion that (25) provides an approximation of the LCEs in the vicinity of the anti-resonance close enough to include all relevant effects.

6 Conclusion

An approximate formulation of the LCEs in the vicinity of an anti-resonance was derived and verified in limit cases and numerically. With the approximate solution, it can be easily seen that an increase in the amplitude of parametric excitation leads to a widening of the anti-resonance. This effect can be utilized to significantly increase robustness against parameter variation when realizing anti-resonance in real-world systems.

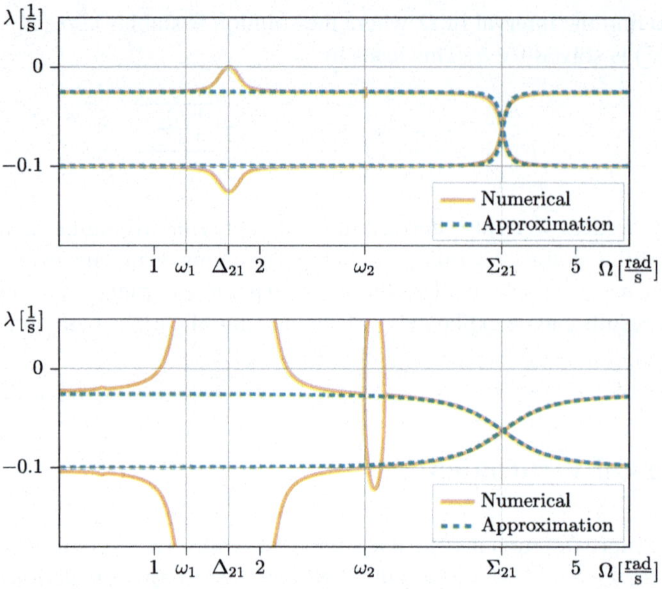

Fig. 3 LCEs calculated with numerical (Floquet) and approximate solution for $\varepsilon^2 = 0.16\,\mathrm{s}^{-4}$ (top) and $4\,\mathrm{s}^{-4}$ (bottom)

References

1. V.V. Bolotin, *The Dynamic Stability of Elastic Systems*. Holden-Day (1964)
2. L. Cesari, *Sulla stabilità delle soluzioni dei sistemi di equazioni differenziali lineari a coefficienti periodici (On the stability of systems of linear differential equations with periodic coeficients)* (Consiglio nazionale delle ricerche, 1940)
3. L. DeBroeck, "Untersuchung des Stabilitätsverhaltens erweiterter parametererregter Systeme mit der Normalformtheorie". Bachelor Thesis (AG Dynamik und Schwigungen, TU Darmstadt, Darmstadt, 2017)
4. F. Dohnal, "A contribution to the mitigation of transient vibrations: parametric anti-resonance; theory, experiment and interpretation" (Habilitation Thesis, 2012)
5. F. Dohnal, Experimental studies on damping by parametric excitation using electromagnets. Proc. Inst. Mech. Eng., Part C: J. Mech. Eng. Sci. **226**(8), 2015–2027 (2012)
6. F. Dohnal, General parametric stiffness excitation—anti-resonance frequency and symmetry. Acta Mechanica **196**(1), 15– 31 (2008)
7. F. Dohnal, A. Chasalevris, Inducing modal interaction during run-up of a magnetically supported rotor, in *13th international conference in dynamical systems theory and applications DSTA* (2015)
8. N. Eicher, *Einführung in die Berechnung parametererregter Schwingungen* (TUB-Dokumentation / Weiterbildung, TU Berlin, 1981)
9. A. Karev, Asynchronous Parametric Excitation in Dynamical Systems. Ph.D. thesis (TU Darmstadt, Darmstadt, 2020)
10. A. Karev, P. Hagedorn, Asynchronous parametric excitation: validation of theoretical results by electronic circuit simulation. Nonlinear Dyn. **102**(1), 555–565 (2020)
11. A. Karev, P. Hagedorn, Global stability effects of parametric excitation. J. Sound Vib. **448**, 34–52 (2019). (May)

12. Z. Kraus et al., Enhancing vibration mitigation in a Jeffcott rotor with active magnetic bearings through parametric excitation. Nonlinear Dyn. **109**(2), 393–400 (2022)
13. E. Mettler, Allgemeine Theorie der Stabilität erzwungener Schwingungen elastischer Körper". Ingenieur-Archiv **17**(6), 418–449 (1949)
14. A.H. Nayfeh, D.T. Mook, *Nonlinear oscillations*. Physics textbook. (Wiley, Weinheim 2004), p. 704
15. B. Pfau, Ein verstellbares Zweiflächengleitlager zur Schwingungsminderung flexibler Rotoren. Ph.D. thesis (TU Darmstadt, Darmstad, 2018)
16. E. Schmidt et al., Design of an electromagnetic actuator for parametric stiffness excitation. COMPEL—Int. J. Comput. Math. Electr. Electron. Eng. 26(3), 800–813 (2007). Ed. by O. Biro
17. H. Schmieg, Kombinationsresonanz bei Systemen mit allgemeiner harmonischer Erregermatrix (Combination resonance in systems with general harmonic excitation matrix). Ph.D. thesis. (University of Karlsruhe, Karlsruhe, 1976)
18. A. Tondl, On the interaction between self-excited and parametric vibrations, in *Monographs and Memoranda 25. Special Printing of the National Research Institute for Machine Design*, vol. 60 (1978), pp. 61–61
19. A. Tondl, To the problem of quenching self-excited vibrations. Acta technica ČSAV **43**(1), 109–116 (1998)
20. A. Tondl, To the problem of self-excited vibration suppression. Engineering Mechanics (2008)
21. F. Verhulst, Parametric and autoparametric resonance. Acta Applicandae Mathematica **70**, 231–264 (2002)

A Study of Mathematical Model of AFM with Casimir Force by Means of Averaging Method

G. A. Kurina, J. M. Balthazar, M. A. Ribeiro, and A. M. Tusset

Abstract The paper deals with a mathematical model of atomic force microscopy with a microcantilever containing an analysis tip. This model takes into account the Casimir force. The averaging method is applied to determine an asymptotic solution of the equation for the microcantilever tip displacement. Explicit expressions of problems for finding asymptotic solution of the fourth order are derived. The results of numerical experiment are given.

1 Introduction

Atomic force microscopy (AFM) is a powerful method of analysis in modern materials science and technology. Because it allows surface analysis at the atomic level of synthetic or biological structures [2, 4, 12, 14–16].

An essential component of the atomic force microscope is a microcantilever with a tip attached to its free end that touches the surface of the material that undergoes vibrations during reading, these vibrations are collected by a laser that strikes the microcantilever and is reflected to a photodiode that collects surface imperfection data and so, generating a three-dimensional image of the surface of the material. The interaction forces between the atoms existing between the tip and the surface to be measured, cause the deflection of the microcantilever [1, 4, 11, 13, 14].

G. A. Kurina (✉)
Voronezh State University, Universitetskaya pl., 1, 394018 Voronezh, Russia
e-mail: kurina@math.vsu.ru

Federal Research Center "Computer Science and Control" of Russian Academy of Sciences, Vavilova ul., 44/2, 119333 Moscow, Russia

J. M. Balthazar · M. A. Ribeiro · A. M. Tusset
Federal University of Technology - Paraná, R. Doutor Washington Subtil Chueire 330-Jardim Carvalho, Ponta Grossa - PR 84017-220, Brazil

J. M. Balthazar
São Paulo State University, av. Eng. Luís Edmundo Carrijo Coube, 14-01 - Vargem Limpa, Bauru - SP 17033-360, Brazil

M. Belhaq (ed.), *Advances in Nonlinear Dynamics and Control of Mechanical and Physical Systems*, Springer Proceedings in Physics 301,
https://doi.org/10.1007/978-981-99-7958-5_6

The Casimir effect is discussed in many publications (see, for instance, [3, 9, 10]). The authors of [11] analyzed not only the influence of the Casimir force on nonlinear dynamics of AFM but also demonstrated the influence on fractional dynamics, and presented control techniques that can suppress the chaotic behavior.

In the present paper, a dimensionless nonlinear mathematical model of an atomic force microscopy with the Casimir force from [11] is studied. That is, the microcantilever tip displacement with the influence of the Casimir force is considered. For the study of this displacement, it is proposed to use the averaging method (see, for instance, [8]) which is very effective in research of oscillatory processes.

The first-order approximation of the averaging method is often close to an exact solution. Higher order approximations are sometimes required, which have been applied, for example, in a nonlinear theory of a cyclotron resonance maser with a Fabry–Perot resonator [5, 7], a theory of micro-electromechanical systems [6], and a theory of atomic force microscopy without the Casimir force [13].

The manuscript is organized as follows: in Sect. 2, the considered mathematical model is reduced to the system in a standard form and the algorithm for finding solution approximations by averaging method is briefly described. Section 3 is devoted to obtaining relations for the first, second, third, and fourth orders of approximations of the averaging method for the studied mathematical model of AFM with the Casimir force. Section 4 is dedicated to the results of numerical experiments and, finally, Sect. 5 is the conclusion of the work.

2 Mathematical Modeling and Algorithm of Averaging Method

Figure 1 illustrates a simplified atomic force microscopy (AFM) scheme. In this scheme, the simplest devices of the AFM composed of the microcantilever, the analysis tip, the laser, and the photodetector can be seen. The mathematical modeling is adapted from [11].

We consider the equation of displacement for analysis tip of AFM in the form

$$m\frac{d^2y}{dt^2} = -k_l y - k_{nl} y^3 - c\frac{dy}{dt} - \frac{\mu_e \eta^3 l}{(z_0+y)^3}\frac{dy}{dt} + \frac{A_1 R}{180(z_0+y)^8} - \frac{A_2 R}{6(z_0+y)^2} + \frac{hc_l \pi^2}{240(z_0+y)^4} + f_0 \cos(\omega t), \tag{1}$$

where y is the microcantilever tip displacement. This equation and the physical sense of its coefficients are given, for instance, in [11, 13].

Rewriting (1) in the dimensionless form of the system of two first-order differential equations, we have:

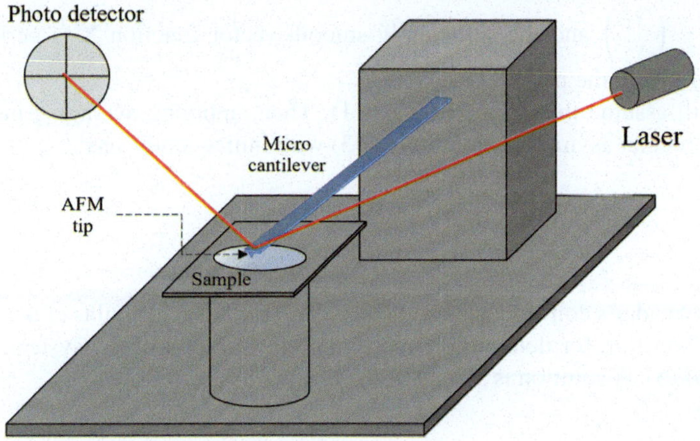

Fig. 1 Scheme of the atomic force microscope composed of the microcantilever, the analysis tip, laser, and the photodetector. Adapted [11]

$$\begin{aligned}
\frac{dx_1}{dt} &= x_2,\\
\frac{dx_2}{dt} &= -rx_2 - bx_1 - cx_1^3 - \frac{d}{(a+x_1)^2} - \frac{p}{(a+x_1)^3}x_2\\
&\quad + \frac{\beta}{(a+x_1)^4} + \frac{e}{(a+x_1)^8} + \Gamma\cos(\omega t),
\end{aligned} \tag{2}$$

where meaning of the dimensionless coefficients is presented in [11].

If $\beta = 0$, then the Casimir force is not taken into account.

Introduce the new notation

$$\tau = \omega t, \quad \varepsilon = \frac{1}{\omega}$$

and write (2) in the form of (3):

$$\begin{aligned}
\frac{dx_1}{d\tau} &= \varepsilon x_2,\\
\frac{dx_2}{d\tau} &= \varepsilon\Bigg[-rx_2 - bx_1 - cx_1^3 - \frac{d}{(a+x_1)^2} - \frac{p}{(a+x_1)^3}x_2\\
&\quad + \frac{\beta}{(a+x_1)^4} + \frac{e}{(a+x_1)^8} + \Gamma\cos(\tau)\Bigg].
\end{aligned} \tag{3}$$

Considering (3) in its vector form, we obtain the standard form

$$\frac{d\mathbf{x}}{d\tau} = \varepsilon\mathbf{X}(\mathbf{x}, \tau), \tag{4}$$

where $\mathbf{x} = \begin{pmatrix} x_1 \\ x_2 \end{pmatrix}$ and the sufficiently smooth vector function $\mathbf{X}$ is periodic with respect to τ with the period 2π.

We will assume that $\varepsilon << 1$ ($\omega >> 1$). Then, applying averaging method, we can construct an asymptotic solution of (3) with initial conditions

$$x_1(0) = x_1^0,\\ x_2(0) = x_2^0. \tag{5}$$

For the reader's convenience, we present here the basic formulas of the averaging method algorithm for determining an asymptotic solution of the system described by (4) and (5). A solution is sought in the form

$$\mathbf{x} = \overline{x} + \varepsilon w_1(\overline{x}, \tau) + \varepsilon^2 w_2(\overline{x}, \tau) + \varepsilon^3 w_3(\overline{x}, \tau) + \cdots , \tag{6}$$

where the new variable $\overline{x}$ satisfies the equation

$$\frac{d\overline{x}}{d\tau} = \varepsilon A_1(\overline{x}) + \varepsilon^2 A_2(\overline{x}) + \varepsilon^3 A_3(\overline{x}) + \cdots . \tag{7}$$

Assuming that the mean value of functions $w_j(\overline{x}, \tau)$ with respect to the interval $[0, 2\pi]$ is equal to zero, using the relations

$$\frac{\partial w_j}{\partial \tau}(\overline{x}, \tau) = g_j(\overline{x}, \tau) - A_j(\overline{x}), \tag{8}$$

where $g_j(\overline{x}, \tau)$ depends on approximations of orders smaller than j, and

$$A_j(\overline{x}) = \frac{1}{2\pi} \int_0^{2\pi} g_j(\overline{x}, \tau) d\tau, \tag{9}$$

we can uniquely define unknown terms in (6) and (7).

Often a good approximation to an exact solution gives the first approximation of the averaging method. Sometimes higher order approximations are needed.

3 Approximations of Averaging Method for a Mathematical Model of AFM with the Casimir Force

Applying the algorithm of the previous section, namely (8) and (9), we obtain for the problem with the Casimir force, defined by (3) and (5) the following:

$$\begin{aligned} A_1^1(\overline{x}) &= \overline{x}_2, \\ A_1^2(\overline{x}) &= -r\overline{x}_2 - b\overline{x}_1 - c\overline{x}_1{}^3 - \frac{d}{(a+\overline{x}_1)^2} - \frac{p}{(a+\overline{x}_1)^3}\overline{x}_2 \\ &\quad + \frac{\beta}{(a+\overline{x}_1)^4} + \frac{e}{(a+\overline{x}_1)^8}, \end{aligned} \tag{10}$$

i.e., the first-order approximation of averaging method for a solution $x_1(\tau)$ and $x_2(\tau)$ is the functions $\overline{x}_1(\tau)$ and $\overline{x}_2(\tau)$ satisfying the equations:

$$\begin{aligned} \frac{d\overline{x}_1}{d\tau} &= \varepsilon\overline{x}_2, \\ \frac{d\overline{x}_2}{d\tau} &= \varepsilon\Bigg[-r\overline{x}_2 - b\overline{x}_1 - c\overline{x}_1{}^3 - \frac{d}{(a+\overline{x}_1)^2} - \frac{p}{(a+\overline{x}_1)^3}\overline{x}_2 \\ &\quad + \frac{\beta}{(a+\overline{x}_1)^4} + \frac{e}{(a+\overline{x}_1)^8}\Bigg] \end{aligned} \tag{11}$$

with the initial conditions

$$\begin{aligned} \overline{x}_1(0) &= x_1^0, \\ \overline{x}_2(0) &= x_2^0. \end{aligned} \tag{12}$$

Then we consecutively get

$$\begin{aligned} w_1^1(\overline{x},\tau) &= 0, \\ w_1^2(\overline{x},\tau) &= \Gamma\sin\tau, \\ A_2^1(\overline{x}) &= A_2^2(\overline{x}) = 0. \end{aligned}$$

The second-order approximation for a solution $x_1(\tau)$ and $x_2(\tau)$ is defined, respectively, by the equalities

$$\begin{aligned} x_1(\tau) &= \overline{x}_1(\tau), \\ x_2(\tau) &= \overline{x}_2(\tau) + \varepsilon\Gamma\sin\tau, \end{aligned}$$

where $\overline{x}_1(\tau), \overline{x}_2(\tau)$ are defined by (11) and (12). Therefore, the first- and second-order approximations for the solution $x_1(\tau)$ are the same.

Further, using formulas from the preceding section, we have

$$\begin{aligned} w_2^1(\overline{x},\tau) &= -\Gamma\cos\tau, \\ w_2^2(\overline{x},\tau) &= \Gamma\left(r + \frac{p}{(a+\overline{x}_1(\tau))^3}\right)\cos\tau, \\ A_3^1(\overline{x}) &= A_3^2(\overline{x}) = 0. \end{aligned}$$

Hence, differential equations for the first-, second-, and third-order approximations are the same. The third-order approximation for a solution $x_1(\tau)$ and $x_2(\tau)$ is defined, respectively, by the expressions

$$x_1(\tau) = \overline{y}_1(\tau) - \varepsilon^2\Gamma \cos\tau,$$
$$x_2(\tau) = \overline{y}_2(\tau) + \varepsilon\Gamma \sin\tau + \varepsilon^2\Gamma \left[r + \frac{p}{(a+\overline{y}_1(\tau))^3} \right] \cos\tau,$$

where the pair $\overline{y}_1(\tau), \overline{y}_2(\tau)$ is a solution for (11) with the initial values

$$\overline{y}_1(0) = x_1^0 + \varepsilon^2\Gamma,$$
$$\overline{y}_2(0) = x_2^0 - \varepsilon^2\Gamma \left[r + \frac{p}{(a+x_1^0)^3} \right].$$

From the previous section, it also follows that

$$w_3^1(\overline{x}, \tau) = \Gamma \left(r + \frac{p}{(a+\overline{y}_1(\tau))^3} \right) \sin\tau, \tag{13}$$

$$\begin{aligned} w_3^2(\overline{x}, \tau) = -\Gamma \Bigg\{ & \left[r + \frac{p}{(a+\overline{y}_1(\tau))^3} \right]^2 - b - 3c\overline{y}_1^2(\tau) + \frac{2(d-4e)}{(a+\overline{y}_1(\tau))^3} \\ & + \frac{3p\overline{y}_2(\tau)}{(a+\overline{y}_1(\tau))^4} - \frac{4\beta}{(a+\overline{y}_1(\tau))^5} \Bigg\} \sin\tau, \end{aligned} \tag{14}$$

$$A_4^1(\overline{x}) = A_4^2(\overline{x}) = 0.$$

In view of (13) and (14),

$$w_3^1(\overline{x}, 0) = w_3^2(\overline{x}, 0) = 0,$$

hence, the system for finding the fourth-order approximation coincides with the system for $\overline{y}_1(\tau)$ and $\overline{y}_2(\tau)$. Thus, the fourth-order approximation is defined by the relations

$$\begin{aligned} x_1(\tau) &= \overline{y}_1(\tau) + \varepsilon^3 w_3^1(\overline{x}, \tau), \\ x_2(\tau) &= \overline{y}_2(\tau) + \varepsilon^3 w_3^2(\overline{x}, \tau), \end{aligned} \tag{15}$$

where $w_3^1(\overline{x}, \tau)$ and $w_3^2(\overline{x}, \tau)$ are given by (13) and (14), respectively. Note, that the functions $\overline{y}_1(\tau)$ and $\overline{y}_2(\tau)$ in (13) and (14) can be replaced by $\overline{x}_1(\tau)$ and $\overline{x}_2(\tau)$ without loss of the accuracy order.

Sometimes, only the third approximation for a solution has the form of the exact solution. See, for instance, Fig. 5 below. Here, the first-order approximation coincides with the second one and the form of them differs from the exact solution $x_1(\tau)$.

4 Results of Calculations

The following values of parameters were considered: $r = 0.1,\ a = 1.6,\ b = 0.05,\ c = 0.35,\ d = 4/27,\ p = 0.009,\ e = 0.0001,\ \Gamma = 0.2,\ x_1^0 = x_2^0 = 0.$

Figs. 2 and 3 show the influence of the Casimir force on the behavior of trajectories $x_1(\tau)$ and $x_2(\tau)$ with $\varepsilon = 1$, $\beta = 0$ (red line) and $\beta = 0.01818$ (blue line), the parameter β is related to the Casimir force.

The results of calculating solutions $x_1(\tau)$ and $x_2(\tau)$ of (3) and (5) with $\beta = 0.01818$, $\varepsilon = 0.2$ and their approximations due to the averaging method are presented in Figs. 4, 5, 6 and 7. In these figures, the red, black, dotted green, and dotted blue lines denote the exact solution and the first, second, and third approximations of averaging method for the solution, respectively.

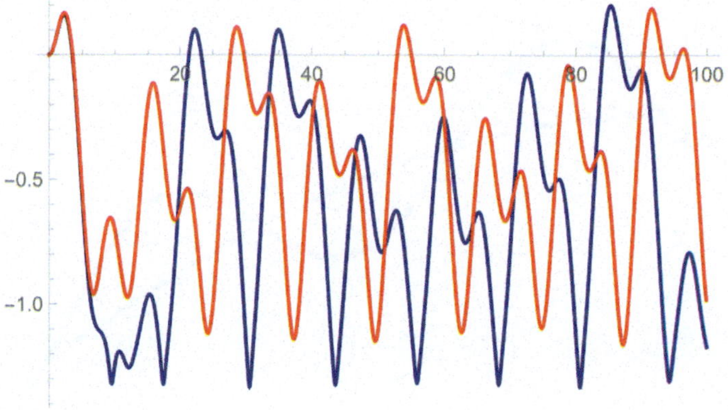

Fig. 2 Trajectories $x_1(\tau)$ with $\beta = 0$ (blue) and $\beta = 0.01818$ (red), $\tau \in [0, 100]$

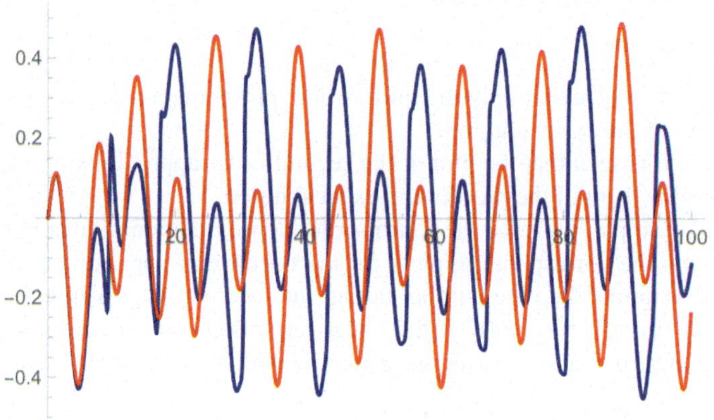

Fig. 3 Trajectories $x_2(\tau)$ with $\beta = 0$ (blue) and $\beta = 0.01818$ (red), $\tau \in [0, 100]$

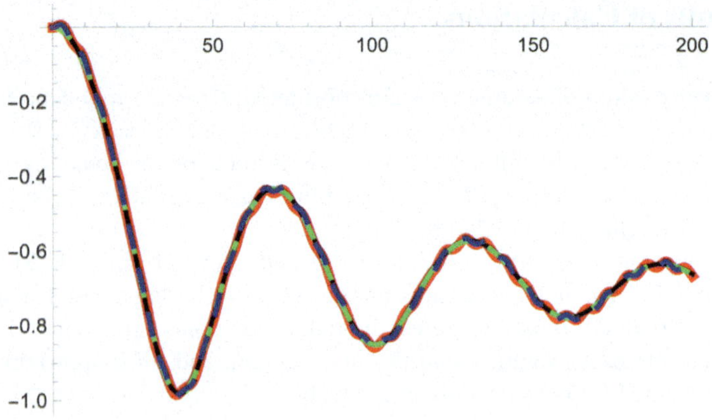

Fig. 4 Trajectories $x_1(\tau), \tau \in [0, 200]$

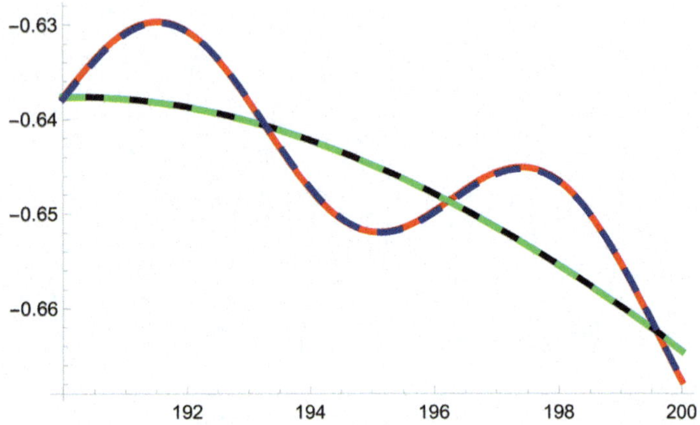

Fig. 5 Trajectories $x_1(\tau), \tau \in [190, 200]$

In Fig. 5, it is seen that only the third-order approximation of averaging method has the form of the exact solution.

In Fig. 7, it is seen that the plot of the second-order approximation of averaging method has the form of the plot for the exact solution.

For this example, plots for the fourth-order approximation of averaging method coincide with the corresponding plots for the third-order approximation. It should be noted that unlike the case without taking into account the Casimir force (see in [13] Fig. 11 for $\omega = 5$ $(\varepsilon = 0.2)$) there is a very good coincidence of plots for the third-order approximation of averaging method with the corresponding plots for the exact solution.

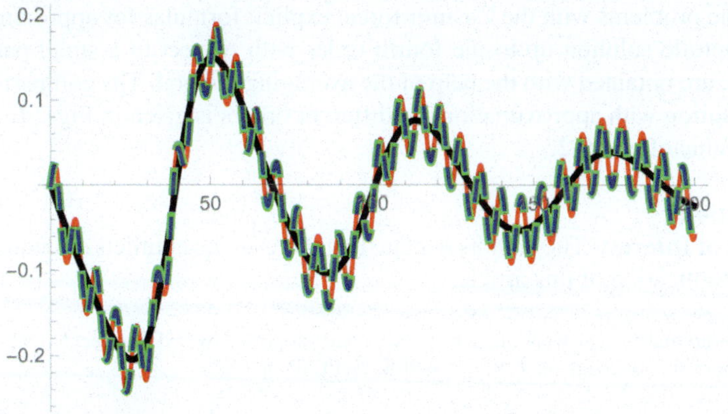

Fig. 6 Trajectories $x_2(\tau), \tau \in [0, 200]$

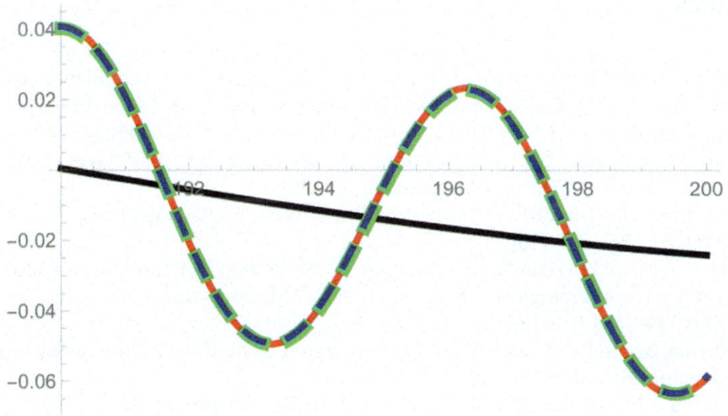

Fig. 7 Trajectories $x_2(\tau), \tau \in [190, 200]$

5 Conclusion

In this manuscript, we explore the atomic force microscopy (AFM) system considering the Casimir force interaction. We use the averaging method to find approximate solutions to the equation of the microcantilever tip displacement. These approximate solutions are extremely important for technical applications of AFM, where a chaotic behavior is also possible [11]. Figures 2 and 3 present the behavior of the considered problem solution in the dependence on the parameter β that relates the Casimir force to the model, for $\beta = 0$, there is no presence of the Casimir force, however, for $\beta = 0.01818$, there is a Casimir force contribution.

For the problems with the Casimir force, explicit formulas for approximations of an asymptotic solution up to the fourth order with respect to a small parameter ε inclusive are obtained with the help of the averaging method. The comparison of the exact solution with approximations of different orders is given in Figs. 4, 5, 6 and 7 for the value of $\varepsilon = 0.2$.

Declarations:
Conflict of Interest: The authors declare that there are no conflicts of interest regarding the publication of this article.

Acknowledgements The work of the first author was supported by RSF (Project No. 21-11-00202), other authors acknowledge the CNPq, CAPES, FAPESP, and FA.

References

1. J.M. Balthazar, A.M. Tusset, S.L.T. De Souza, A.M. Bueno, Microcantilever chaotic motion suppression in tapping mode atomic force microscope. Proc. Instit. Mech. Eng. Part C: J. Mech. Eng. Sci. **227**(8), 1730–1741 (2013). https://doi.org/10.1177/0954406212467933
2. G. Binnig, C.F. Quate, Ch. Gerber, Atomic force microscope. Phys. Rev. Lett. **56**(9), 930 (1986). https://doi.org/10.1103/PhysRevLett.56.930
3. H.B.G. Casimir, On the attraction between two perfectly conducting plates. Proc. Kon. Nederl. Akad. Wet. **51**, 793–795 (1948)
4. S.V. Kontomaris, A. Stylianou, G. Chliveros, A. Malamou, Determining spatial variability of elastic properties for biological samples using AFM. Micromachines **14**(1), 182 (2023). https://doi.org/10.3390/mi14010182
5. G.A. Kurina, Second approximation of the averaging method in electron maser theory. Radiophys. Quantum Electron. **19**(7), 747–751 (1976)
6. G.A. Kurina, J.M. Balthazar, A.M. Tusset, Using different approximations of averaging method in theory of micro electromechanical systems (MEMS), in *Vibration Engineering and Technology of Machinery: Proceedings of VETOMAC XV 2019* (Springer, 2021), pp. 333–342
7. A.F. Kurin, G.A. Kurina, V.V. Novikov, Nonlinear theory of a cyclotron-resonance maser with a Fabry-Perot resonator. Radiophys. Quantum Electron. **19**(7), 742–747 (1976)
8. N.N. Moiseev, *Asymptotic Methods of Nonlinear Mechanics* (Nauka, Moscow, 1981)
9. V.M. Mostepanenko, New experimental results on the Casimir effect. Braz. J. Phys. **30**(2), 309 (2000). https://doi.org/10.1590/S0103-97332000000200012
10. V.M. Mostepanenko, N.N. Trunov, The Casimir effect and its applications. Sov. Phys. Usp. **31**(11), 965–987 (1998). https://doi.org/10.1070/PU1988v031n11ABEH005641
11. M.A. Ribeiro, J.M. Balthazar, W.B. Lenz, R.T. Rocha, A.M. Tusset, Numerical exploratory analysis of dynamics and control of an atomic force microscopy in tapping mode with fractional order. Shock Vib. Hindawi Article ID 4048307 (2020). https://doi.org/10.1155/2020/4048307
12. M.A. Ribeiro, J.M. Balthazar, A.M. Tusset, A.M. Bueno, H.H. Daum, MEMS-based atomic force microscope: nonlinear dynamics analysis and its control. Vib. Monit. Anal.—Recent. Adv. (2022). https://doi.org/10.5772/intechopen.108880
13. M.A. Ribeiro, G.A. Kurina, A.M. Tusset, J.M. Balthazar, Nonlinear numerical analysis and averaging method applied atomic force microscopy with viscoelastic term. Arch. Appl. Mech. **92**(12), 3817–3827 (2022). https://doi.org/10.1007/s00419-022-02264-5
14. M.A. Ribeiro, A.M. Tusset, J.M. Balthazar, On nonlinear fractional dynamics analysis of atomic force microscopy system in trolling mode with viscoelasticity term. Braz. J. Phys. **52**, Article number: 186 (2022). https://doi.org/10.1007/s13538-022-01188-3

15. M.A. Ribeiro, A.M. Tusset, W.B. Lenz, I. Kirrou, J.M. Balthazar, Numerical analysis of fractional dynamical behavior of Atomic Force Microscopy. Eur. Phys. J. Spec. Top. **230**(18–20), 3655–3661 (2021)
16. P.S. Yamaguchi, A.M. Tusset, M.A. Ribeiro, J.M. Balthazar, An atomic force microscopy (AFM) modelling in fractional order: nonlinear control system design. Braz. J. Phys. **52**(5), Article number: 151 (2022). https://doi.org/10.1007/s13538-022-01155-y

On Dynamical Behavior of Atomic Force Microscopy: A Short Review

M. A. Ribeiro, J. M. Balthazar, Ilham Kirrou, A. M. Tusset, Átila M. Bueno, and Hilson H. Daum

Abstract Atomic force microscopy is an experimental analysis widely used in recent years to determine the topology of biological and synthetic samples. This experimental analysis has been widely applied because it is not a destructive technique, that is, there is no destruction of the samples to be analyzed. Due to this wide applicability, we explore a mini-review of two systems in a mathematical way to determine the behavior of nonlinear dynamics (Lyapunov exponent, Bifurcation Diagrams, and phase portraits) and of basins of attraction. Such results corroborate the literature and allow establishing the parameters in which the nonlinear dynamics is in a chaotic or periodic regime, thus opening possibilities for the application of techniques that can suppress chaos and thus determine more accurate images of the analyzed samples.

1 Introduction

In recent years, technological advances have allowed for more detailed analysis of the composition of synthetic or biological materials. Following this line of thought, atomic force microscopy (AFM) has been gaining ground. Therefore, it allows not only to analyze the composition of the materials that constitute a certain sample, but also allows the surface analysis of samples. As one of the classes of microscopy

M. A. Ribeiro (✉) · J. M. Balthazar · A. M. Tusset
Universidade Tecnológica Federal do Paraná - Campus Ponta Grossa, Ponta Grossa, PR, Brazil
e-mail: mau.ap.ribeiro@gmail.com

J. M. Balthazar
School of Engineering, São Paulo State University, Bauru, SP, Brazil

I. Kirrou
ENSA-Agadir, University Ibn Zohr, Agadir, Morocco

Á. M. Bueno
Polytechnic School of the University of São Paulo, So Paulo, SP, Brazil

H. H. Daum
Universidade Tecnológica Federal do Paraná - Campus Guarapuava, Guarapuava, PR, Brazil

M. Belhaq (ed.), *Advances in Nonlinear Dynamics and Control of Mechanical and Physical Systems*, Springer Proceedings in Physics 301,
https://doi.org/10.1007/978-981-99-7958-5_7

analysis, the AFM is derived from the tunneling microscope, which allowed the analysis of interaction forces between the analysis tip and the surface [1–17].

There are numerous applications of AFM in the most diverse areas of science, such as lipid bilayer analysis. From this perspective, the authors [2] analyze the behavior of ruptures in these lipid bilayers, as such applications aim to analyze ruptures in vesicles. Another application, which the same authors explore is the measurement of electrochemical components of these cell membranes.

However, other applications of AFM are in the analysis of synthetic samples, such as the analysis of the particle size of $LiFePO_4$ for electric car batteries since the size of the particles influences the state of charge of the batteries. These results are observed in [9] that use AFM to control particle size.

The AFM device consists of a microbeam with one end attached to the device and the other containing the nanopoint for analysis. This nano analysis point interacts with the surface of the materials and a laser is reflected on it, which is reflected its vibrations and is collected by a photodiode that reconstructs the surface of the sample. There are countless ways to analyze the AFM, however, in this review, we will show the dynamics in two ways [1–10].

The first is the contact mode, presenting the viscoelasticity term and the Casimir force. This modeling is approached numerically. The viscoelastic term determines the behavior of soft samples, that is, an approximation for biological samples. Works, such as [11–13], explore the fractional behavior of the term and propose strategies to suppress chaotic behavior using control techniques.

The second mode is the intermittent oscillation close to its resonant frequency by a small piezoelectric element mounted on the tip of the AFM. The signal obtained from the photodetectors measures the oscillation amplitude, which must be kept constant. Piezoelectric actuators control the height of the console above the sample. In this case, the tip gently touches the sample surface and eliminates the frictional force present in contact mode. This method is used to circumvent the restrictions present in the contact and non-contact modes, the sample surface is less modified in this mix of methods. The intermittent contact mode is used in biological materials, polymers, and very rough samples, as these are malleable and deformable by the tip [11–17].

Therefore, in this manuscript we will address the review of the nonlinear dynamics of two mathematical models for atomic force microscopy. First, we consider the tapping mode considering the Casimir force and the viscoelasticity term. The second case is using the intermittent mode considering the viscoelasticity term. We demonstrate the basins of attraction and their possible attractors, the maximum Lyapunov exponents, bifurcation diagrams, phase maps, and Poincaré maps [19–25].

2 Mathematical Modeling and Numerical Nonlinear Dynamics

The mathematical modeling that we will discuss is the one-dimensional mathematical modeling of two AFM structures, one considers the tapping mode and the other considers the intermittent mode system. Both are assigned the term of viscoelasticity to the system that allows approximations of the behavior of soft samples when analyzed. The numerical analyses in this manuscript were performed using Julia Lang and Python scripts, all organized at: https://sciencemau.github.io/Olivia-Project/.

2.1 Case One: Tapping Mode with Casimir Force

Figure 1 shows a schematic of the AFM system in moto tapping. The system basically consists of a microcantilever attached to the device support, however, at the free end of the microcantilever there is the analysis tip that is in contact with the sample to be analyzed. In this way, a laser is fired at the end with the analysis tip and reflected to a photodiode that collects the vibrations caused by the irregularities of the surface of the sample [11].

The one-dimensional equation of motion of the AFM tapping mode considered in this modeling is attributed to the Van der Waals force with a combination of an attractive and a repulsive portion, as follows:

$$U_{VdW} = \frac{A_1}{180(z_0 + x)^8} - \frac{A_2 R}{6(z_0 + x)^2} \tag{1}$$

where z_0 is the distance between the equilibrium point of the microcantilever and the analyzed sample, R is the radius of the sphere of the tip of the AFM, $U(x, z_0)$ is the

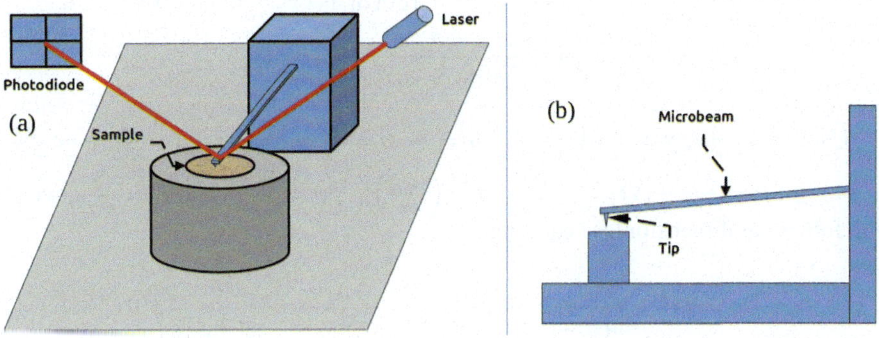

Fig. 1 **a** Scheme of the AFM device and **b** Microbeam containing the micro point of analysis

Lennard-Jones potential, A_1 is the Hamaker constant to the attractive potential and A_2 is the Hamaker constant to the repulsive potential.

For an approximate one-dimensional mathematical model of the displacement of the microcantilever, we will consider a damping defined by

$$F_{damp} = C_d \dot{x} \tag{2}$$

where c_d is the structural damping of the microcantilever.

For an approximate one-dimensional mathematical model of the displacement of the microcantilever, we will consider a damping defined by

$$F_{sf} = \frac{\mu_{eff} \eta^3 l}{(z_0 + x)} \tag{3}$$

where μ_{eff} is the coefficient of effective viscosity, and η and l are the width and length of the microcantilever, respectively.

As a conservative force of the system, a nonlinear spring can be considered, defined by

$$F_{nls} = k_l x + k_{nl} x^3 \tag{4}$$

where k_l is the linear stiffness of the AFM and k_{nl} is the nonlinear stiffness of the AFM microcantilever.

We also attribute the Casimir strength, being defined by [11]

$$F_{cas} = \frac{\hbar c_l \pi^2}{240(x + z_0)^4}, \tag{5}$$

where $\hbar$ is Planck's constant and c_l is the speed of light.

Therefore, the equation of motion of the microcantilever is defined by

$$\begin{aligned} m\ddot{x} &= -c_d \dot{x} - k_l x - k_{nl} x^3 + \frac{A_1 R}{180(z_0 + x)^8} - \frac{A_2 R}{6(z_0 + x)^2} \\ &- \frac{\mu_{eff} \eta^3 l}{(z_0 + x)^3} \dot{x} + \frac{\hbar c_l \pi^2}{240(x + z_0)^4} + f_0 \cos(\omega t) \end{aligned} \tag{6}$$

Using $\tau = \omega_1 t$, $y = \frac{x}{z_s}$, $\dot{y} = \frac{\dot{x}}{\omega_1 z_s}$, $a = \frac{z_0}{z_s}$, $b = \frac{k_l}{k_{at}}$, $c = \frac{k_{nl}}{k_{at}} z_s^2$, $d = \frac{4}{27}$, $e = \frac{2a^6}{405 z_s^6}$, $\Gamma = \frac{f_0}{k_{at} z_s}$, $p = \frac{\mu_{eff} \eta^3 l}{m \omega_1 z_s^3}$, $\Omega = \frac{\omega}{\omega_1}$, $z_s = \left(\frac{2}{3}\right)\left(\frac{A_2 R}{3k_1}\right)^{1/3}$, $\psi = \frac{\hbar \pi^2 c_l}{240}$ and $\beta = \frac{\psi}{\omega_1^2 z_s^5}$, we get the following dimensionless equation:

$$\ddot{x} = -r\dot{x} - bx - cx^3 - \frac{d}{(a + x)^2} + \frac{e}{(a + x)^8} - \frac{p\dot{x}}{(a + x)^3} + \frac{\beta}{(a + x)^4} + \Gamma \cos(\Omega \tau) \tag{7}$$

Table 1 Properties and dimensions of the AFM system considered [5]

Parameter	Values
d_1	0.01
B_1	−0.148967
C_1	-4.59118×10^{-5}
C_2	0.149013
ζ	0.99
Ω	1
E_1	1.57367

The values of the physical constants and dimensionless parameters are given by Table 1. These dimensionless parameters are approximated values of those performed during an experimental analysis that were obtained in [18].

Description	Value
Length	449 nm
Width	46 nm
Thickness	1.7 nm
Tip radius	150 nm
Material density	2,330 kg/m^3
Young's modulus	176 GPa
Bending stiffness	0.11 Nm^{-1}
First-order resonant frequency	11804 kHz
Quality factor	100
Hamaker constant (Repulsive)	1.3596×10^{-70} J m^6
Hamaker constant (Attractive)	1.865×10^{-19} J

Therefore, the basin of attraction was calculated using the dimensionless parameters used are $r = 0.1, a = 1.6, b = 0.05, c = 0.35, d = 4/27, e = 0.0001, \Gamma = 0.2$, $p = 5.757 \times 10^{-5}$ and $\Omega = 1.0$. The Fig. 2 shows the emergence of six attractors that form the basin of attraction for a value $p = 2.503 \times 10^{-5}$, in Fig. 2a–f are the attractive points found and Fig. 2g shows the basin of attraction, the colors represent the regions where the initial conditions converge for the attractors. To determine these basins of attraction, we use the following intervals $x_0 \times \dot{x}_0 \in [-0.95, 0.95] \times [-0.95, 0.95]$.

Thus, the dimensionless parameters used are: $r = 0.1$, $a = 1.6$, $b = 0.05$, $c = 0.35$, $d = 4/27$, $e = 0.0001$, $\Gamma = 0.2$, $p \in [0.0, 0.02]$, $\Omega = 1.0$. In Fig. 3 we have the bifurcation diagram in Fig. 3a and the Lyapunov exponent in Fig. 3b. We can observe the periodic and chaotic nodes obtained with the equation of motion of the microcantilever considering the viscoelasticity term of the sample. This viscoelasticity term corresponds to the force caused by soft samples as described in [11]. The range for $p \in [0; 0.00399]$ determine the chaotic region (see (7))

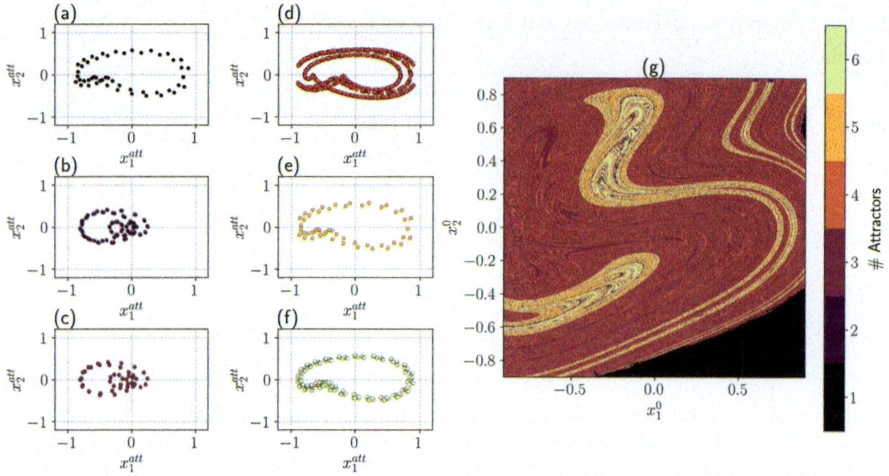

Fig. 2 Basins of attraction and attractor. **a** #1 attractor, **b** #2 attractor, **c** #3 attractor, **d** #4 attractor, **e** #5 attractor, **f** #6 attractor and **g** Basins of attraction

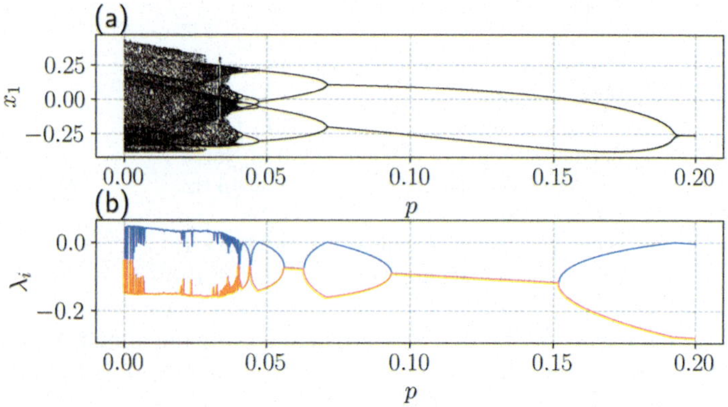

Fig. 3 **a** Bifurcation diagram for $p \in [0, 0.02]$ and **b** Lyapunov exponent $p \in [0, 0.02]$

However, we consider $p = 0.005$ (Fig. 4a), $p = 0.01$ (Fig. 4b), $p = 0.01465$ (Fig. 4c) and $p = 0.02$ (Fig. 4d) for the construction of phase maps (gray line) and Poincaré Maps (dots in black).

2.2 *Case Two: Mathematical Modeling Intermittent*

The intermittent mode is seen in Fig. 5, in which the analysis point behavior is caused by a curvature during the sample surface reading process. Works such as [12, 13] explore the intermittent AFM mechanism considering a mathematical modeling of

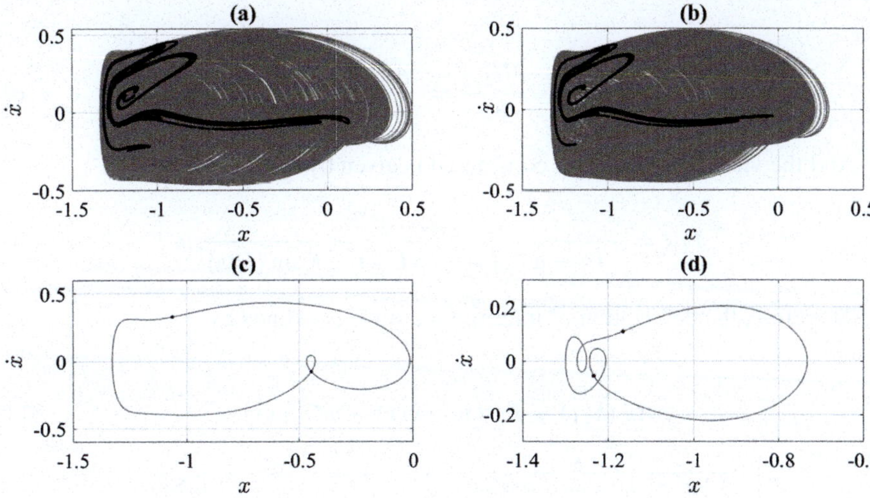

Fig. 4 Portrait Phase (Gray line) and Poincare Maps (Black dots). **a** $p = 0.005$, **b** $p = 0.01$, **c** $p = 0.01465$ and **d** $p = 0.02$

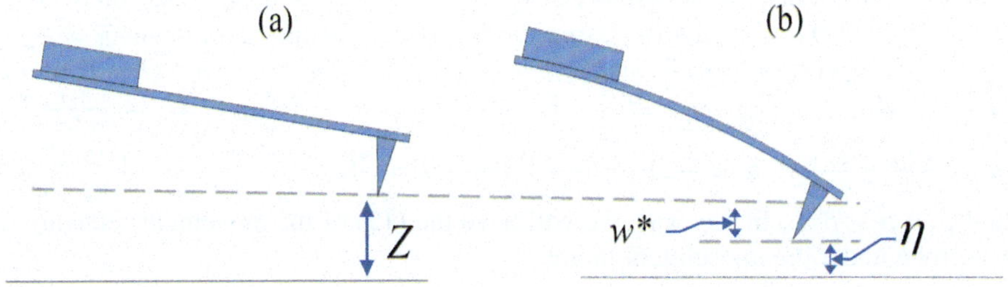

Fig. 5 Simplify schemes of AFM in intermittent mode. **a** System in its initial start and **b** intermittent mode

a microbeam with deflection and under the action of an external force that describes the irregularities of the sample surface [12, 18].

Figure 5 shows a simplified diagram of the deformation of the microbeam that has the AFM system analysis point. Figure 5a shows z the distance of tip between sample, and Fig. 5b shows the deflection of the microbeam where η is equilibrium between the probe tip and the sample $\eta^* = Z - w^*(L)$, where L is the length of the microbeam and w^* constant deflection toward the sample of the linear rod interacting with nonlinear analysis tip forces.

For the analysis of the equations of motion, the equations of the potentials described by the equations below are also assigned:

$$U_{L,j} = \frac{A_1 R}{1260 z^7} - \frac{A_2 R}{6z} \tag{8}$$

$$P_{L,j} = -\frac{\partial U_{L,j}}{\partial z} = \frac{A_1 R}{180 z^8} - \frac{A_2 R}{6 z^2} \tag{9}$$

And the viscoelasticity term considered is given by

$$F_{CS} = -\frac{\mu_{eff}\ \beta^3 l}{\left[z - w^*(L) - u(L,t) - Y\sin(\Omega t)\right]^3}\dot{u} \tag{10}$$

Therefore, the description of intermittent mode is defined by

$$\rho A\ddot{u}(x,t) + EI\left(u^{''''}(x,t) + w^{*''''}(x)\right) = \tag{11}$$
$$\left[-\frac{A_1 R}{180(z-w^*(L)-u(L,t)-Y\sin(\Omega t))^8} + \frac{A_2 R}{6(z-w^*(L)-u(L,t)-Y\sin(\Omega t)^2} - \frac{\mu_{eff}\beta^3 L\dot{u}}{z-w^*(L)-u(L,t)-Y\sin(\Omega t)}\right]\delta(x-L) + \rho A\Omega^2 Y\sin(\Omega t)$$

where R is tip radius, z is the distance of tip, A_1 and A_2 are Hamarker constant is, $A_1 = \pi^2\rho_1\rho_2 c_1$ and $A_1 = \pi^2\rho_1\rho_2 c_2$. Using the following change of variables $\overline{\eta} = \frac{x_1(\tau)}{\eta^*}$, $x_1 = \phi_1(L)\,q_1(\tau)$, $\eta^* = z - w^* L$, $x_1 = \phi_1(L)\,q_1(\tau)$, $\eta^* = z - w^*(L)$, $\tau = \omega_1 t$, $\overline{\Omega} = \frac{\Omega}{\omega_1}$ $d_1 = \frac{C_1}{\omega_1\rho A\ \int_0^L \phi_1^2 dx}$, $B_1 = (1-\overline{\alpha})\,\Gamma_1$, $C_1 = \frac{A_1 R}{180\ k\ (\eta^*)^9}\Gamma_1, C_2 = \frac{A_2 R}{6k\ (\eta^*)^3}\Gamma_1$, $\omega_1^2 = \frac{EI\int_0^L \phi_1\phi_1^{''''} dx}{\rho A\int_0^L \phi_1^2 dx}$, $\overline{\alpha} = \frac{z}{\eta^*}, \Gamma_1 = \frac{k\ \phi_1^2(L)}{\omega_1^2\rho A\int_0^L \phi_1^2 dx}, k = \frac{3EI}{L^3}$, $\zeta = \frac{Y}{\eta^*}$, $E_1 = \frac{\phi_1(L)\int_0^L \phi_1 dx}{\int_0^L \phi_1^2 dx}$ and $p = \mu_{eff}\beta^2 L$. In this way, we will have the (12) of the movement referring to the probe under the intermittent mode,

$$\ddot{x} + \dot{x} + d_1\dot{x} + B_1 - \frac{C_1}{(1-x-\zeta\sin(\Omega\tau))^8} - \frac{C2}{(1-x-\zeta\sin(\Omega\tau))^2} + \frac{p\dot{x}}{(1-x-\zeta\sin(\Omega\tau))^3} = \zeta\Omega^2 E_1\sin(\Omega\tau) \tag{12}$$

For numerical analysis we use the following parameters described in Table 1.

Figure 6 shows the emergence of six attractors that form the basin of attraction for a value $p = 5.757\times 10^{-5}$. In Fig. 6a–e are the attractive points found and Fig. 2f shows the basin of attraction, the colors represent the regions where the initial conditions converge for the attractors. To determine these basins of attraction, we use the following regions $x_0 \times \dot{x}_0 \in [-0.95, 0.95] \times [-0.95, 0.95]$.

Thus, considering $x_0 = 0.0$ and $\dot{x}_0 = 0.0$ and $p \in [0.0, 0.2]$, the Lyapunov maximum exponent and the bifurcation diagram were calculated, in which we can observe that the values of $p \in [0.0004; 0.0088]$, $[0.0088; 0.0108]$, $[0.0156; 0.0170]$ and $[0.0184; 0.0019]$ the system has a chaotic behavior, as can be seen in Fig. 7a representing the bifurcation diagram and confirmed by the Lyapunov maximum exponent as shown in Fig. 7b.

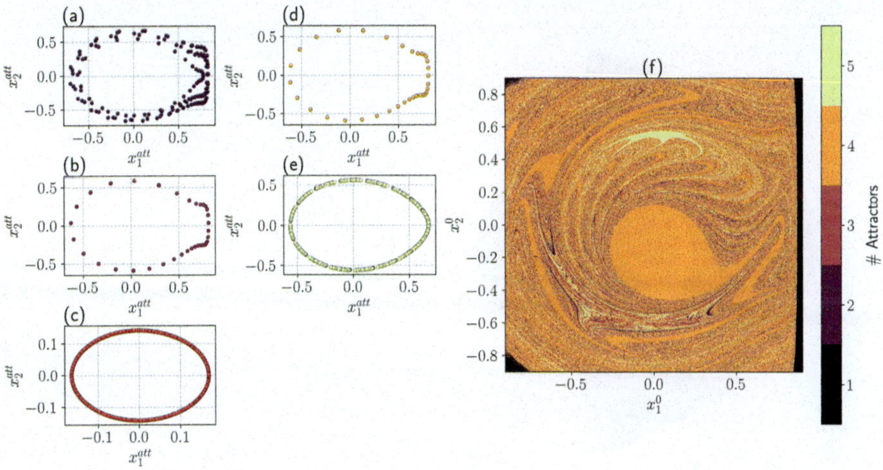

Fig. 6 Basins of attraction and attractor. **a** #1 attractor, **b** #2 attractor, **c** #3 attractor, **d** #4 attractor, **e** #5 attractor and **f** Basins of attraction

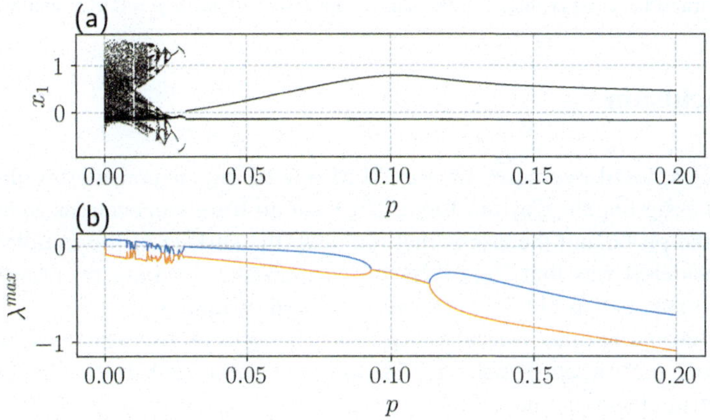

Fig. 7 **a** Diagram bifurcation and **b** maximum Lyapunov exponent. Both for $p \in [0.0, 0.2]$

Therefore, we consider the values and $p = 0.005$, Fig. 8a shows the chaotic behavior in the phase map (gray line) and the Poincaré maps (black dots) and $p = 0.01$ in Fig. 8b shows the periodic behavior of the phase map (gray line) and the Poincaré map (black dot).

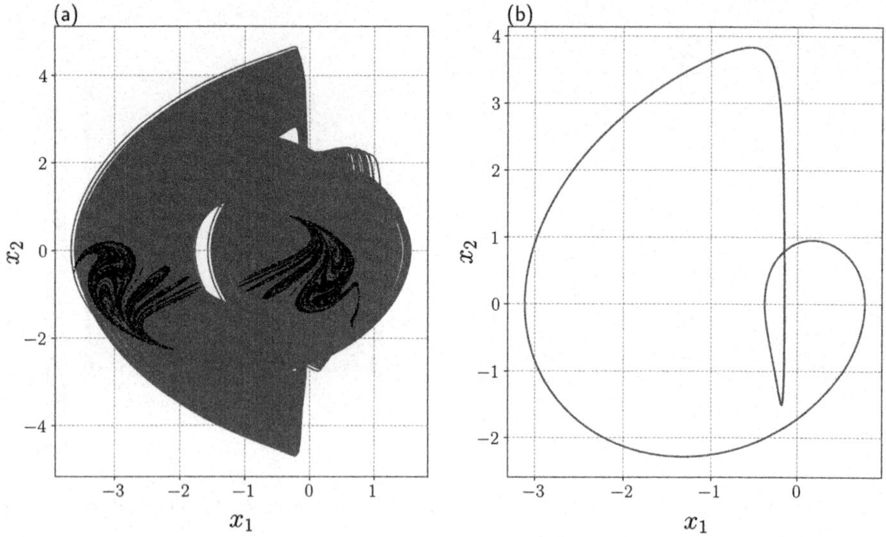

Fig. 8 Portrait Phase (Gray line) and Poincare Maps (Black dots). **a** $p = 0.005$, and **b** $p = 0.02$

3 Conclusion

Our work explored two cases for the AFM device, one is considering the tapping mode and assigning the Casimir force. The Casimir force can occur as an interaction between the probe and the atoms that make up the surface of the sample. Another force considered was that caused by the viscoelastic damping resulting from the sample, works such as [11, 12] explore fractional dynamics.

The intermittent mode was also exploring the term of viscoelasticity, works like [14] explore the fractional behavior of this term and establish the chaotic regions resulting from the dynamics of the system.

Therefore, our work showed a brief review of the nonlinear dynamics of these mathematical models, in which we show the basins of attraction and the attractor points for both cases. The basins of attraction show the behavior of the ensemble of initial conditions on a 1200×1200 grid in $x_0 \times \dot{x}_0 \in [-0.95, 0.95] \times [-0.95, 0.95]$. The low values ($p = 2.503 \times 10^{-5}$) of the viscoelastic term the emergence of 6 attractors for the intermittent mode and 5 attractors for the tapping mode. Analysis of the basins of attraction brings topological consequences to the phase space that allow, indirectly, to carry out a global analysis, in the parameter space, of the stability of the equilibrium points. Therefore, we explore the maximum Lyapunov exponent and the Bifurcation diagrams that show the chaotic and periodic windows. For confirmation, it is shown with the Phase portrait and the Poincaré maps.

Establishing nonlinear dynamic quantities in electromechanical systems such as AFM not only allow a better understanding of the mathematical model, but also determine methodologies for control projects that suppress the chaotic behavior of these devices. The suppression of chaos in these devices makes it possible to work better and thus establish a clearer image of the sample surfaces.

References

1. J. Nader, K. Laxminarayana, Mechatr. **14**, 8 (2004)
2. M.H. Korayem, S. Zafari, A. Amanati, M. Damircheli, N. Ebrahimi, Int. J. Adv. Manuf. Technol. **50** (2010)
3. A.M. Tusset, M.A. Ribeiro, W.B. Lenz, R.T. Rocha, J.M. Balthazar, J. Vib. Eng. Technol. (2019)
4. K. dos Santos Rodrigues, J.M. Balthazar, A.M. Tusset, B.R. de Pontes, Á.M. Bueno, Int. J. Control Autom. Syst. **25** (2014)
5. S. Rützel, S.I. Lee, A. Raman, Proc. R. Soc. Lond. **459** (2003)
6. M.A. Ribeiro, J.M. Balthazar, W.B. Lenz, R.T. Rocha, A.M. Tusset, Shock. Vib. (2020)
7. A.M. Tusset, J.M. Balthazar, R.T. Rocha, M.A. Ribeiro, W.B. Lenz, Nonlinear Dyn. **99** (2020)
8. I. Podlubny, *Fractional Differential Equations: An Introduction to Fractional Derivatives, Fractional Differential Equations, to Methods of Their Solution and Some of Their Applications*, 2nd edn. (Elsevier, Slovak Republic, 1998)
9. R. Arvind, J. Melcher, R. Tung, Nano Today **3** (2008)
10. F. Jamitzky, M. Stark, W. Bunk, W.M. Heckl, R.W. Stark, Nanotechnology **17** (2016)
11. M.A. Ribeiro, J.M. Balthazar, W.B. Lenz, R.T. Rocha, A.M., Tusset, Shock vib. (2020)
12. M.A. Ribeiro, A.M. Tusset, W.B. Lenz, I. Kirrou, J.M. Balthazar, Eur. Phys. J. Plus. **230** (2021)
13. M.A. Ribeiro, A.M. Tusset, J.M. Balthazar, Braz. J. Phys. **52** (2022)
14. M.A. Ribeiro, G.A. Kurina, A.M. Tusset, J.M. Balthazar, Arch. Appl. Mech. **92**, 3817–3827 (2022)
15. A.M. Tusset, M.A. Ribeiro, W.B. Lenz, R.T. Rocha, J.M. Balthazar, J. Vib. Eng. **8** (2020)
16. A.M. Tusset, J.M. Balthazar, M.A. Ribeiro, W.B. Lenz, R.T. Rocha, Eur. Phys. J. Plus. **230** (2021)
17. P.S. Yamaguchi, A.M. Tusset, M.A., Ribeiro, J.M. Balthazar, Braz. J. Phys. **52** (2022)
18. M.A. Ribeiro, J.M. Balthazar, A.M. Tusset, A.M. Bueno, H.H. Daum, *MEMS-Based Atomic Force Microscope: Nonlinear Dynamics Analysis and Its Control, Vibration Monitoring and Analysis, Recent Advances* [Working Title] (IntechOpen, 2022)
19. R.W. Stark, Mater. Today **13** (2010)
20. C. Wang, N. Pai, H. Yau, Commun. Nonlinear Sci. **15** (2010)
21. A. Wolf, J.B. Swift, H.L. Swinney, J.A. Vastano, Physica D **16** (1985)
22. E. Hairer, C. Lubich, M. Schlichte, SIAM J. Sci. Statist. Comput. **6**(1985)
23. R. Garrappa, Int. J. Comput. Math. **87** (2010)
24. M.F. Danca, M. Fečkan, N. Kuznetsov, G. Chen, Nonlinear Dyn. **98** (2019)
25. A. Bonfanti, J.L. Kaplan, G. Charras, A. Kabla, Soft Matter **16**, 26 (2020)

Methodological Description of Obtaining and Stabilizing Solitons in Macromechanical Damped Resonators Subjected to a Base Acceleration

Arthur Barbosa, Najib Kacem, and Noureddine Bouhaddi

Abstract This study focuses on the phenomenon of Intrinsic Localized Modes (ILMs) and solitons in nonlinear systems. Unlike linear systems, nonlinear ones can exhibit solitons even when the periodicity hypothesis is considered. The Nonlinear Schrödinger equation (NLSE) is used to describe the phenomenon in the damped Duffing oscillator chain analyzed in this article. The study presents a general methodology for finding and stabilizing solitonic solutions in macromechanical resonators under base acceleration where a mathematical treatment is presented for an equivalent discrete model and the solution is obtained using the method of multiple scales. The stability of the nonlinear wave is assessed by analyzing physical parameters (stiffness, acceleration, and damping) in order to demonstrate numerically the methodology's ability to reproduce stationary waves in the coupled oscillators even considering the presence of damping.

1 Introduction

The phenomenon of energy localization in coupled oscillators has gained significant interest from researchers in recent years due to its practical advantages. This strategy has a wide range of applications, from detecting masses in microcantilevers [1] to modal analysis of structures [2]. Numerical analyses reported in the literature indicate that both linear and nonlinear oscillators can exhibit this phenomenon. However, only nonlinear structures can reproduce localized vibration modes while respecting the property of periodicity [3].

A. Barbosa (✉) · N. Kacem · N. Bouhaddi
University of Franche-Comté, 24 Rue de l'Epitaphe, 25000 Besançon, France
e-mail: arthur.barbosa@femto-st.fr

N. Kacem
e-mail: najib.kacem@femto-st.fr

N. Bouhaddi
e-mail: noureddine.bouhaddi@femto-st.fr

M. Belhaq (ed.), *Advances in Nonlinear Dynamics and Control of Mechanical and Physical Systems*, Springer Proceedings in Physics 301,
https://doi.org/10.1007/978-981-99-7958-5_8

A promising approach to obtaining energy localization in mechanical systems is the study of solitonic waves. This technique involves obtaining nonlinear partial differential equations that describe the behavior of the chain of oscillators through soliton wave solutions. The methodology in question has been utilized in the domain of microelectromechanical systems and is supported by both theoretical [4] and empirical [5] evidence. Nonetheless, few works in this area have been presented from a macromechanical point of view (such as in [6–8]), especially in the presence of damping.

The architecture of a macromechanical design that reproduce solitons varies according to the desired application. Examples in the literature include circular [9], unidirectional [10], and pendular structures [11]. These architectures, in contrast to micromechanical systems, typically do not have a high number of oscillators. For comparison, [12] experimentally analyzed the propagation of solitons in 60 coupled pendulums, while [5] presented experimental analyses in 152 microcantilevers. Not only the number of oscillators and the architecture but also the nature of the system excitation is a fundamental design choice. Some studies reported external [13], parametric [14], and both excitations [15] as possible methods.

Considering the lack of works that propose to design damped macromechanical systems that could reproduce stationary solitons, the aim of this work is to introduce a methodology for obtaining localized vibration patterns by the NLSE equation in coupled oscillators. Despite the vast literature on the subject, there remains a need for more comprehensive experimental studies on the effects of damping in architectures under external excitation. In this context, this study also stands out from the literature by elucidating potential design parameters that ensure the stability of nonlinear waves in experimental setups.

The methodology involves evaluating the linear properties of each oscillator so that, after verifying the conditions of stability and existence of solutions for the obtained NLSE equation, the parameters of coupling and nonlinear stiffness can be designed. According to the presented numerical example, this approach proved to be effective in locating the vibration amplitude of the coupled oscillators.

2 System Description and Mathematical Model

As mentioned in the introduction, the physical architecture of the oscillator chain is designed based on the chosen application. In order to generalize the methodology, the results of this work will be based on equivalent systems that are independent of the chosen architecture (see Fig. 1). This mathematical approach can be found in several studies (as an example in [16]).

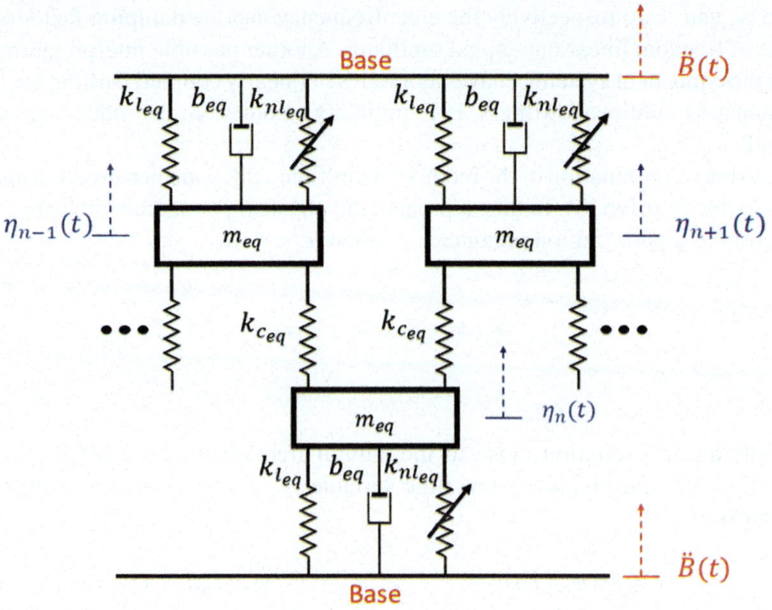

Fig. 1 Graphical representation of the system using equivalent physical variables: m_{eq} (mass), $k_{l_{eq}}$ (linear stiffness between oscillators and base), $k_{c_{eq}}$ (linear coupling stiffness), $k_{nl_{eq}}$ (nonlinear stiffness between oscillators and base), b_{eq} (damping), $\eta_n(t)$ (relative displacement of oscillator n with respect to the base displacement), and $\ddot{B}(t)$ (base acceleration, whose oscillation frequency is ω)

2.1 *Development of a Mathematical Model for Nonlinear Oscillator Chains*

The discrete model represented in Fig. 1 can be obtained by the Lagrangian approach, where the following relationship is derived:

$$m_{eq}\ddot{\eta}_n + b_{eq}\dot{\eta}_n + k_{l_{eq}}\eta_n - k_{nl_{eq}}\eta_n^3 + k_{c_{eq}}(2\eta_n - \eta_{n+1} - \eta_{n-1}) = -m_{eq}\ddot{B}. \quad (1)$$

For:

$$\omega_0^2 = \frac{k_{l_{eq}}}{m_{eq}}, \quad \frac{b_{eq}}{m_{eq}} = 2\zeta\omega_0, \quad \ddot{B} = B_A\left(0.5e^{i\omega t} + 0.5e^{-i\omega t}\right) \quad (2)$$

the new dynamic equation can then be written in the form:

$$\begin{gathered}\ddot{\eta}_n + 2\zeta\omega_0\dot{\eta}_n + \omega_0^2\eta_n - \frac{k_{nl_{eq}}}{m_{eq}}\eta_n^3 + \frac{k_{c_{eq}}}{m_{eq}}(2\eta_n - \eta_{n+1} - \eta_{n-1}) = \\ = -0.5B_A\left(e^{i\omega t} + e^{-i\omega t}\right),\end{gathered} \quad (3)$$

where ω_0 and ζ are respectively the eigenfrequency and the damping factor of the 1 degree of freedom linear uncoupled oscillator. Another possible interpretation of (3) is that the equivalent system behaves as a series of linearly coupled Duffing oscillators with negative nonlinear stiffness. For simplicity of notation, the index "*eq*" will be omitted.

By using a combination of the method of multiple scales and perturbation analysis, it is possible to solve (3). In this approach, the physical parameters are expressed as a function of a perturbation parameter ϵ, where $\epsilon << 1$:

$$\begin{aligned} B_A &= \varepsilon B_{Ae}, \; k_c = \varepsilon k_{ce}, \\ k_{nl} &= \varepsilon k_{nle}, \quad \zeta = \varepsilon \zeta_e. \end{aligned} \tag{4}$$

Applying an excitation close to the natural frequency of the oscillators ($\omega = (1-\varepsilon)\omega_0$) and introducing a new time variable ($T = \varepsilon t$), the system solution can be expressed as:

$$\eta_n = -(\psi_n(T) e^{-it\omega_0(1-\varepsilon)} + e^{it\omega_0(1-\varepsilon)} \overline{\psi}_n(T)) + O(\epsilon), \tag{5}$$

where, ψ_n denotes the amplitude of oscillation of the oscillator n, the overbar indicates the complex conjugate of the variable, and $O(\epsilon)$ refers to terms whose order of magnitude is related to the perturbation factor ε.

After the inclusion of (5) and Relations (4) in (3), an expression that relates the physical terms and the perturbation factor is obtained. Due to the high number of algebraic steps required to derive this relationship, it is recommended the use of symbolic programming to obtain an equation in the following format:

$$O(\epsilon^0) + O(\epsilon^1) + O(\epsilon^2) = O(\epsilon^1) + O(\epsilon^2) = 0. \tag{6}$$

It should be noted that the independent terms $O(\epsilon^0)$ are canceled during the algebraic steps, whereas the terms $O(\epsilon^1)$ and $O(\epsilon^2)$ are not. This procedure leads to the emergence of secular terms that should be canceled (terms proportional to $e^{\omega_0 t}$ or $e^{-\omega_0 t}$). This cancellation results in the discretized nonlinear Schrödinger Equation:

$$\begin{gathered} \frac{0.25 m B_{Ae}}{k_l} + i\zeta_e \psi_n(T) + \frac{3k_{nle}}{2k_l} \psi_n^2(T) \overline{\psi}_n(T) \\ + \\ \left(\frac{k_{ce}}{k_l} \psi_n(T) - \frac{k_{ce}}{2k_l} \psi_{n+1}(T) - \frac{k_{ce}}{2k_l} \psi_{n-1}(T) \right) - \psi_n(T) \\ = \\ -\frac{i}{\omega_0} \frac{d\psi_n(T)}{dT}. \end{gathered} \tag{7}$$

This strategy is well documented in the literature, for instance, the studies [6, 17] employed different orders of magnitude for the perturbation factor, resulting in different magnitudes among the physical parameters. Moreover, the studies [9, 18] used distinct formats for (5), taking into account different wave numbers and, therefore, different ways of canceling the secular terms.

2.2 *Extraction of the Continuous Nonlinear Schrödinger Equation*

In our investigation, two procedures will be implemented: the adimensionalization and the approximation of the discrete to the continuous domain. As will be detailed in Sect. 3, there is an analytical advantage of fundamental interest for this study when these approaches are adopted: when following this format, the stability of the nonlinear Schrödinger equation is well known in the literature, which enables the use of previously reported results in the selection of the design parameters.

Intuitively, the continuity hypothesis becomes more accurate as the number of oscillators in the chain increases and/or the distance between them decreases. Nonetheless, the validation of this assumption will not be presented. By incorporating this procedure into the model, the following expressions can be defined:

$$\psi_n(T) = \psi\left(n\Delta s, T\right) = \psi\left(s_n, T\right), \tag{8}$$

where the variable s is related to the position of the oscillators in the lattice, n the oscillator count variable and Δs the distance between them (see Fig. 2). A direct consequence of (8) is the approximation of the oscillation amplitude of adjacent oscillators as follows:

$$\psi_{n\pm1} \approx \psi\left(s_n, T\right) \pm \Delta s \frac{\partial}{\partial s}\psi\left(s_n, T\right) + 0.5\Delta s^2 \frac{\partial^2}{\partial s^2}\psi\left(s_n, T\right). \tag{9}$$

To accomplish the nondimensionalization of the (7), the following relationships will be considered:

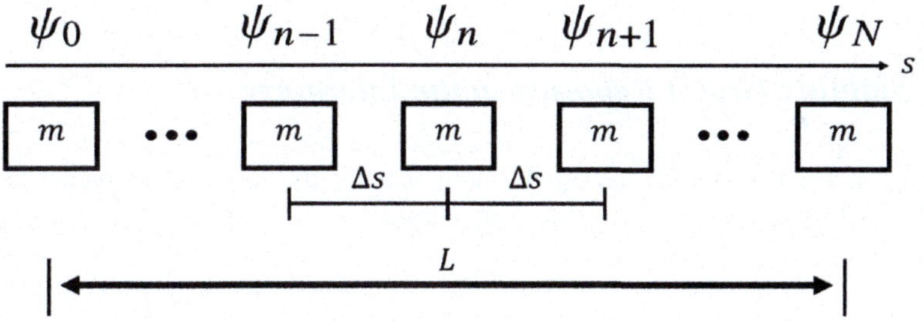

Fig. 2 The positioning of the oscillators amplitude along the chain

$$\begin{aligned} \hat{\psi} &= \sqrt{\tfrac{3k_{nle}}{4k_l}}\psi(s,T), & \hat{T} &= T\omega_0, & \hat{s} &= \tfrac{s}{\delta}, \\ \Delta\hat{s} &= \tfrac{\Delta s}{\delta}, & k_{ce} &= \tfrac{2\delta^2}{\Delta s^2}k_l, \quad h = 0.125\sqrt{3}m B_{Ae}\sqrt{\tfrac{k_{nle}}{k_l^3}}, \end{aligned} \tag{10}$$

where the superscript notation above variables indicates their dimensionless version. The parameter δ is a design variable that connects the physical space L with the dimensionless domain L/δ (see Fig. 2).

After employing the (8), (9), and (10) in (7), the NLSE under external excitation in its dimensionless form is derived:

$$-i\frac{\partial\hat{\psi}}{\partial\hat{T}} = h + (2|\hat{\psi}|^2 + i\zeta_e - 1)\hat{\psi} + \frac{\partial^2\hat{\psi}}{\partial\hat{s}^2}. \tag{11}$$

The (11) balances the damping ζ_e with an excitation h of external origin, being a particular case of mechanical systems that also consider excitations of parametric origin or forms of nonlinear damping [19].

3 Conditions of Existence and Stability of Solitonic Solutions

The (11) plays a crucial role in describing a plurality of natural phenomena. Consequently, numerous studies have reported numerical and analytical solutions for similar versions of this Equation that vary according to the specific application, such as alternative forms of damping or excitation. Recent examples of research apply the NLSE in fields such as biology [20], optics [21], photonics [22], plasma physics [23], and metamaterials [24]. To the best of our knowledge, no analytical solution has been reported for (11) when a linear damping is considered ($\zeta_e > 0$). However, for a certain range of parameters (h and ζ_e), the NLSE has numerical stable time solutions, which enable the construction of physical parameters that enable the existence of solitons.

3.1 Stability Graph Reported in the Literature

The search for time-stable $\hat{\psi}$ can be achieved by the scheme proposed by [13]. The methodology presented by this author allows us to analyze stationary solutions $\left(\frac{\partial\hat{\psi}}{\partial\hat{T}} = 0\right)$ that could serve as initial points for numerical integrations in time [19].

According to [13], the (11) for $\frac{\partial\hat{\psi}}{\partial\hat{T}} = 0$ and $\zeta_e = 0$ has two solutions, but just one that could be used in the search of time-stable solitons in the presence of damping:

$$\psi(x) = \psi_0 \left(1 + \frac{2\sinh^2\alpha}{1 - \cosh\alpha \cosh(Ax)}\right), \tag{12}$$

where

$$h = \frac{\sqrt{2}\cosh^2\alpha}{(1+2\cosh^2\alpha)^{3/2}}, \quad \psi_0 = \frac{1}{\sqrt{2(1+2\cosh^2\alpha)}}, \quad A = 2\psi_0 \sinh\alpha. \tag{13}$$

Starting from the undamped solution, the damped ones could be obtained by the continuous analog of Newton's method, while considering the numerical continuation among solutions with close parameters. For a better understanding of this implementation, it is recommended the reading of the suggested bibliography.

The work [13] introduced the existence/stability graph for generic values of h and ζ_e, which will be utilized in this study (see Fig. 3). It is noteworthy that this chart reveals the existence of regions in which numerical solutions for (11) can be obtained, but they are not stable over time. Another non-trivial characteristic that can be observed is the fact that the solitary wave solutions become increasingly constrained as damping increases.

The derivation of the curves that generated the boundaries of the stability diagram was deduced by the author. Meanwhile, in this paper, the calculation of the Hopf bifurcation curve (in purple) was approximated using a polynomial function based on the results presented in [13] employing the least squares method:

- Lower limit (curve in black): $h = \frac{2\zeta_e}{\pi}$,
- First upper limit (blue curve): $h = \left(\frac{1}{3}\left(\frac{1}{3}\left(-\gamma^2 + \frac{1}{3}\right)^3\right)^{0.5} + \frac{1}{3}\left(\gamma^2 + \frac{1}{9}\right)\right)^{0.5}$,
- Second upper limit (orange curve): $h = \left(\zeta_e^3 - \zeta_e^2 + \frac{\zeta_e}{2}\right)^{0.5}$,
- Hopf bifurcation (purple curve): $h = 1.8728\zeta_e^3 + 1.6822\zeta_e^2 + 0.0997\zeta_e + 0.0777$.

A remark should be made: just as there is an approximation involving the Hopf bifurcation, there is also an approximation for the Lower limit (curve in black). In spite of that, according to [13], this approximation is highly effective in delineating the stability area.

3.2 *Design Methodology for Stability*

It becomes possible to propose a design methodology for the solitonic problem with the obtained relationships, where, discriminating the disturbance factor, can be summarized as follows:

$$\zeta = \varepsilon\zeta_e, \quad \hat{L} = \frac{L}{\delta}, \quad k_c = \frac{2\varepsilon\delta^2}{\Delta s^2} k_l, \quad h = 0.125\sqrt{3}m\frac{B_A}{\varepsilon}\sqrt{\frac{k_{nl}}{\varepsilon k_l^3}}, \tag{14}$$

As can be inferred from Fig. 3, h is approximately located between $0 \leq h \leq 0.5$, while ζ_e is located about between $0 \leq \zeta_e \leq 0.7$. This information must be taken into consideration when determining the parameters.

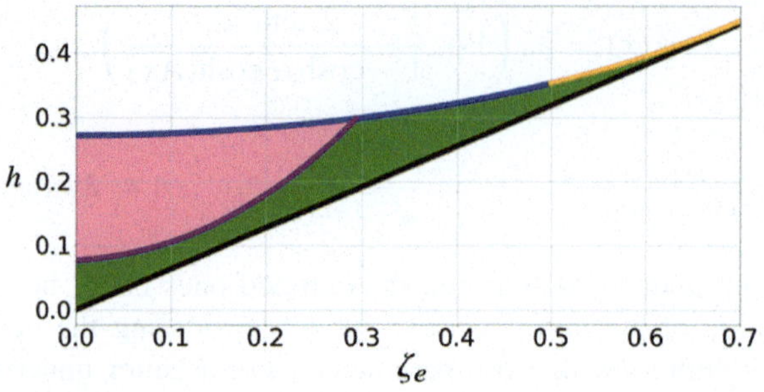

Fig. 3 Diagram of existence and stability of solitons (11) based on the parameter pairs h x ζ_e. The green region indicates the area of numerical existence and stability, whereas the pink region represents the area where the solutions start to gain instability as go further from the Hopf bifurcation (curve in purple). Based on [13]

According to (14), low values of ε would result in high values of h, while high values of ε would result in high values of ζ_e, making the emergence of solitons unfeasible. Hence, it is the designer's task to evaluate the physical range of possibilities that specific applications permit for the stiffness, the damping, and the base acceleration amplitude.

To illustrate this problem, we will consider $\epsilon = 0.008$ and $\zeta = 0.24\%$ (consequently $\zeta_e = 0.3$). The stability diagram (Fig. 3) reveals that there are values of h that ensure the existence of solitonic solutions for (11). In the present example, the value of h will be taken as $h = 0.243450$.

Once the damping has been assessed, the next step involves seeking an appropriate δ that can ensure numerically favorable boundary conditions. As depicted in Fig. 4, for these chosen h and ζ_e, the energy is localized around $-5 < \hat{s} < 5$. This particularity is critical for the selection of δ, due to the fact that if the physical space is nondimensionalized so that its endpoints are located inside $-5 < \hat{s} < 5$ (see Fig. 2), the localized vibration modes will not occur. As a consequence, a low value for δ is desirable. However, this choice implies a low value for the coupling stiffness between the oscillators, as shown in (14). Thus, in order to achieve greater energy localization, the designer should opt for the Minimum viable value of coupling stiffness for each application.

Not only δ but also the space between the oscillators is also responsible for the coupling stiffness value. For a given L, the number of oscillators determines the value of Δs by the expression:

$$\Delta s = \frac{L}{N - 1}. \tag{15}$$

Therefore, for a given δ, the hypothesis of continuity will be more respected the greater the coupling stiffness, which would imply in low values of δ and high number of oscillators.

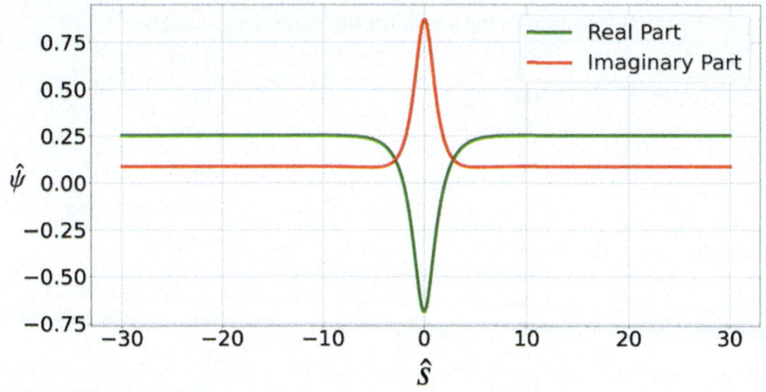

Fig. 4 Numerical solution of the time-invariant (11) $\left(\dfrac{\partial \hat{\psi}}{\partial \hat{T}} = 0\right)$ for $h = 0.243450$ and $\zeta_e = 0.3$. Obtained using the continuous analog Newton's method [13, 19]

Finally, the nonlinearity k_{nl} can be determined by balancing the amplitude of acceleration B_A, the perturbation factor ε and the linear stiffness k_l to achieve the chosen value of h. For a given choice of perturbation factor ε, high values of k_l must be accompanied by high values of k_{nl} or high values of B_A, as can be inferred from (14).

To summarize the analyses discussed in this section, the methodology for obtaining localized modes can be understood by the following steps:

1. Identify the physical variables of the oscillators: natural frequency ω_0, linear stiffness k_l, and damping ζ;
2. Assess the perturbation factor ε based on ζ_e, as indicated on the stability diagram;
3. Analyze the relationship between δ, Δs and the coupling stiffness k_c of the system;
4. Select an appropriate value of h that ensures that stable solutions are achievable, based on the stability chart;
5. Finally, balance the physical quantities among k_{nl} and B_A for the particular value of h.

3.3 Numerical Example

As highlighted in [4], the dynamic formation of solitons from resonators in a motionless configuration is not obvious. Due to the nonlinearity of the system, it is necessary to consider possible stable basins of attraction that would lead the system with zero initial conditions to different solutions of the solitonic equation. To address this issue, a possible strategy is to enforce stabilization conditions just after an unstable configuration. For the simulation, the parameters listed in Table 1 will be used.

The numerical technique adopted for integrating (3) is the fourth-order Runge–Kutta method (see Table 2). Initially, the system starts at $\eta_n = 0$ and $\dot{\eta}_n = 0$, under an

Table 1 Physical parameters used in the chain for the numerical example

Parameter	Value	Unit
m	0.5	kg
k_l	24180.530	N/m
ω_0	70 π	rad/s
ζ	0.3%	Dimensionless
B_A	2.943	m/s^2
N	41	Dimensionless
L	2	m
Δs	0.05	m
ε	0.01	Dimensionless
δ	0.118	Dimensionless
k_c	2693.518	N/m

Table 2 Simulation parameters

Parameter	Value	Unit
Total simulation time	25	s
Time step between iterations	0.0005	s
h for time $< 5s$	0.4	Dimensionless
k_{nl} for time $< 5s$	22287207.593	N/m^3
h for time $\geq 5s$	0.25	Dimensionless
k_{nl} for time $\geq 5s$	8705940.466	N/m^3

unstable condition for values of h and ζ_e. After 5 s, the system is then subjected to a new set of parameters h and ζ_e, but under a condition of solitonic stability. Following the numerical integration, the amplitudes of the oscillators are plotted as a function of the space, and the formation of the stationary soliton is analyzed.

The results obtained after the numerical procedure are presented in Fig. 5. As can be seen, the vibration amplitude of the oscillators is concentrated at the origin, where some oscillators have high amplitudes, while increasing the distance from the origin leads to a decrease in the amplitude.

4 Discussion and Conclusion

This work introduces a methodology for obtaining localized vibration modes using solitonic equations in damped macromechanical structures. The main contribution of this work is a set of five steps that could be followed in the design of a structure to ensure that it has the conditions to reproduce solitonic standing waves.

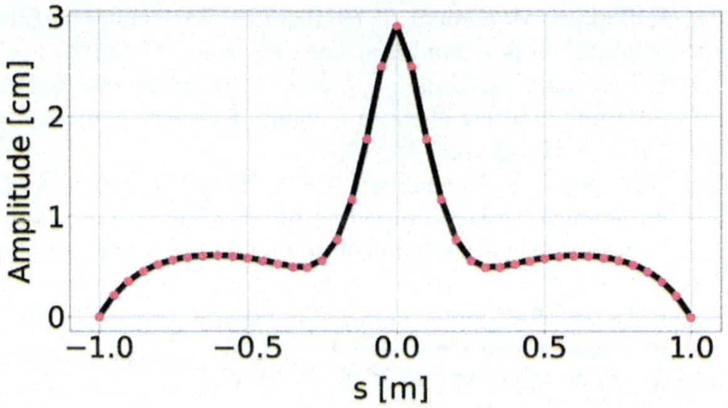

Fig. 5 Result of the numerical integration of (3). Pink dots representing the position and oscillation amplitude of the resonators at the end of the integration. Black curve connecting the dots for a better visualization of the curve

The damping plays a critical role in solitons and can considerably limit the range of possible design options. To overcome this limitation, it is essential to utilize a parameter chart that depicts the existence and stability of different configurations. This analysis can assist engineers in selecting an appropriate design that meets the desired performance.

An example of a possible physical configuration is provided, along with a suggested strategy for experimental implementation. Based on the numerical simulations, the methodology has demonstrated its ability to reproduce stationary waves in the oscillator chain, nevertheless, more study must be done regarding the numerical simulation algorithm, especially from the perspective of searching initial conditions for the integration in time.

Future work involves evaluating the continuity hypothesis through numerical analysis of the optimal number of oscillators and experimental validation of the presented approach. Furthermore, experimental validation is critical to confirm the results obtained by the analytical results.

References

1. M. Spletzer, A. Raman, H. Sumali, J.P. Sullivan, Highly sensitive mass detection and identification using vibration localization in coupled microcantilever arrays. Appl. Phys. Lett. **92**(11), 114102 (2008)
2. M. Manav, A.S. Phani, E. Cretu, Mode localization and sensitivity in weakly coupled resonators. IEEE Sens. J. **19**(8), 2999–3007 (2018)
3. Z. Zergoune, N. Kacem, N. Bouhaddi, On the energy localization in weakly coupled oscillators for electromagnetic vibration energy harvesting. Smart Mater. Struct. **28**(7), 07LT02 (2019)
4. E. Kenig, B.A. Malomed, M.C. Cross, R. Lifshitz, Intrinsic localized modes in parametrically driven arrays of nonlinear resonators. Phys. Rev. E **80**(4), 046202 (2009)
5. M. Sato, A.J. Sievers, Visualizing intrinsic localized modes with a nonlinear micromechanical array. Low Temp. Phys. **34**(7), 543–548 (2008)
6. M. Kimura, T. Hikihara, Coupled cantilever array with tunable on-site nonlinearity and observation of localized oscillations. Phys. Lett. A **373**(14), 1257–1260 (2009)
7. M. Kimura, T. Hikihara, Experimental manipulation of intrinsic localized modes in macro-mechanical system. Nonlinear Theory Appl. IEICE **3**(2), 233–245 (2012)

8. E. Perkins, M. Kimura, T. Hikihara, B. Balachandran, Effects of noise on symmetric intrinsic localized modes. Nonlinear Dyn. **85**, 333–341 (2016)
9. F. Fontanela, A. Grolet, L. Salles, A. Chabchoub, N. Hoffmann, Dark solitons, modulation instability and breathers in a chain of weakly nonlinear oscillators with cyclic symmetry. J. Sound Vib. **413**, 467–481 (2018)
10. A.D. Adile, F. Kenmogne, A.K.S. Tewa, H. Simo, A.M. Tahir, S. Kumar, Dynamics of a mechanical network consisting of discontinuous coupled system oscillators with strong irrational nonlinearities: resonant states and bursting waves. Int. J. Non-Linear Mech. **137**, 103812 (2021)
11. E. Destyl, S.P. Nuiro, D.E. Pelinovsky, P. Poullet, Coupled pendula chains under parametric PT-symmetric driving force. Phys. Lett. A **381**(46), 3884–3892 (2017)
12. X. Liu, X. Wang, H. Liu, S. Wang, M. Wen, Experimental demonstration of emission of solitons from a resonant localized wave. Phys. Rev. E **102**(5), 052201 (2020)
13. I.V. Barashenkov, Y.S. Smirnov, Existence and stability chart for the ac-driven, damped nonlinear Schrödinger solitons. Phys. Rev. E **54**(5), 5707 (1996)
14. N.V. Alexeeva, I.V. Barashenkov, G.P. Tsironis, Impurity-induced stabilization of solitons in arrays of parametrically driven nonlinear oscillators. Phys. Rev. Lett. **84**(14), 3053 (2000)
15. D. Bitar, N. Kacem, N. Bouhaddi, M. Collet, Collective dynamics of periodic nonlinear oscillators under simultaneous parametric and external excitations. Nonlinear Dyn. **82**, 749–766 (2015)
16. P. Li, S. Gao, H. Cai, L. Wu, Theoretical analysis and experimental study for nonlinear hybrid piezoelectric and electromagnetic energy harvester. Microsyst. Technol. **22**, 727–739 (2016)
17. D. Bitar, N. Kacem, N. Bouhaddi, Intrinsic localized modes in periodic nonlinear lattices under simultaneous excitations, in *International Congress of Sound and Vibration* (2017)
18. Y. Bromberg, M.C. Cross, R. Lifshitz, Response of discrete nonlinear systems with many degrees of freedom. Phys. Rev. E **73**(1), 016214 (2006)
19. D. Bitar, Collective dynamics of weakly coupled nonlinear periodic structures. Doctoral dissertation (Université Bourgogne Franche-Comté, 2017)
20. C. Boopathy, L. Kavitha, R. Ravichandran, Nonlinear modelling of soliton collision dynamics for blood flow in a stenotic artery. Mater. Today: Proc. **51**, A1–A13 (2022)
21. K. Shehzad, A.R. Seadawy, J. Wang, M. Arshad, Multi peak solitons and btreather types wave solutions of unstable NLSEs with stability and applications in optics. Opt. Quantum Electron. **55**(1), 7 (2023)
22. S.M. Hernández, J. Bonetti, N. Linale, D.F. Grosz, P.I. Fierens, Soliton solutions and self-steepening in the photon-conserving nonlinear Schrödinger equation. Waves Random Complex Media **32**(5), 2533–2549 (2022)
23. S.A. El-Tantawy, A.H. Salas, H.A. Alyousef, M.R. Alharthi, Novel approximations to a non-planar nonlinear Schrödinger equation and modeling nonplanar rogue waves/breathers in a complex plasma. Chaos, Solitons Fractals **163**, 112612 (2022)
24. A.S. Ali, A. Govindarajan, M. Lakshmanan, Filamentation and stabilization of vortex solitons in nonlinear metamaterial waveguides. Phys. Lett. A **451**, 128416 (2022)

Global Dynamical Analysis of a Boost Converter with a Constant Power Load and Actively Damped by a Series Loss-Free Resistor

Luis Benadero, Abdelali El Aroudi, Max Sebastiá-Rullo, Hugo Valderrama-Blavi, Angel Cid-Pastor, and Luis Martínez-Salamero

Abstract Two-stage cascaded DC–DC converters appear in many power electronics engineering applications. In this cascaded scheme, the first stage is called the source or upstream converter while the second one is called the load or downstream converter. If the downstream converter is controlled to maintain a tightly regulated output voltage on the load, it behaves as a constant power load (CPL) for the source converter. The overall dynamics of this arrangement tends to be unstable, so it can be stabilized by means of a suitable feedback, or instead in open loop operation, by using an additional resistor, which can be a loss free resistor (LFR), thus avoiding efficiency losses. This work deals with the dynamical analysis of an open-loop DC—DC boost converter with a stabilizing resistor in series with the inductor and a constant power load by means of an averaged model. Fold and Hopf bifurcations explain the existence and stability conditions of the two possible equilibrium points. Moreover, an appropriate design requires not only that the equilibrium point must be stable but also that its basin of attraction must be large enough to include normal initial conditions. Then, we make an extended analysis taking into account the global dynamics of the system, mainly related to the saddle stable manifold or to an unstable large limit cycle, whenever it exists.

L. Benadero
Universitat Politècnica de Catalunya, Barcelona, Spain
e-mail: luis.benadero@upc.edu

A. El Aroudi (✉) · M. Sebastiá-Rullo · H. Valderrama-Blavi · A. Cid-Pastor · L. Martínez-Salamero
Universitat Rovira i Virgili, Tarragona, Spain
e-mail: abdelali.elaroudi@urv.cat

M. Sebastiá-Rullo
e-mail: max.sebastia@urv.cat

H. Valderrama-Blavi
e-mail: hugo.valderrama@urv.cat

A. Cid-Pastor
e-mail: angel.cid@urv.cat

L. Martínez-Salamero
e-mail: luis.martinez@urv.cat

M. Belhaq (ed.), *Advances in Nonlinear Dynamics and Control of Mechanical and Physical Systems*, Springer Proceedings in Physics 301,
https://doi.org/10.1007/978-981-99-7958-5_9

1 Introduction

Two-stage cascaded DC–DC converters are widely used in various applications such as electric vehicles and ships, microgrids, more electric power aircrafts and more electric ships among others [1–5]. In these systems, the first stage is called the source or the upstream converter and the second stage is the load or the downstream converter. When the second stage is controlled to maintain a tightly regulated output voltage on the load, it behaves as a constant power load (CPL) for the source converter [3–5]. An electrical machine in an electric vehicle with tight velocity control and under constant torque operation will absorb constant power from the three-phase DC–AC inverter hence resulting in a CPL to the upstream DC–DC converter [4]. With a constant power load, the output voltage decreases as the load current increases to maintain the constant power demand. In such a scenario, the load converter exhibits a negative incremental impedance to the source converter leading to a high risk of instability [1, 2, 4, 6, 7]. Since the overall dynamics of the converter with CPL tends to be unstable, it can be stabilized by means of a suitable feedback.

The stabilization and control of DC–DC converters with CPLs have been extensively studied, and various control strategies have been proposed to achieve stable performance [6–11]. Instead of using feedback, to make stable a DC–DC converter with CPL in open loop operation, damping filters can be used [12]. Another way to stabilize these systems is by aggregating parasitic resistances to some key elements such as inductors and switching elements such as transistors and diodes [5]. Because this strategy would imply a reduction in the efficiency of the converter, a convenient alternative solution would be the use of a loss-free resistor (LFR). Namely, an LFR is a two-port element with resistive behaviour at its input port and transfers the absorbed power at the input port to its output port without losses. This LFR can be synthesized using power converters by imposing the proportionality between the voltage and current at its input port [13, 14].

This work deals with the analysis of the dynamics of an open loop DC–DC boost converter with a stabilizing resistor in series with the inductor and a CPL. To simplify the analysis and to provide algebraic explicit mathematical results, an averaged model of the circuit is used with a reduced three-dimensional parameter space. This model has two possible equilibrium points, a node and a saddle, whose existence and stability are linked to codimension one smooth fold and Hopf bifurcations, both emerging from a codimension-two Bogdanov–Takens bifurcation. Actually, the node equilibrium, which can be a focus or not, is the approximation of the desired limit cycle of the real switched system. Then, we get parameter constraint for a correct operation from this analysis. Moreover, an appropriate design requires not only a stable limit cycle with small amplitude, but also that its basin of attraction being wide enough to include normal initial conditions. Then we make an extended analysis taking into account the global dynamics of the system, mainly related to the saddle stable manifold or to an unstable large limit cycle (ULLC) whenever it exists. The boundaries for existence in the parameter space of this unstable large amplitude cycle are those corresponding to Hopf bifurcation and a global homoclinic bifurcation, which also emerges from the Bogdanov–Takens bifurcation.

An averaged model of a bidirectional boost DC–DC converter with constant power load (CPL) is introduced in Sect. 2 making use of normalized variables and parameters. In Sect. 3, different limit sets and related bifurcations are justified. The accessibility of the desired equilibrium point under reasonable initial conditions is extensively studied in Sect. 4. Finally, in the last section, the main conclusions are summarized.

2 Circuit: Diagram, Expressions and Normalization

A simplified scheme of the boost converter is shown in Fig. 1. Let τ be the physical time and v_C and i_L the capacitor voltage and the inductor current, which are the state variables of the system. Physical parameters are C and L, that is the capacitance and inductance of corresponding elements, R_S is the resistance of the LFR, P is the constant power load and V_i is the main voltage source. The following differential equations are obtained by applying standard Kirchhoff laws to the circuit shown in Fig. 1:

$$\begin{aligned} C\frac{dv_C}{d\tau} &= \quad -\frac{P}{v_C} + (1-u)i_L \\ L\frac{di_L}{d\tau} &= -(1-u)v_C - R_S i_L + V_i \end{aligned} \quad . \tag{1}$$

The two switches S and $\overline{\text{S}}$ are activated in a complementary way, that is $\overline{u} = 1 - u$. Then if $u = 1$, S is ON (closed) and $\overline{\text{S}}$ is OFF (open). The opposite action is done when $u = 0$. Concerning the driving signal applied to the switches, T_S and d are its period and duty ratio (fraction for which $u = 1$), respectively. From this set of variables and parameters, the following normalized variables are defined,

$$t = \omega_0 \tau, \quad x = \frac{v_C}{V_i}, \quad y = \frac{i_L Z_0}{V_i}, \tag{2}$$

where t, x, y are the normalized time and state variables and

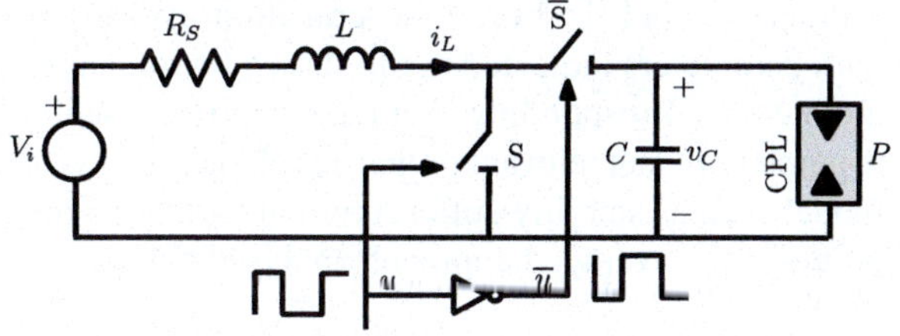

Fig. 1 Circuit schematic diagram of a bidirectional boost converter loaded by a CPL

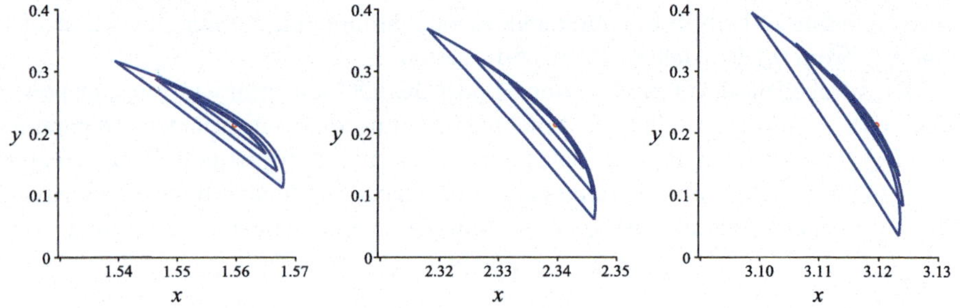

(a) $d = 0.4$ (b) $d = 0.6$ (c) $d = 0.7$

Fig. 2 Limit cycles (periodic orbits) with parameters $r = 0.3$, $p = 0.2$, $T \in \{0.25, 0.4, 0.55\}$ and d in the caption. Red points correspond to equilibrium in the averaging approach

$$\omega_0 = \sqrt{\frac{1}{LC}}, \quad Z_0 = \sqrt{\frac{L}{C}}.$$

Besides, normalized period, power and equivalent resistance parameters are

$$T = \omega_0 T_S, \quad p = \frac{P Z_0}{V_i^2}, \quad r = \frac{R_S}{Z_0}. \tag{3}$$

Finally, the averaged vector field, that is the normalized differential averaged equations for state variables (x, y) with parameters (p, r, d) can be summarized as

$$\dot{\mathbf{x}} = \mathbf{F}(\mathbf{x}) \quad \text{with} \quad \left\{ \begin{array}{l} F_x(x, y) = \dot{x} = \quad -\dfrac{p}{x} + (1-d)y \\ F_y(x, y) = \dot{y} = -(1-d)x - ry + 1 \end{array} \right\}, \tag{4}$$

where $\mathbf{x} = (x, y)^\mathsf{T}$ is the state vector. Notice that the binary variable $u \in \{0, 1\}$ in the switched model has been substituted by the duty ratio d, which is valid in the open interval $(0, 1)$. In the averaged approach, a continuous time and vector field are assumed for the evolution of the system, such that the state variables are in fact their averaged values within a switching cycle, thus replacing the exact switched model by the simplified averaged one. Note that the small amplitude desired limit cycle corresponding to the switched model becomes an equilibrium point for the averaged model. In Fig. 2, several limit cycles are represented in the state plane with parameters indicated in the shown figure. Note that cycles grow as T increases around the corresponding averaged value represented by a red point. It is also worth to emphasize that relevant properties of the limit cycles corresponding to the switched model such as stability suffer minimal variations regarding those of the equilibrium points of the averaged approach analysed below.

3 Stability and Bifurcation Analysis

3.1 Equilibrium Points and Associated Bifurcations

Two equilibrium points exist for the averaged system (4), whenever $4pr < 1$. These will be called here $\mathbf{x}_N = (x_N, y_N)^\mathsf{T}$ and $\mathbf{x}_S = (x_S, y_S)^\mathsf{T}$ and their expressions are straightforward from $\mathbf{F}(\mathbf{x}) = \mathbf{0}$.

$$\mathbf{x}_N = \begin{pmatrix} \dfrac{1+\sqrt{1-4rp}}{2(1-d)} \\ \dfrac{1-\sqrt{1-4rp}}{2r} \end{pmatrix}, \quad \mathbf{x}_S = \begin{pmatrix} \dfrac{1-\sqrt{1-4rp}}{2(1-d)} \\ \dfrac{1+\sqrt{1-4rp}}{2r} \end{pmatrix}. \tag{5}$$

The eigenvalues pairs λ_N and λ_S corresponding to $\mathbf{x}_N$ and $\mathbf{x}_S$, respectively, are obtained by means of the Jacobian matrix of the vector field $\mathbf{F}(\mathbf{x})$, that is the equivalent linear system of (4) around equilibrium (5)

$$\lambda_N = \frac{p}{2x_N^2} - \frac{r}{2} \pm \sqrt{W_N} \quad \text{where} \quad W_N = \left(\frac{p}{2x_N^2} + \frac{r}{2}\right)^2 - (1-d)^2, \tag{6}$$

$$\lambda_S = \frac{p}{2x_S^2} - \frac{r}{2} \pm \sqrt{W_S} \quad \text{where} \quad W_S = \left(\frac{p}{2x_S^2} + \frac{r}{2}\right)^2 - (1-d)^2. \tag{7}$$

Note that under the critical parameter condition $4rp = 1$, the two equilibrium collapse at the common fold point $\mathbf{x}_F$ with eigenvalues λ_{F1} and λ_{F2},

$$\mathbf{x}_F = \left(\frac{1}{2(1-d)}, \frac{1}{2r}\right)^\mathsf{T}, \quad \lambda_{F1} = 0, \quad \lambda_{F2} = \frac{(1-d)^2}{r} - r = \delta\left(2 + \frac{\delta}{r}\right), \tag{8}$$

where $\delta = 1 - r - d$. These results can be summarized in the following Remark.

Remark 1 In system (4), for the critical parameter condition $p = p_F = (4r)^{-1}$, the two equilibrium collapse, being null one of the common eigenvalues thus resulting in a smooth fold, also called saddle-node, codimension-one bifurcation.

Note that for this condition, the nonzero eigenvalue $\lambda_{F2} > 0$ has the same sign than δ. In the boundary, $\delta = 0$, we have

Remark 2 In system (4), for the two critical parameter conditions $p = p_F = (4r)^{-1}$ and $d = 1 - r$, in addition to the equilibrium extinction, their two common eigenvalues become null, thus resulting a smooth codimension-two Bogdanov–Takens bifurcation with two zero eigenvalues.

Moreover, a codimension-three parameter point defined as $\mathcal{P}_1 \equiv \{r = 1, p = 1/4, d = 1\}$, such that the Bogdanov–Takens bifurcation is located at the extreme of the duty ratio domain ($d = 1$), is associated to a relevant change in the bifurcation pattern. As a consequence the following remark results

Remark 3 If $r > 1$, the Hopf bifurcation cannot be given. This bifurcation only exists if $r < 1$, thus starting at the codimension-two Bogdanov–Takens, according to Remark 2.

In order to check the evolution of the eigenvalues λ_N and λ_S for the two equilibrium $\mathbf{x}_N$ and $\mathbf{x}_S$, with parameters $r > 0$, $0 < p < p_F = (4r)^{-1}$ and $d \in (0, 1)$, we evaluate W_N and W_S in (6) and (7) respectively, taking into account x_N and x_S in (5). Noticing that either W_N or W_S are expressed as a difference of two squared terms, both of them positive, they can only permute its sign when the difference of that two terms, called here Σ_N or Σ_S vanishes. These magnitudes are

$$\Sigma_N = \frac{(1-d)^2}{\sigma_N} + r - 2(1-d), \text{ where } \sigma_N = \frac{1 - 2rp + \sqrt{1-4rp}}{2p} \tag{9}$$

$$\Sigma_S = \frac{(1-d)^2}{\sigma_S} + r - 2(1-d), \text{ where } \sigma_S = \frac{1 - 2rp - \sqrt{1-4rp}}{2p} \tag{10}$$

We check that $\Sigma_S > 0$ in our parameter domain, so $\mathbf{x}_S$ is always a saddle.

However, concerning $\mathbf{x}_N$, $\Sigma_N = 0$ if $d = d_W = 1 - \sigma_N \left(1 \pm \sqrt{1 - r/\sigma_N}\right)$. Besides, $\Sigma_N = 0$ if $d = 1 - r/2$ with $p = 0$ and also, $\Sigma_N = 0$ if $d = 0$ with $p = p_0 = (2-r)(1 + 2r - r^2)^{-2}$. Therefore, if $r < 1$, the positive branch is valid in the range $d_W \in (0, 1-r)$ corresponding to p from p_0 to p_F and the negative branch is valid in the range $d_W \in (1-r, 1-r/2)$ corresponding to p from p_F to 0. If $1 < r < 2$, the positive branch for d_W is not valid and the negative one is given for $d_W \in (0, 1-r/2)$ associated to p from p_0 to 0. If $r > 2$, no branch applies for d_W, so the node is always non-spiral and stable.

Crossing the positive d_W branch implies a transition for the equilibrium node $\mathbf{x}_N$ between spiral or focus and non-spiral behaviour. The positive branch stands for unstable node and the negative one for the stable node.

The next and relevant question concerns stability. The spiral node $\mathbf{x}_N$, with an associated frequency $\sqrt{-W_N}$, becomes unstable when the real part of λ_N vanishes, which occurs at the critical value

$$d_H = 1 - \sqrt{r\sigma_N} = 1 - \frac{1 + \sqrt{1-4rp}}{2}\sqrt{\frac{r}{p}}, \tag{11}$$

where σ_N is defined in (9). Actually, this is a smooth subcritical Hopf bifurcation, at which an ULLC is born. More details about the existence and significance of this large limit cycle, which coexists with the stable spiral equilibrium $\mathbf{x}_N$, will be given below.

In Fig. 3, the continuous black and red lines stand for the smooth fold (emergence of the saddle and stable node pairs) and the subcritical Hopf (stability boundary of the spiral node) bifurcations, respectively. The continuous grey line is associated to the saddle and unstable node pair appearance. The sparse dotted line corresponds to the transition between spiral and non-spiral behaviour of the node.

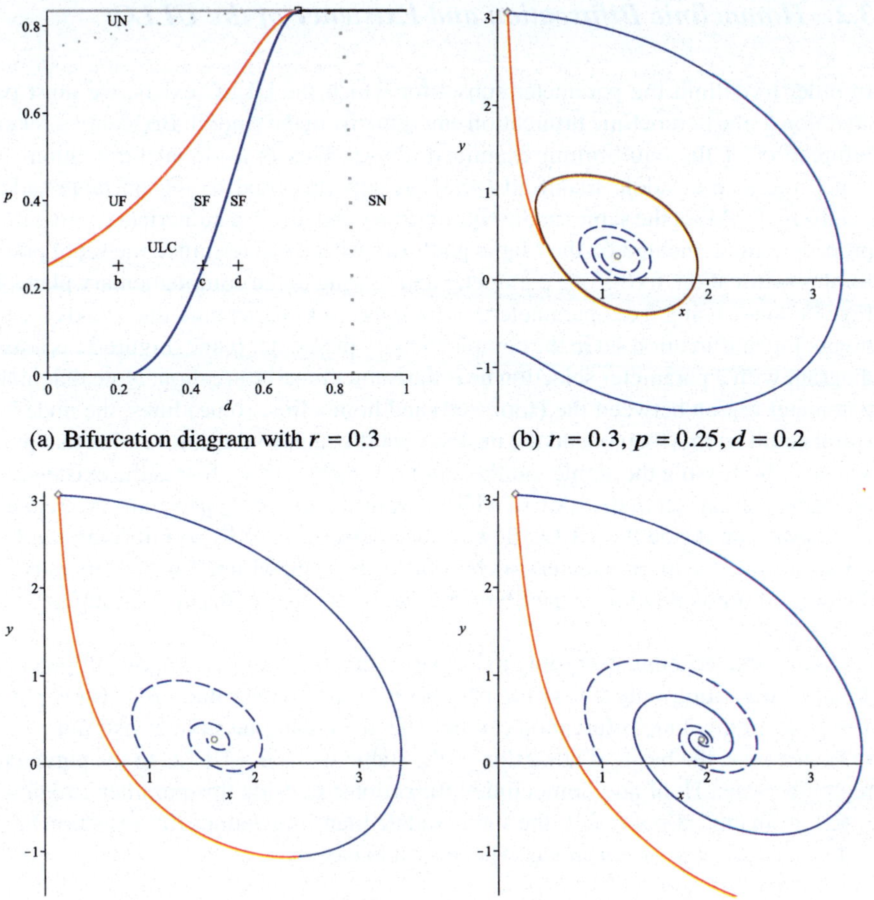

(a) Bifurcation diagram with $r = 0.3$

(b) $r = 0.3$, $p = 0.25$, $d = 0.2$

(c) $r = 0.3$, $p = 0.25$, $d = 0.432796$

(d) $r = 0.3$, $p = 0.25$, $d = 0.53$

Fig. 3 **a** Diagram using fixed parameter $r = 0.3$ in the complementary (d, p) parameter plane, including bifurcation lines: fold (black), Hopf (red), homoclinic (blue) and also the spiral to non-spiral transition (dotted grey line). The possible equilibrium, SF or UF and SN or UN (stable or unstable focus and stable or unstable node) are indicated in their existence regions. Besides, the ULLC exists in the parameter region between Hopf and homoclinic bifurcation. Notice also that the saddle equilibrium exists for $p < (4r)^{-1}$ and the codimension two bifurcation is given at $p = (4r)^{-1}$, $d = 1 - r$ (squared point), from which the bifurcating lines emerge. Panels (**b**–**d**) stand for dynamics in the state plane with parameters in the caption (see cross dots in panel (**a**) standing for these values), including relevant invariant sets: the stable spiral node and the saddle equilibrium (circle and diamond points), the stable and unstable saddle manifolds (red and blue lines), and the orbit from $\mathbf{x}_1 = (1, 0)^\mathsf{T}$ state (dashed blue line). Case (**b**) includes also the ULLC (maroon closed line) and case (**c**) close to the homoclinic bifurcation for which the stable and unstable saddle manifolds merge together

3.2 Homoclinic Bifurcation and Existence of the ULLC

In order to delimit the parameter subset for which the ULLC exists, we must pay attention to the homoclinic bifurcation emerging from the smooth Bogdanov–Takens bifurcation of the equilibrium examined above. This is a global bifurcation, for which the stable saddle manifold (SSM) is superimposed to the unstable saddle manifold (USM) of the same saddle equilibrium point. From a numerical continuation procedure implemented on the Maple platform, we have determined the set of points for this bifurcation, fixing one parameter and solving in the complementary plane. In Fig. 3a, made using fixed parameter $r = 0.3$, the continuous blue line stands for the homoclinic bifurcation set in the complementary parameter plane. Figure 3c is a state diagram with a parameter selection near the homoclinic bifurcation. Note that in the parameter region between the Hopf (red) and homoclinic (blue) lines, the node is a spiral equilibrium (focus), and the unstable saddle manifold cannot end at that point as it avoids crossing the stable saddle manifold (SSM). Thus, forces the existence of an extra invariant set in the form of a ULLC splitting the state space into two regions. Orbits starting inside the ULLC flow to the stable focus (SF) equilibrium, see Fig. 3b. Conversely, with parameters such in Fig. 3d, the SSM itself is the separatrix of the desired orbits flowing to the SF or the undesired ones collapsing as approaching to $x = 0$.

Taking into account, according to Remark 3 that the codimension two Bogdanov–Takens bifurcation only exists for parameter $r < 1$, with values $p = (4r)^{-1}$, $d = 1 - r$, the homoclinic bifurcation can only be given with the same constraint $r < 1$. In such a case, the basin of attraction of the stable focus is a ULLC in the parameter region between Hopf and homoclinic bifurcations. Outside this parameter region if $r < 1$, or in any case if $r > 1$, the stable saddle manifold defines the attraction basin of the stable (spiral or not) node, whenever it exists.

4 Parameter Sets Associated to the Boundary of the Basin of Attraction of the Stable Equilibrium

This section deals with the feasibility of reaching the equilibrium point whenever it exists and is stable, thus concerning the boundary of its basin of attraction. We will concentrate on two conditions. As a stricter one, the inclusion in that basin of the state $\mathbf{x}_1 = (1, 0)^\mathsf{T}$ as a likely starting point of the dynamics, is considered. As a second more relaxed condition, we impose that at least one point of the abscissa, that is $\mathbf{x} = (x, 0)^\mathsf{T}$, belongs to the basin. In order to find critical parameter values for which these conditions apply, two excluding cases are possible. One refers to the ULLC, and the stable saddle manifold (SSM) is involved in the second case. As it is well known that these two invariant sets can be boundaries of an attraction basin. If the ULLC exists, this set will determine the critical condition; otherwise, the SSM will do. Let us define some codimension-three bifurcation points concerning the basin of attraction of the stable equilibrium.

A first (defined above) codimension-three bifurcation point corresponds to the parameter set $\mathcal{P}_1 \equiv \{r = r_1 = 1, p = 1/4, d = 1\}$. For this set, the Bogdanov–Takens bifurcation is done at one extreme of the parameter d domain. Therefore, Hopf and homoclinic bifurcation cannot exist if $r > r_1$.

A second codimension-three point is obtained numerically by imposing the following constraints: homoclinic bifurcation with tangency at $\mathbf{x}_1$ at the extreme value $d = 0$. Approximately, this is $\mathcal{P}_2 \equiv \{r = r_2 \simeq 0.695540, p \simeq 0.262139, d = 0\}$. If $r > r_2$, the state $\mathbf{x}_1$ cannot be attained by the ULLC, then the critical boundary condition at $\mathbf{x}_1$ is limited exclusively by the SSM. If $r < r_2$, the homoclinic bifurcation with tangency at $\mathbf{x}_1$ is given at a codimension-two point with $d = d_C > 0$. If $d < d_C$, the ULLC determines the critical boundary of the basin at $\mathbf{x}_1$. Otherwise, if $d > d_C$, the SSM must be used instead. Summarizing, if the ULLC exists, this set will determine the boundary of the basin of attraction, otherwise only the SSM defines this boundary.

4.1 Boundary of the Basin of Attraction Includes State $\mathbf{x}_1$

We will first determine the parameter regions for which orbits starting at the state $\mathbf{x}_1 = (1, 0)^\mathsf{T}$ flows to the stable equilibrium. In order to determine the frontier of this region, two scenarios are possible. If $r < r_2 \simeq 0.695540$ (parameter r at $\mathcal{P}_2$) and inside the parameter region between the Hopf and the homoclinic bifurcation, the frontier line should be determined by forcing the ULLC passing through $\mathbf{x}_1$. This case is represented in Fig. 4 with long dashed green lines in several panels for fixed values of parameter r. Otherwise, the frontier line, if exists, is determined by forcing the SSM passing through $\mathbf{x}_1$. Dot–dash green lines, also in Fig. 4, correspond to this second occurrence.

In order to give a more global view of the parameter region with appropriate starting conditions, some alternative representations are given in Fig. 5. In panels (a–c) of this figure, the critical $p(r)$ condition such that $\mathbf{x}_1$ belongs to the boundary is plotted for some fixed values of parameter d. Herein, dashed and dot-dashed green lines stand for ULLC and SSM cases, respectively, and the solid black line represents the fold existence limit, because in these cases, the state $\mathbf{x}_1$ is always inside the basin of attraction. The connection between those cases is represented by squared points. In panel (d), the boundary (green lines in Fig. 4) is plotted all together. From these diagrams, we can appreciate that certain parameter combinations are more suitable in order to optimize the basin of attraction of equilibrium, thus avoiding undesired dynamics if $\mathbf{x}_1$ is assumed as the starting state. In the panel (e), we represent the optimal value of r (red line) in front of the duty ratio d, thus providing a maximum available value of parameter p with safe conditions. The blue line in this diagram indicates this maximum value of p. Note that this critical situation corresponds to the ULLC as the basin of attraction including the state $\mathbf{x}_1$.

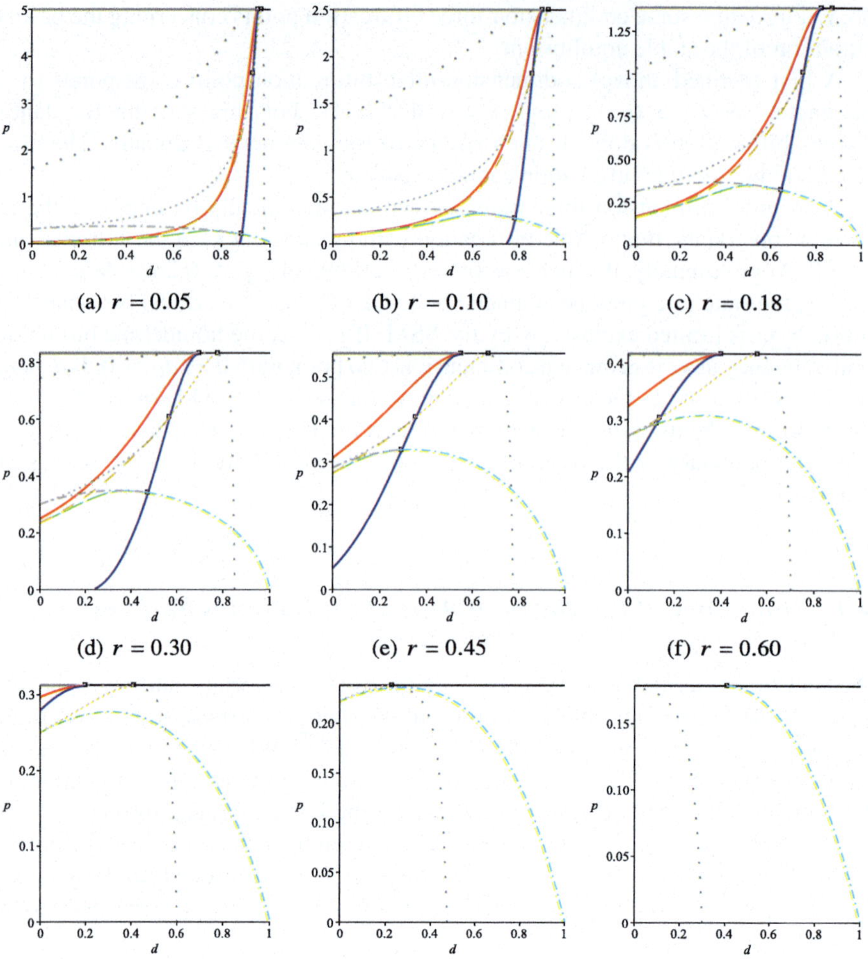

(a) $r = 0.05$ (b) $r = 0.10$ (c) $r = 0.18$

(d) $r = 0.30$ (e) $r = 0.45$ (f) $r = 0.60$

(g) $r = 0.80$ (h) $r = 1.05$ (i) $r = 1.40$

Fig. 4 Diagrams in (d, p) parameter plane with parameter r fixed given in the caption

4.2 Boundary of the Basin of Attraction Tangent to x-Axis

As a second more relaxed condition, we impose that at least one point of the abscissa, that is $\mathbf{x} = (x, 0)^\mathsf{T}$ belongs to the basin. Actually, $\mathbf{x}$ is tangent to the abscissa if it is an isolated point. If that condition were not satisfied, dynamics will not flow to the stable equilibrium point with a zero initial inductor current condition for any value of the initial capacitor voltage. As in the above cases for boundary at $\mathbf{x}_1$, this condition should be applied either to the ULLC or to the SSM. Orange lines in Fig. 4 stand for these more relaxed cases, dashed for ULLC and dotted for SSM cases.

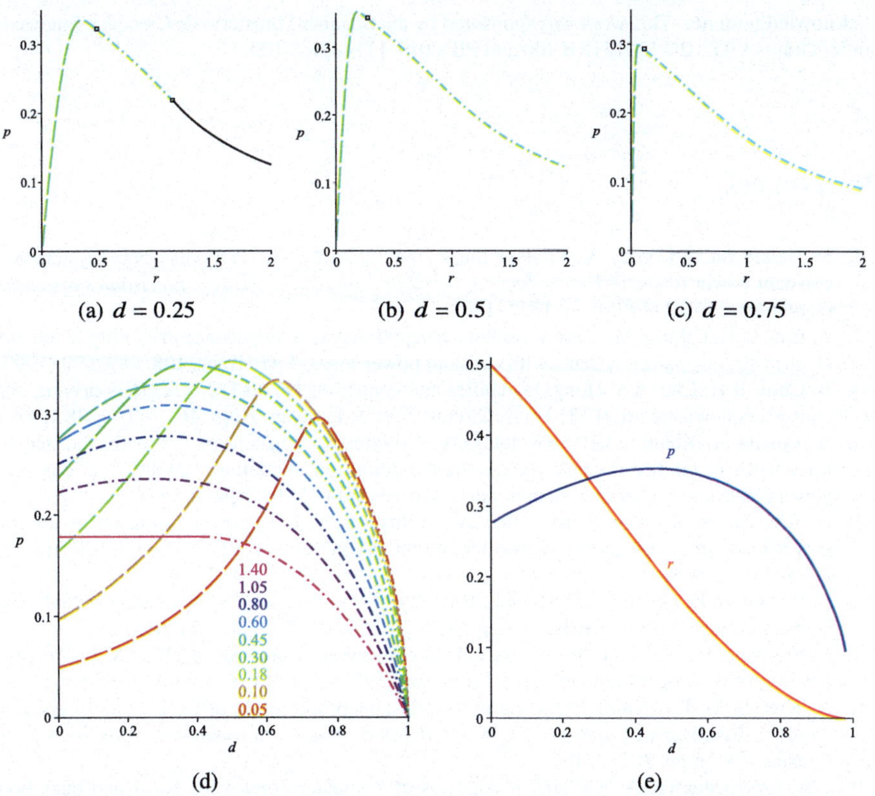

Fig. 5 Diagrams related to basin of attraction boundary at $\mathbf{x}_1 = (1, 0)^\top$. Panels (**a**–**c**) are for $p(r)$ relation with fixed d shown in the caption. Dashed, dot-dashed and black lines refer to ULLC, SSM and no-bound cases respectively. In panel (**d**), $p(d)$ is represented for several values of parameter r from 0.05 to 1.4 (see legend). In panel (**e**), $r(d)$ parameter relation is such that gives a maximum value of p that optimizes the boundary and in the same diagram $p(d)$ indicates that maximum value

5 Overview

The dynamics of a DC–DC boost open loop controlled converter, with a CPL and an actively damped loss-free resistor, has been analysed by means of nonlinear tools in an averaged model, that is a smooth approximation of the exact switched model. Invariant sets such an equilibrium, saddle manifolds and unstable large limit cycles are found, and some related transitions, such as fold, Hopf and homoclinic bifurcations are elucidated. Parameter optimization has been studied focusing on the accessibility of the desired periodic orbit (equilibrium in the averaging approach), by considering the inclusion or not of reasonable starting condition in the equilibrium basin of attraction. Future research will deal with changes due to the discontinuous conduction mode, if a diode acts as complementary switch.

Acknowledgements This work was sponsored by the Spanish Ministerio de Ciencia e Innovación under Grants PID2020-120151RB-I00 and PID2019-111443RB-I00.

References

1. M. Belkhayat, R. Cooley, A. Witulski, Large signal stability criteria for distributed systems with constant power loads, in *Proceedings of the 26th Annual IEEE Power Electronics Specialists Conference*, vol. 2 (1995), pp. 1333–1338
2. A. Emadi, B. Fahimi, M. Ehsani, On the concept of negative impedance instability in the more electric aircraft power systems with constant power loads. SAE Trans. **108**, 689–699 (1999)
3. B. Choi, B.H. Cho, S.S. Hong, Dynamics and control of DC-to-DC converters driving other converters downstream. IEEE Trans. Circuits Syst. I: Fundam. Appl. **46**, 1240–1248 (1999)
4. A. Emadi, A. Khaligh, C.H. Rivetta, G.A. Williamson, Constant power loads and negative impedance instability in automotive systems: definition, modeling, stability, and control of power electronic converters and motor drives. IEEE Trans. Veh. Technol. **55**, 1112–1125 (2006)
5. A. Khaligh, Realization of parasitics in stability of DC-DC converters loaded by constant power loads in advanced multiconverter automotive systems. IEEE Trans. Ind. Electron. **55**, 2295–2305 (2008)
6. L. Benadero, R. Cristiano, D. Pagano, E. Ponce, Nonlinear analysis of interconnected power converters: a case study. IEEE J. Emerg. Sel. Top. Circuits **5**, 326–335 (2015)
7. R. Cristiano, D. Pagano, L. Benadero, E. Ponce, Bifurcation analysis of a DC-DC bidirectional power converter operating with constant power loads. Int. J. Bifurc. Chaos **26** (2016)
8. A. Kwasinski, P.T. Krein, Stabilization of constant power loads in Dc-Dc converters using passivity-based control, in *INTELEC 07—29th International Telecommunications Energy Conference* (2007), pp. 867–874
9. C.N. Onwuchekwa, A. Kwasinski, Analysis of boundary control for boost and buck-boost converters in distributed power architectures with constant-power loads, in *Twenty-Sixth Annual IEEE Applied Power Electronics Conference and Exposition (APEC)* (2011), pp. 1816–1823
10. A. Kwasinski, C.N. Onwuchekwa, Dynamic behavior and stabilization of DC microgrids with instantaneous constant-power loads. IEEE Trans. Power Electron. **26**, 822–834 (2011)
11. B.A. Martinez-Treviño, A. Aroudi, A. Cid-Pastor, L. Martinez-Salamero, Nonlinear control for output voltage regulation of a boost converter with a constant power load. IEEE Trans. Power Electron. **34**, 10381–10385 (2019)
12. R.W. Erickson, D. Maksimovic, *Input Filter Design. Fundamentals of Power Electronics* (Springer International Publishing, 2020), pp. 675–723
13. A. Cid-Pastor, L. Martinez-Salamero, A. El Aroudi, R. Giral, J. Calvente, R. Leyva, Synthesis of loss-free resistors based on sliding-mode control and its applications in power processing. Control. Eng. Pract. **21**, 689–699 (2013)
14. R. Haroun, A. El Aroudi, A. Cid-Pastor, E. Vidal-Idiarte, H. Valderrama-Blavi, L. Martinez-Salamero, Modelling and control of modular DC-nanogrids based on loss-free resistors. IEEE Access **8**, 33305–33317 (2020)

Exact Numerical Method for Predicting Instabilities of Switching Converters with Constant Power Load

Kuntal Mandal and Abdelali El Aroudi

Abstract Switching converters are nonlinear dynamical systems irrespective of the load and can exhibit different types of instabilities with the variation of the parameters. Due to the ON/OFF nature of the periodically controlled switches, this kind of system is modeled by a set of differential equations (i.e., subsystem) after satisfying different switching conditions within a switching period. Therefore, their steady-state behavior shows periodic waveforms in the time domain and periodic orbit in the state space. For resistive load, the stability analysis of the periodic orbit can be done by calculating the fundamental solution matrix or monodromy matrix using exponential matrices for evolution within the linear subsystems, and the saltation matrices for transition across subsystems through switching conditions. However, with Constant Power Load (CPL), the subsystems are nonlinear and the state matrices are state dependent. Therefore, the state transition matrices in each subsystem are not analytically available and have to be calculated numerically by solving the system differential equations and variational equations simultaneously. The presented numerical method for stability analysis is applicable to different operating conditions with no limitation in problem size i.e., dimensions and number of subsystems. A higher number of subsystems in a period only increases the number of matrices to be multiplied for the calculation of the monodromy matrix, and not the dimension of it. Apart from switching converters, this method can be used in general for any hybrid systems including mechanical impacting systems, robotic systems, power systems, and neuron models.

K. Mandal (✉) · A. El Aroudi
Universitat Rovira i Virgili, Tarragona, Spain
e-mail: dr.kuntal.mandal@gmail.com

A. El Aroudi
e-mail: abdelali.elaroudi@urv.cat

M. Belhaq (ed.), *Advances in Nonlinear Dynamics and Control of Mechanical and Physical Systems*, Springer Proceedings in Physics 301,
https://doi.org/10.1007/978-981-99-7958-5_10

1 Introduction

During the last three decades, different types of complex nonlinear phenomena in switching converters have been reported with theoretical and experimental results [3, 21]. These phenomena impose low-frequency oscillations (slow-timescale) as well as subharmonic and chaotic oscillations (fast-timescale) in the system behavior. These have harmful effects on the system operation such as system failure, large ripple, and poor efficiency. Therefore, unfolding different types of instabilities in such systems is important both from theoretical and practical points of view. Recently, power electronic converters and inverters loaded by Constant Power Load (CPL) have emerged as a major topic of research [7] due to its different industrial applications such as microgrids [8] and electrified transportation [20]. In these applications, the input source voltage and output load current may vary widely. Moreover, the converter can operate in Continuous Conduction Mode (CCM) and Discontinuous Conduction Mode (DCM) based on the different applications [19].

The advanced modeling technique for stability analysis could unfold the exact onset of the slow- and fast-timescale instabilities under different operating conditions of the system compared to the averaged or approximate design-oriented models [5, 6, 22]. Therefore, finding the stability boundaries in the parameter space helps the designer to select the parameters for reliable and efficient operation of the system [1, 16]. Though some efforts have been put to quantify the stability of switching converters with CPL [10, 23], to the best of authors' knowledge there is no general automated tool that can detect stable as well as unstable periodic orbit along with quantifying their stability. The detection of the unstable periodic orbit is important for the understanding of the underlying mechanism of different types of instabilities.

In the past, rigorous efforts have been put into developing a general automated tool [14] for stability analysis of switching systems. Most of the examples were mainly taken for switching converters with linear subsystems [15, 17, 18] and switching inverters with linear subsystems [2]. A few examples with nonlinear subsystems such as Alpazur oscillator [4], DC series motor [11] were also analyzed. In this paper, the algorithm [14] has been used for extensive stability analysis of a DC-DC boost converter with CPL. The available methods in the literature have limitations on the problem size with many subsystems in a period [13].

2 System Description and Operation

Consider a boost converter loaded by a CPL depicted in Fig. 1. The system is under a Peak Current Mode Control (PCMC) with a voltage loop closed by a PI compensator. The ON/OFF signal u driving the switch S is generated by a comparator and an S-R latch. The duty ratio of the square signal u is decided according to the comparison between the sensed signal $R_s i_L$ and its desired reference $R_s i_{\text{ref}}$ which is provided by the output voltage mode control (VMC) for regulating the CPL voltage v_o to a

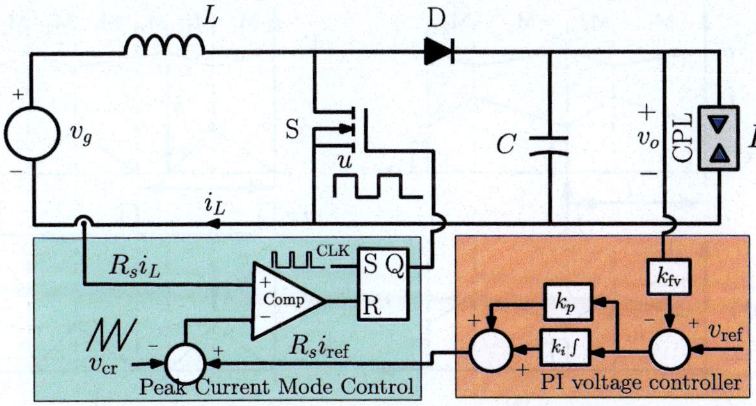

Fig. 1 Circuit diagram of a boost converter loaded by CPL. A two-loop control method is used where the inner loop is PCMC and the outer loop is PI compensated VMC

desired voltage reference v_{ref}. The outer voltage loop has a PI compensator with a proportional gain k_p and integral gain k_i. The switch S is closed periodically at the starting of each switching cycle. A compensating ramp signal $v_{cr} = V_M \left(\frac{t}{T} \bmod 1\right)$ with amplitude V_M and period T is substracted from the reference signal $R_s i_{ref} = k_p(v_{ref} - k_{fv}v_o) + \int k_i(v_{ref} - k_{fv}v_o)$ and the resulting signal is compared with the sensed signal $R_s i_L$. The intersection of these two signals makes the switch S open. Therefore, the switching condition is given as

$$\mathrm{h}_1 : (R_s i_{ref} - v_{cr}) - R_s i_L = 0$$

The other two switching conditions are determined by the inductor current reaching zero (in DCM) and at the end of each clock period as shown in Fig. 2.

These are given by the following equations

$$\mathrm{h}_2 : R_s i_L = 0, \quad \mathrm{h}_3 : \left(\frac{t}{T} \bmod 1\right) = 0$$

Therefore, within a switching period (or clock period) there are two subsystems ($\mathrm{M}_1 - \mathrm{M}_2$) in CCM (Fig. 2a) and three subsystems ($\mathrm{M}_1 - \mathrm{M}_2 - \mathrm{M}_3$) in DCM (Fig. 2b) for a typical steady-state at nominal operating condition. As these subsystems repeat in sequence for each clock period, this steady-state behavior is known as period-1. The transitions between subsystems occur after satisfying one of the switching conditions. Both in CCM and DCM, the closed-loop system can be expressed as a set of differential equations given in Table 1.

Now these differential equations will be represented by state equations for the ease of the analysis. By considering the state as $\boldsymbol{x} = [x_1 \ x_2 \ x_3]^\top = [v_o \ i_L \ v_i]^\top$, and vector field as $\boldsymbol{f} = [f_1(x_1, x_2, x_3) \ \ f_2(x_1, x_2, x_3) \ \ f_3(x_1, x_2, x_3)]^\top$, the differential equations for each subsystem are given in Table 2.

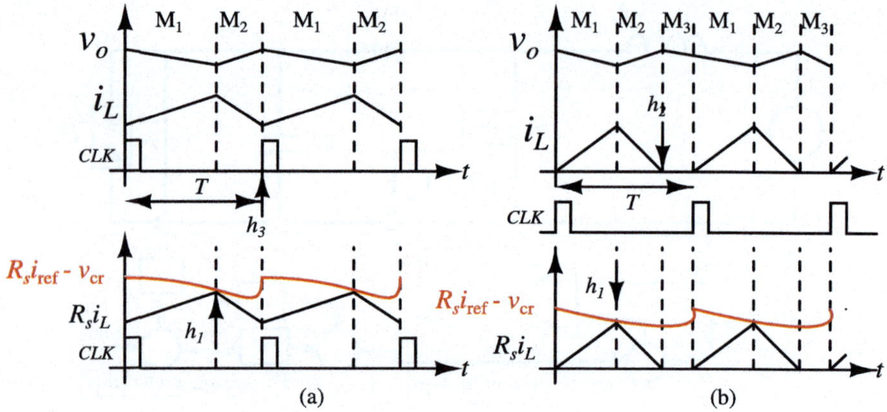

Fig. 2 The inductor current, capacitor voltage and control waveforms of the two-loop controlled closed-loop boost converter in **a** CCM and **b** DCM. The switching conditions are given by h_1, h_2, h_3

Table 1 Differential equations in different subsystems M_i for CCM and DCM operation

M_1: S ON D OFF	M_2: S OFF D ON	M_3: S OFF D OFF
$\frac{dv_o}{dt} = -\frac{P}{Cv_o}$	$\frac{dv_o}{dt} = -\frac{P}{Cv_o} + \frac{i_L}{C}$	$\frac{dv_o}{dt} = -\frac{P}{Cv_o}$
$\frac{di_L}{dt} = \frac{v_g}{L}$	$\frac{di_L}{dt} = -\frac{v_o}{L} + \frac{v_g}{L}$	$\frac{di_L}{dt} = 0$
$\frac{dv_i}{dt} = k_i(v_{\text{ref}} - k_{\text{fv}}v_o)$	$\frac{dv_i}{dt} = k_i(v_{\text{ref}} - k_{\text{fv}}v_o)$	$\frac{dv_i}{dt} = k_i(v_{\text{ref}} - k_{\text{fv}}v_o)$

Table 2 State equations in different subsystems M_i for CCM and DCM operation

M_1: S ON D OFF	M_2: S OFF D ON	M_3: S OFF D OFF
$\dot{x_1} = f_1 = -\frac{P}{Cx_1}$	$\dot{x_1} = f_1 = -\frac{P}{Cx_1} + \frac{x_2}{C}$	$\dot{x_1} = f_1 = -\frac{P}{Cx_1}$
$\dot{x_2} = \frac{v_g}{L}$	$\dot{x_2} = f_2 = -\frac{x_1}{L} + \frac{v_g}{L}$	$\dot{x_2} = f_2 = 0$
$\dot{x_3} = f_3 = k_i(v_{\text{ref}} - k_{\text{fv}}x_1)$	$\dot{x_3} = f_3 = k_i(v_{\text{ref}} - k_{\text{fv}}x_1)$	$\dot{x_3} = f_3 = k_i(v_{\text{ref}} - k_{\text{fv}}x_1)$

For different subsystems (M_1, M_2, M_3) the vector fields are denoted by $\boldsymbol{f}_1, \boldsymbol{f}_2, \boldsymbol{f}_3$. Note that the vector fields are represented in bold letter.

3 Stability Analysis Method for Periodic Orbit

Let us first consider a general nonlinear system which has no switching but exhibits periodic orbit in state-space. For the purpose of stability analysis, one has to find the state transition matrix. To find out this matrix along the periodic orbit we have to solve the system differential equations and variational equations simultaneously. The system equation is given in the form

$$\dot{\boldsymbol{x}} = \boldsymbol{f}(\boldsymbol{x}, t) \tag{1}$$

If $\boldsymbol{x}$ is the position of a particle in the state-space then the direction of the particle's velocity $\dot{\boldsymbol{x}}$ is given by the unique vector function $\boldsymbol{f}(\boldsymbol{x})$. In the state space, at any point $\boldsymbol{x}$, the particle follows a path at the tangent direction parallel to $\boldsymbol{f}(\boldsymbol{x})$. Therefore, a flow of particle will be created which can be represented by flow function $\boldsymbol{\Psi}(t, t_0, \boldsymbol{x}_0)$. In general nonlinear case, the flow depends on initial time t_0 and initial state $\boldsymbol{x}_0$. Therefore, the flow corresponds to the point at time t on the trajectory that passes through $\boldsymbol{x}_0$ at time t_0. The flow function satisfies

$$\frac{\partial \boldsymbol{\Psi}(t, t_0, \boldsymbol{x}_0)}{\partial t} = \boldsymbol{f}(\boldsymbol{\Psi}(t, t_0, \boldsymbol{x}_0), t), \quad \boldsymbol{\Psi}(t_0, t_0, \boldsymbol{x}_0) = \boldsymbol{x}_0 \tag{2}$$

For equilibrium point, $\boldsymbol{\Psi}(t, t_0, \boldsymbol{x}^*) = \boldsymbol{x}^*$ and $\boldsymbol{f}(\boldsymbol{x}^*) = 0$. In case of periodic solutions, $\boldsymbol{\Psi}(T + t_0, t_0, \boldsymbol{x}^*) = \boldsymbol{x}^*$ for $T > 0$.

The solution of the system equation $\dot{\boldsymbol{x}} = \boldsymbol{f}(\boldsymbol{x}, t)$ is given by

$$\boldsymbol{\Psi}(t, t_0, \boldsymbol{x}_0) = \boldsymbol{x}_0 + \int_0^t \boldsymbol{f}(\boldsymbol{\Psi}(\tau, t_0, \boldsymbol{x}_0), \tau) d\tau \tag{3}$$

As the state transition matrix (STM) is defined as the partial derivative of the solution with respect to the initial condition, we take partial derivative of (3)

$$\begin{aligned} \frac{\partial \boldsymbol{\Psi}(t, t_0, \boldsymbol{x}_0)}{\partial \boldsymbol{x}_0} &= \boldsymbol{I}_n + \int_0^t \frac{\partial \boldsymbol{f}(\boldsymbol{\Psi}(\tau, t_0, \boldsymbol{x}_0), \tau)}{\partial \boldsymbol{x}_0} d\tau \\ &= \boldsymbol{I}_n + \int_0^t \frac{\partial \boldsymbol{f}(\boldsymbol{\Psi}(\tau, t_0, \boldsymbol{x}_0), \tau)}{\partial \boldsymbol{\Psi}(\tau, t_0, \boldsymbol{x}_0)} \frac{\partial \boldsymbol{\Psi}(\tau, t_0, \boldsymbol{x}_0)}{\partial \boldsymbol{x}_0} d\tau \end{aligned} \tag{4}$$

where $\boldsymbol{I}_n$ is $n \times n$ identity matrix. By differentiating (4), the evolution of the STM (or sensitivity function [9]) can be given by the following matrix differential equation (also called variational equation)

$$\frac{d}{dt}\left[\frac{\partial \boldsymbol{\Psi}(t, t_0, \boldsymbol{x}_0)}{\partial \boldsymbol{x}_0}\right] = \left.\frac{\partial \boldsymbol{f}(\boldsymbol{x}, t)}{\partial \boldsymbol{x}}\right|_{x=\boldsymbol{\Psi}(t, t_0, \boldsymbol{x}_0)} \left[\frac{\partial \boldsymbol{\Psi}(t, t_0, \boldsymbol{x}_0)}{\partial \boldsymbol{x}_0}\right] \tag{5}$$

At $t = t_0$, $\boldsymbol{\Psi}(t_0, t_0, \boldsymbol{x}_0) = \boldsymbol{x}_0$, therefore $\left[\frac{\partial \boldsymbol{\Psi}(t_0, t_0, \boldsymbol{x}_0)}{\partial \boldsymbol{x}_0}\right] = \boldsymbol{I}_n$.

With this initial condition x_0 and using $\frac{\partial f(x,t)}{\partial x} = \begin{bmatrix} \frac{\partial f_1}{\partial x_1} & \frac{\partial f_1}{\partial x_2} & \frac{\partial f_1}{\partial x_3} \\ \frac{\partial f_2}{\partial x_1} & \frac{\partial f_2}{\partial x_2} & \frac{\partial f_2}{\partial x_3} \\ \frac{\partial f_3}{\partial x_1} & \frac{\partial f_3}{\partial x_2} & \frac{\partial f_3}{\partial x_3} \end{bmatrix}$, the solution of (5) gives the STM.

Moreover we can use the STM by denoting it as $\boldsymbol{\Phi}(t, t_0, \boldsymbol{x}_0) = \left[\frac{\partial \boldsymbol{\Psi}(t, t_0, \boldsymbol{x}_0)}{\partial \boldsymbol{x}_0}\right]$ to find out the evolution of the perturbation (earlier we use STM for the evolution of the trajectory $\boldsymbol{\Psi}(t, t_0, \boldsymbol{x}_0)$) as follows:

$$\frac{d}{dt}[\Delta\boldsymbol{\Psi}(t, t_0, \boldsymbol{x}_0)] = \boldsymbol{\Phi}(t, t_0, \boldsymbol{x}_0)[\Delta\boldsymbol{\Psi}(t, t_0, \boldsymbol{x}_0)] \tag{6}$$

If we have initial perturbation $\Delta\boldsymbol{\Psi}(t_0, t_0, \boldsymbol{x}_0)$ and final perturbation $\Delta\boldsymbol{\Psi}(t, t_0, \boldsymbol{x}_0)$ then applying linearization around the periodic trajectory up to first order

$$\Delta\boldsymbol{\Psi}(t, t_0, \boldsymbol{x}_0) = \boldsymbol{\Phi}(t, t_0, \boldsymbol{x}_0)\Delta\boldsymbol{\Psi}(t_0, t_0, \boldsymbol{x}_0) \tag{7}$$

where the STM ($\boldsymbol{\Phi}(t, t_0, \boldsymbol{x}_0)$) can be found by solving (5).

After evaluating the above perturbation at $t = T + t_0$, we get

$$\Delta\boldsymbol{\Psi}(T + t_0, t_0, \boldsymbol{x}_0) = \boldsymbol{\Phi}(T + t_0, t_0, \boldsymbol{x}_0)\Delta\boldsymbol{\Psi}(t_0, t_0, \boldsymbol{x}_0) \tag{8}$$

The matrix $\boldsymbol{\Phi}(t_0 + T, t_0, \boldsymbol{x}_0)$ is called monodromy matrix (M) of the periodic orbit. Therefore monodromy matrix (M) is a special case of STM calculated over the time period T of the periodic orbit or limit cycle.

The variational equation (6) can be used to describe how perturbation (or state-space volumes) evolve along a trajectory by the influence of STM [4]. For subsystem M_1 of the boost converter it looks like

$$\begin{bmatrix} \dot{\delta}_{x_1x_1} & \dot{\delta}_{x_2x_1} & \dot{\delta}_{x_3x_1} \\ \dot{\delta}_{x_1x_2} & \dot{\delta}_{x_2x_2} & \dot{\delta}_{x_3x_1} \\ \dot{\delta}_{x_1x_3} & \dot{\delta}_{x_2x_3} & \dot{\delta}_{x_3x_3} \end{bmatrix} = \begin{bmatrix} \frac{\partial f_1}{\partial x_1} & \frac{\partial f_1}{\partial x_2} & \frac{\partial f_1}{\partial x_3} \\ \frac{\partial f_2}{\partial x_1} & \frac{\partial f_2}{\partial x_2} & \frac{\partial f_2}{\partial x_3} \\ \frac{\partial f_3}{\partial x_1} & \frac{\partial f_3}{\partial x_2} & \frac{\partial f_3}{\partial x_3} \end{bmatrix} \begin{bmatrix} \delta_{x_1x_1} & \delta_{x_2x_1} & \delta_{x_3x_1} \\ \delta_{x_1x_2} & \delta_{x_2x_2} & \delta_{x_3x_2} \\ \delta_{x_1x_3} & \delta_{x_2x_3} & \delta_{x_3x_3} \end{bmatrix}$$
$$= \begin{bmatrix} \frac{P}{Cx_1^2} & 0 & 0 \\ 0 & 0 & 0 \\ -k_i k_{\mathrm{fv}} & 0 & 0 \end{bmatrix} \begin{bmatrix} \delta_{x_1x_1} & \delta_{x_2x_1} & \delta_{x_3x_1} \\ \delta_{x_1x_2} & \delta_{x_2x_2} & \delta_{x_3x_2} \\ \delta_{x_1x_3} & \delta_{x_2x_3} & \delta_{x_3x_3} \end{bmatrix} \tag{9}$$

The $\delta_{x_1x_2}$ is the component of the x_1 variation that came from the previous x_2 variation, and so on.

As the variations depend on the slope at each point along the trajectory, to find out the STM, we solve the system equation (1) and the variational equation (6) simultaneously. For this purpose, we define an augmented state vector by putting the system state on top of the variation state as

$$\boldsymbol{x}_{\mathrm{aug}} = [x_1, x_2, x_3, \delta_{x_1x_1}\delta_{x_1x_2}, \delta_{x_1x_3}, \delta_{x_2x_1}, \delta_{x_2x_2}, \delta_{x_3x_2}, \delta_{x_3x_2}, \delta_{x_3x_2}, \delta_{x_3x_3}]^{\top}$$

Note that for n-dimensional system, $(n + n^2)$ differential equations are solved. After using the system state equation from Table 2 and variational equation (9) for subsystem M_1, we get the final form of the differential equations as

$$
\dot{\boldsymbol{x}}_{\text{aug}} = \begin{bmatrix} f_1(x_1, x_2, x_3) \\ f_2(x_1, x_2, x_3) \\ f_3(x_1, x_2, x_3) \\ \frac{\partial f_1}{\partial x_1}\delta_{xx} + \frac{\partial f_1}{\partial x_2}\delta_{xy} + \frac{\partial f_1}{\partial x_3}\delta_{xz} \\ \frac{\partial f_1}{\partial x_1}\delta_{yx} + \frac{\partial f_1}{\partial x_2}\delta_{yy} + \frac{\partial f_1}{\partial x_3}\delta_{yz} \\ \frac{\partial f_1}{\partial x_1}\delta_{zx} + \frac{\partial f_1}{\partial x_2}\delta_{zy} + \frac{\partial f_1}{\partial x_3}\delta_{zz} \\ \frac{\partial f_2}{\partial x_1}\delta_{xx} + \frac{\partial f_2}{\partial x_2}\delta_{xy} + \frac{\partial f_2}{\partial x_3}\delta_{xz} \\ \frac{\partial f_2}{\partial x_1}\delta_{yx} + \frac{\partial f_2}{\partial x_2}\delta_{yy} + \frac{\partial f_2}{\partial x_3}\delta_{yz} \\ \frac{\partial f_2}{\partial x_1}\delta_{zx} + \frac{\partial f_2}{\partial x_2}\delta_{zy} + \frac{\partial f_2}{\partial x_3}\delta_{zz} \\ \frac{\partial f_3}{\partial x_1}\delta_{xx} + \frac{\partial f_3}{\partial x_2}\delta_{xy} + \frac{\partial f_3}{\partial x_3}\delta_{xz} \\ \frac{\partial f_3}{\partial x_1}\delta_{yx} + \frac{\partial f_3}{\partial x_1}\delta_{yy} + \frac{\partial f_3}{\partial x_2}\delta_{yz} \\ \frac{\partial f_3}{\partial x_1}\delta_{zx} + \frac{\partial f_3}{\partial x_1}\delta_{zy} + \frac{\partial f_3}{\partial x_2}\delta_{zz} \end{bmatrix} = \begin{bmatrix} -\frac{P}{Cx_1} \\ \frac{v_g}{L} \\ k_i(v_{\text{ref}} - k_{\text{fv}}x_1) \\ \left(\frac{P}{Cx_1^2}\right)\delta_{xx} + 0.\delta_{xy} + 0.\delta_{xz} \\ \left(\frac{P}{Cx_1^2}\right)\delta_{yx} + 0.\delta_{yy} + 0.\delta_{yz} \\ \left(\frac{P}{Cx_1^2}\right)\delta_{zx} + 0.\delta_{zy} + 0.\delta_{zz} \\ 0.\delta_{xx} + 0.\delta_{xy} + 0.\delta_{xz} \\ 0.\delta_{yx} + 0.\delta_{yy} + 0.\delta_{yz} \\ 0.\delta_{zx} + 0.\delta_{zy} + 0.\delta_{zz} \\ (-k_i k_{\text{fv}})\delta_{xx} + 0.\delta_{xy} + 0.\delta_{xz} \\ (-k_i k_{\text{fv}})\delta_{yx} + 0.\delta_{yy} + 0.\delta_{yz} \\ (-k_i k_{\text{fv}})\delta_{zx} + 0.\delta_{zy} + 0.\delta_{zz} \end{bmatrix} \tag{10}
$$

Note that there are 12 differential equations with initial condition

$$
\boldsymbol{x}_0 = [x_{10}\ x_{20}\ x_{30}\ 1\,0\,0\,0\,1\,0\,0\,0\,1]^{\top}.
$$

The final solution of the variational equations (at the event point $t = t_1$ i.e., after satisfying a switching condition $h_1(\boldsymbol{x}, t) = 0$) gives the state transition matrix for the subsystem M_1 in the form

$$
\boldsymbol{\Phi}_1 = \begin{bmatrix} \delta_{x_1x_1} & \delta_{x_2x_1} & \delta_{x_3x_1} \\ \delta_{x_1x_2} & \delta_{x_2x_2} & \delta_{x_3x_1} \\ \delta_{x_1x_3} & \delta_{x_2x_3} & \delta_{x_3x_3} \end{bmatrix}_{|t=t_1} \tag{11}
$$

Note that there are two and three subsystems within a periodic orbit in CCM and DCM respectively. Therefore, after solving the STM for the nonlinear subsystem M_1 we move to other nonlinear subsystems sequentially along with the calculation of saltation matrices at the switching events as shown in Fig. 3. The saltation matrix is also a state transition matrix across the switching event when the transition from one subsystem to the next subsystem happens [12].

The monodromy matrix of the typical period-1 orbit for CCM and DCM operation can be found as follows

$$
M_{\text{cycle}}^{\text{CCM}} = S_2\Phi_2 S_1\Phi_1, \quad M_{\text{cycle}}^{\text{DCM}} = S_3\Phi_3 S_2\Phi_2 S_1\Phi_1 \tag{12}
$$

This method can also be extended for higher periodic orbits. The periodic orbit becomes unstable when any of the eigenvalues of the matrix M_{cycle} goes out of the unit circle. If an eigenvalue becomes equal to -1, a first-timescale subharmonic oscillation sets in; if an eigenvalue becomes equal to $+1$, a saddle-node bifurcation is about to happen; and if a pair of complex conjugate eigenvalues assume a magnitude of unity, a slow-timescale oscillation sets in.

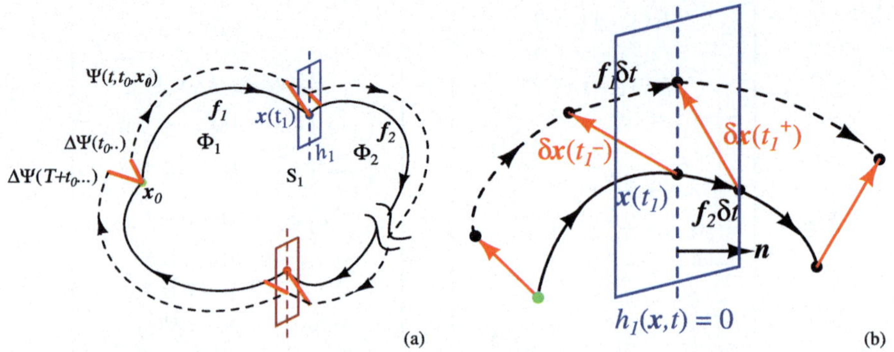

Fig. 3 **a** Evolution of flow $\Psi(t, t_0, x_0)$ and perturbation $\Delta\Psi(t, t_0, x_0)$ (red color) in the state space. **b** Enlarge view of evolution of perturbation across switching at $h_1(\boldsymbol{x}, t) = 0$

4 Quantification of the Instabilities

The parameter values of a practically designed boost converter (Fig. 1) loaded with CPL is given in Table 3.

By choosing the input voltage v_g as the bifurcation parameter, with the variation (decreasing) of it a bifurcation diagram is shown in Fig. 4a. It shows that a stable period-1 orbit bifurcates to a stable period-2 orbit. Then 4-piece chaotic attractors are created in a nonsmooth fashion. For clarity in the diagram, chaotic attractors are not shown here. The unstable period-1, period-2, and period-4 orbits are shown in Fig. 5 to understand the underlying mechanism of different types of bifurcation.

Table 3 Specifications and parameters for CCM and DCM operation

Specification	Value (CCM)	Value (DCM)
Input voltage (v_g)	6–12 V	6–12 V
Output voltage (v_0)	24 V± 1%	24 V± 1%
Output power (P)	48–200 W	10–50 W
Parameter	Value (CCM)	Value (DCM)
Inductance (L)	100 μH	100 μH
Capacitance (C)	47 μF	200 μF
Switching frequency (f_s)	100 kHz	10 kHz
Proportional gain (k_p)	1	1
Integral gain (k_i)	200 s^{-1}	200 s^{-1}
Ramp amplitude (V_M)	0.5	0.5
Current sensor resistance (R_s)	1 Ω	1 Ω
feedback gain (k_{vf})	0.1	0.1
Voltage reference (v_{ref})	2.4 V	2.4 V

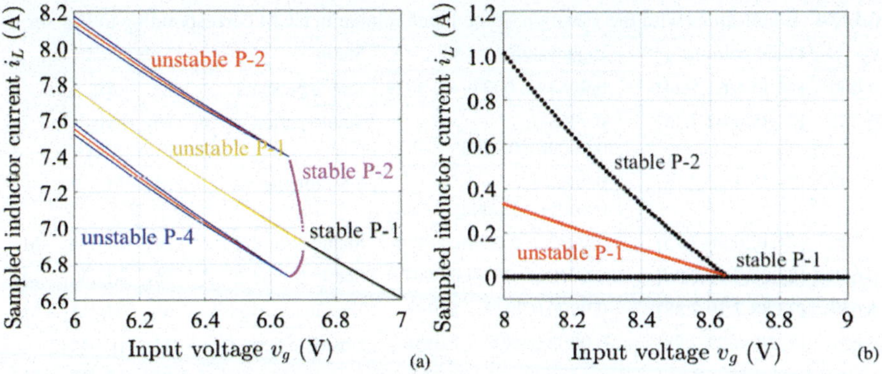

Fig. 4 The bifurcation diagrams with the variation of the input voltage v_g **a** CM ($P = 48\ W$) **b** DCM ($P = 24\ W$). Other parameter values are given in Table 3

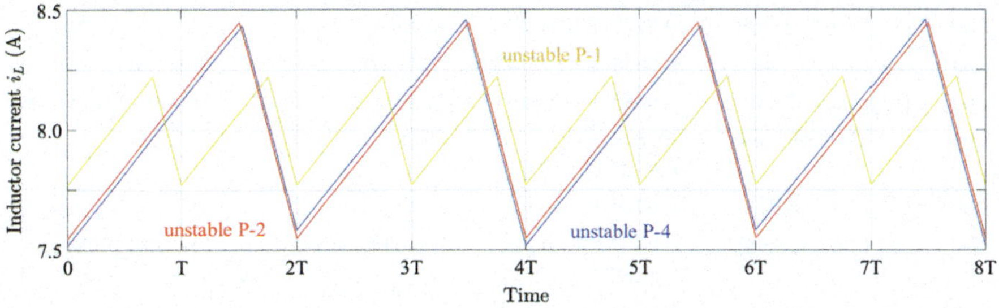

Fig. 5 The unstable periodic waveforms at $v_g = 6.0$ V in CCM corresponding to Fig. 4a

Therefore, a path following a bifurcation diagram (using the combination of shooting method with Newton-Raphson and continuation algorithm) is drawn which shows the stable and the unstable periodic orbits. All the bifurcation points and their types are identified by monitoring the eigenvalues of the monodromy matrix of the fixed point as shown in Table 4.

Similarly, with the parameter values for DCM, another bifurcation diagram is drawn with the variation of input voltage v_g as shown in Fig. 4b. With the decreasing value of the v_g, the system undergoes through a period doubling bifurcation. The bifurcation points and their types are identified by monitoring the eigenvalues of the monodromy matrix of the fixed-point as shown in Table 5.

Using the same stability analysis method, the stability curves have been drawn for CCM and DCM operations as shown in Fig. 6. By monitoring the maximum eigenvalues of the monodromy matrix with the variation of two parameters, the stability curves are constructed using the bifurcation points. In the case of CCM operation, the fast- and slow-timescale instability boundaries are compared with the expressions given in [5]. This approximate design expression for a specific operating condition shows good agreement in fast-timescale boundary with the exact stability

Table 4 Eigenvalues with the variation of the input voltage in CCM corresponding to Fig. 4a

v_g	Fixed point $[\boldsymbol{x}^*]^\top$	Eigenvalues	Status	Subsystem sequence
7.000	[24.15,6.61,7.44]	0.9964±0.0004i,−0.9539	Stable period-1	$M_1 - M_2$
6.701	[24.15,6.92,7.75]	0.9965 ± 0.0009i,−0.9996	Stable period-1	$M_1 - M_2$
6.700	[24.15,6.92,7.75]	0.9965 ± 0.0009i,−1.0002	Unstable period-1	$M_1 - M_2$
6.700	[24.12,6.98,7.75]	0.9932 ± 0.0098i,0.9998	Stable period-2	$M_1 - M_2 \| M_1 - M_2$
6.658	[23.88,7.39,7.88]	0.9910±0.0637i,0.9997	Stable period-2	$M_1 - M_2 \| M_1 - M_2$
6.657	[23.88,7.39,7.88]	0.9900,0.9949,−1.0146	Unstable period-2	$M_1 - M_2 \| M_1 - M_2$
6.657	[24.30,6.73,7.88]	0.9802,0.9898,−1.0156	Unstable period-4	$M_1 \| M_1 - M_2 \|$ $M_1 - M_2 \| M_1 - M_2$

Table 5 Eigenvalues with the variation of the input voltage in DCM corresponding to Fig. 4b

v_g	Fixed point $[\boldsymbol{x}^*]^\top$	Eigenvalues	Status	Subsystem sequence
9.00	[24.11, 0.0, 5.76]	0.9849 ± 0.0109i, 0.0	Stable period-1	$M_1 - M_2 - M_3$
8.65	[24.13,0.0,5.84]	0.9854 ± 0.0114i,0.0	Stable period-1	$M_1 - M_2 - M_3$
8.64	[24.13, 0.0,5.84]	0.9685 ± 0.0206i,0.0	Stable period-2	$M_1 – M_2 \| M_1 – M_2 – M_3$
8.64	[24.13,0.006,5.84]	0.9844 ± 0.0107i,−1.6741	Unstable period-1	$M_1 - M_2$
8.00	[24.14, 0.0,6.01]	0.9789 ± 0.0250i,0.0	Stable period-2	$M_1 – M_2 \| M_1 – M_2 – M_3$
8.00	[24.14,0.33, 5.98]	0.9864 ± 0.0121i, −1.8723	Unstable period-1	$M_1 - M_2 - M_3$

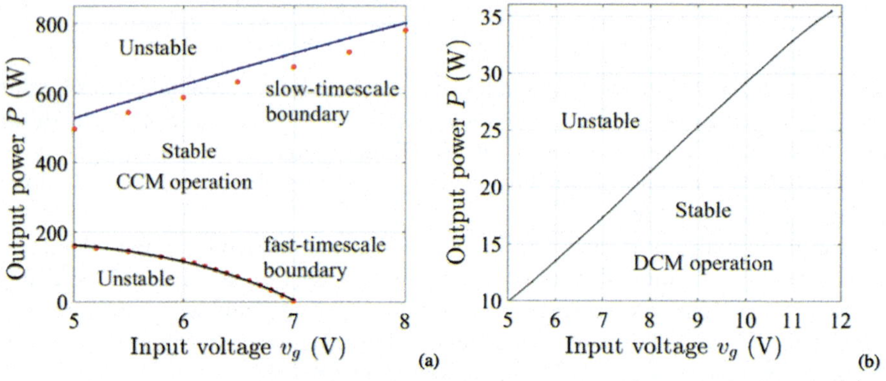

Fig. 6 The stability curve in the parameter space for **a** CCM and **b** DCM operation. In CCM, the solid lines (black and blue) are drawn using the design expression given in [5]

curve given by red dots. Surprisingly, there is some amount of difference between the approximate slow-timescale boundary and the exact stability curve (by red dots). For the slow-timescale oscillation, the bifurcation points and their eigenvalues are also given in Table 6.

Table 6 Eigenvalues for slow-timescale oscillations corresponding to Fig. 6a

v_g	P	Fixed point $[x^*]^\top$	Eigenvalues	Status	Subsystem
5.0	499.9	[25.72,99.78,100.39]	0.9914 ± 0.1271i, 0.9972	Stable period-1	$M_1 - M_2$
5.0	500.0	[25.72,99.80,100.41]	0.9919 ± 0.1271i, 0.9972	Unstable period-1	$M_1 - M_2$
7.0	676.8	[26.08,96.44,97.07]	0.9861 ± 0.1658i, 0.9977	Stable period-1	$M_1 - M_2$
7.0	676.9	[26.09,96.45,97.08]	0.9865 ± 0.1659i, 0.9977	Unstable period-1	$M_1 - M_2$
8.0	762.5	[26.22,95.04,95.68]	0.9832 ± 0.1820i, 0.9978	Stable period-1	$M_1 - M_2$
8.0	762.6	[26.22,95.05,95.69]	0.9835 ± 0.1821i, 0.9978	Unstable period-1	$M_1 - M_2$

5 Conclusion

A generalized method for stability analysis of the switching systems with nonlinear subsystems is presented. It makes use of the monodromy matrix over a complete period. This is calculated by solving the system equation and variational equation simultaneously for evolution within the subsystems, and the saltation matrices for transition across subsystems. Because this method treats each switching separately, it does not have any limitation on the complexity of the system, given by the dimensionality, and the number of subsystems and switching conditions in a periodic orbit. It is expected to be useful in carrying out a number of computations (fixed points, their stability, bifurcation diagram, phase space, stability curve in parameter space) in an efficient way.

Acknowledgements This work is supported and funded by the Agency for Management of University and Research Grants (AGAUR) under the Ministry of Research and Universities of the Government of Catalonia (Grant Reference No. 2020 BP 00260). A. El Aroudi acknowledges the support and funding of the Spanish Ministerio de Ciencia e Innovación under Grant PID2020-120151RB-I00.

References

1. A. Abusorrah, K. Mandal, D. Giaouris, A. El Aroudi, M.M. Al-Hindawi, Y. Al-Turki, S. Banerjee, Avoiding instabilities in power electronic systems: toward an on-chip implementation. IET Power Electron. **10**(13), 1778–1787 (2017)
2. A. Bandyopadhyay, K. Mandal, S. Parui, A Filippov method based analytical perspective on stability analysis of a dc-ac h-bridge inverter with nonlinear rectifier load. Int. J. Circuit Theory Appl. **50**(5), 1686–1708 (2022)
3. S. Banerjee, G.C. Verghese, *Nonlinear Phenomena in Power Electronics—Attractors, Bifurcations, Chaos, and Nonlinear Control* (IEEE Press, New York, 2001)

4. Q. Brandon, T. Ueta, D. Fournier-Prunaret, T. Kousaka, Numerical bifurcation analysis framework for autonomous piecewise-smooth dynamical systems. Chaos, Solitons Fractals **42**, 187–201 (2009)
5. A. El Aroudi, M.S. Al-Numay, R. Haroun, G. Zhang, Analytical determination of fast-scale instability boundaries for current mode controlled dc-dc converters with cpl and closed voltage loop. IEEE J. Emerg. Sel. Top. Circuits Syst. **11**(1), 39–48 (2021)
6. A. El Aroudi, R. Haroun, M. Al-Numay, J. Calvente, R. Giral, Fast-scale stability analysis of a dc-dc boost converter with a constant power load. IEEE J. Emerg. Sel. Top. Circuits Syst. **9**(1), 549–558 (2021)
7. A. El Aroudi, L. Benadero, R. Haroun, L. Martínez-Salamero, C.K. Tse, Bifurcation phenomena in open-loop dcm-operated dc-dc switching converters feeding constant power loads. Electronics **12**, 1–23 (2023)
8. A. Emadi, A. Khaligh, C.H. Rivetta, G.A. Williamson, Constant power loads and negative impedance instability in automotive systems: definition, modeling, stability, and control of power electronic converters and motor drives. IEEE Trans. Veh. Technol. **55**(4), 1112–1125 (2006)
9. I. Hiskens, M. Pai, Trajectory sensitivity analysis of hybrid system. IEEE Trans. Circuits Syst. I **47**(2), 204–220 (2000)
10. L. Huang, Y. Lu, A method for calculating the Lyapunov exponent spectrum of dc-dc converter feeding with a switching constant power load. IEEJ Trans. Electr. Electron. Eng. **15**(7), 1040–1047 (2020)
11. S. Kundu, D. Chatterjee, K. Chakrabarty, Bifurcation behaviour of pwm controlled dc series motor drive. Int. J. Circuit Theory Appl. **10**(3), 279–291 (2017)
12. R. Leine, H. Nijmeijer, *Dynamics and Bifurcations of Non-smooth Mechanical Systems*. (Springer, Berlin, 2004)
13. Y. Ma, J. Ing, S. Banerjee, M. Wiercigroch, E. Pavlovskaia, The nature of the normal form map for soft impacting systems. Int. J. Non-Linear Mech. **43**(6), 504–513 (2008)
14. K. Mandal, *SBSDS—Stability and Bifurcation of Switched Dynamical System* (DSWeb 2019 Tutorials on DS Software Contest, Natick, MA, 2019). https://dsweb.siam.org/Software/pager/225/page/4
15. K. Mandal, A. Abusorrah, M.M. Al-Hindawi, Y. Al-Turki, S. Banerjee, A new software for stability and bifurcation analysis of switched dynamical systems, in *2014 International Symposium on Non Linear Theory and its Applications (NOLTA2014)* (Luzern, Switzerland, 2014)
16. K. Mandal, A. Abusorrah, M.M. Al-Hindawi, Y. Al-Turki, A. El Aroudi, D. Giaouris, S. Banerjee, Control-oriented design guidelines to extend the stability margin of switching converters, in *2017 IEEE International Symposium on Circuits and Systems (ISCAS)* (Baltimore, MD, USA, 2017)
17. K. Mandal, S. Banerjee, A new software for dynamical analysis of nonsmooth systems, in *IEEE International Meeting on Analysis and Applications of Nonsmooth Systems (AANS2014)* (Como, Italy, 2014)
18. K. Mandal, C. Chakraborty, A. Abusorrah, M.M. Al-Hindawi, Y. Al-Turki, S. Banerjee, An automated algorithm for stability analysis of hybrid dynamical systems. Eur. Phys. J. Spec. Top. **222**(3–4), 757–768 (2013)
19. A.M. Rahimi, A. Emadi, Discontinuous-conduction mode dc/dc converters feeding constant-power loads. IEEE Trans. Ind. Electron. **57**(4), 1318–1329 (2010)
20. L.M. Saublet, R. Gavagsaz-Ghoachani, J.P. Martin, B. Nahid-Mobarakeh, S. Pierfederici, Bifurcation analysis and stabilization of dc power systems for electrified transportation systems. IEEE Trans. Transp. Electrification **2**(1), 86–95 (2016)
21. C.K. Tse, *Complex Behavior of Switching Power Converters* (CRC Press, Boca Raton, 2003)
22. H. Wu, V. Pickert, M. Ma, B. Ji, C. Zhang, Stability study and nonlinear analysis of dc-dc power converters with constant power loads at the fast time-scale. IEEE J. Emerg. Sel. Top. Power Electron. **8**(4), 3225–3236 (2020)
23. M.K. Zadeh, R. Gavagsaz-Ghoachani, J.P. Martin, S. Pierfederici, B. Nahid-Mobarakeh, M. Molinas, Discrete-time tool for stability analysis of dc power electronics-based cascaded systems. IEEE Trans. Power Electron. **32**(1), 652–667 (2017)

Multi-scale Dynamics and Nonlinear Eigenvalue Problem of Heterogeneous Metastructures Using a Wave Finite Element Scheme and Modal Strain Energy Method

Dongze Cui, Mohamed Ichchou, Abdel-Malek Zine, and Noureddine Atalla

Abstract Wave Finite Element (WFE) scheme is developed to reveal the multi-scale dynamics and damping characteristics of heterogeneous metastructures. Firstly, a clarification of high contrast and high dissipation in metastructures is addressed, especially on the multi-scale dynamics of Highly Contrasted Structure (HCS) and rheological damping of Polyvinyl Butyral (PVB) in Highly Dissipative Structures (HDS). The Bending-Shear Coupling effect in HCS is introduced through the dynamics of the monolithic and bi-layer limits. The Nonlinear Eigenvalue Problem (NEP) for wavevectors propagating in different heading angle is tackled by the Contour Integral method, the wavenumbers is identified by Weighted Wave Assurance Criteria accounting for the energetic distribution for pairing the numerically derived wavemodes, the WFE scheme is further developed to handle the wave identification and calculate the Damping Loss Factor (DLF) of metastructures using the Modal Strain Energy Method. For validation, the wavenumbers of the WFE scheme show good agreement with results from the Analytical method, General Laminate Model (GLM), and classical model RKU. The agreement of DLF verifies the feasibility of WFE for HDS, the multi-scale behavior is captured by analyzing the wavenumber and wavemodes. Some useful conclusions are discussed.

D. Cui (✉) · M. Ichchou
Vibroacoustics & Complex Media Research Group, École Centrale de Lyon, LTDS-CNRS UMR 5513, 69134 Écully, France
e-mail: dongze.cui@ec-lyon.fr; dongze.cui@usherbrooke.ca

M. Ichchou
e-mail: mohamed.ichchou@ec-lyon.fr

D. Cui · N. Atalla
Groupe d'Acoustique de l, Université de Sherbrooke, Université de Sherbrooke, Québec J1K2R1, Canada
e-mail: noureddine.atalla@usherbrooke.ca

A.-M. Zine
Institute Camille Jordan–CNRS, École Centrale de Lyon, UMR 5208, 69134 Écully, France
e-mail: abdel-malek.zine@ec-lyon.fr

M. Belhaq (ed.), *Advances in Nonlinear Dynamics and Control of Mechanical and Physical Systems*, Springer Proceedings in Physics 301,
https://doi.org/10.1007/978-981-99-7958-5_11

1 Introduction

Novel laminar designs are pursued to handle the shortcomings of classical noise control strategies, for example, heterogeneous metastructures have been researched to manipulate the wave propagation that conventional structures either cannot fulfill the requirements or are infeasible in the manufacturing process.

Among all heterogeneous metastructures, Highly Contrasted Structures (HCS) are commonly applied in transportation engineering where a thick soft core is sandwiched between thin stiff skins. Moreover, HCS are also encountered in other engineering fields where stiff layers are glued with soft cohesive materials, for example, Laminated Glass (LG) made of glass layers bounded by thin viscoelastic polymer is widely used in modern skyscrapers. On the other hand, the application of viscoelastic materials is becoming popular for use. This introduces another important notion, the Highly Dissipative Structures (HDS). Thus, heterogeneous metastructures are recently attracting emerging attention.

Wave Finite Element scheme (WFE) is useful to predict dispersion relations of periodic structures which is usually the case of heterogeneous metastructures. WFE converts the study of the whole periodic structure into a Unit Cell (UC). The mathematics of wave propagation in periodic systems was proposed and discussed by [1–3]. Afterward, [4, 5] employed the FE technique for wave propagation analysis. Mencik and Ichchou [6] have then formulated an approach based on the traditional FE model to efficiently calculate the dispersion curves of complex waveguides.

Moreover, WFE has wide application prospects to tackle the drawback of conventional FE. Despite its sophisticated mathematical derivation, previous research proves that WFE is beneficial in comparison with the FE concepts for smaller model size, enhanced computational efficiency, and higher frequency limitation. However, the main drawbacks of WFE are that:

(a) When varying the wave heading direction, WFE leads to an NEP in wavenumbers, which is the main challenge.
(b) The wave solution takes chaotic sequences along frequency when high contrast occurs, the wave identification is fatal to the calculation of indicators such as wavenumber space and Damping Loss Factor (DLF).

As a result, the application of WFE to HDS and HCS has not received much attention in the past, despite its unquestionable advantage [8].

In Sect. 2, the rheological properties of PVB are introduced; the multi-scale dynamics for HCS are clarified. The NEP and the corresponding Contour Integral solver are developed in Sect. 3, afterward, the wave identification technique and the derivation of DLF based on an energetic approach are detailed. In Sect. 4, the results concerning sandwich and LG are validated with results from GLM [9]. High Order Homogenization Method (HOHM) [10] and the classical RKU model [11]. The discussion regarding the results is fully devoted in Sect. 5.

Table 1 Parameters of the series of spring-dashpot Kelvin-Voigt model describing the shear modulus of pvb in equation

$\tau_k(s)$	10^{-5}	10^{-4}	10^{-3}	10^{-2}	10^{-1}	1	10	10^2	10^3	10^4	10^5	10^6	10^7	10^8
$G_k(MPa)$	2.3	25.2	30.8	39.2	35.2	6.2	1.9	0.7	0.17	0.07	0.08	0.11	0.10	0.14

2 Heterogeneous Metastructures

2.1 *Frequency and Temperature-Dependent Properties of PVB*

The viscoelastic properties of PVB type RB41 at the reference temperature $T_0 = 20°C$ is described in harmonic regime [10], by the shear modulus $\mu(\omega, T_0) = Re(\mu) + Im(\mu)$. The modulus of PVB can be modeled by

$$\mu(\omega, T_0) = \sum_{j=1}^{n} \frac{i\omega\tau_k G_k}{1 + i\omega\tau_k} \tag{1}$$

The series of rheological parameters (τ_k, G_k) identified experimentally for the PVB in [12], are given in Table 1.

2.2 *Multi-scale Dynamics for Highly Contrasted Structure*

The stratified plates with relatively thin core, such as LG with PVB interlayer [13], contain both stiffness and geometric contrasts between layers, the viscoelastic property of the PVB layer facilitates the study of viscoelasticity.

Two variables are defined to evaluate the kinematics of the structure [13]: the differential in-plane motion between two stiff layers D and the differential in-plane motion between the middle planes of the stiff layers d, as shown in Fig. 1, the three different kinematics regimes are therefore summarized:

(a) when the core has the same order of stiffness as the skin, there is weak contrast, the whole structure behaves as a monolithic plate and $D = 0$.
(b) when the core is soft compared with that of the stiff layers, the whole structure shows the general kinematics and $D \neq 0$, $d \neq 0$.
(c) when the core is extremely soft, thus inducing high contrast in the structure, the whole structure behaves as a multi-layer plate and $d = 0$.

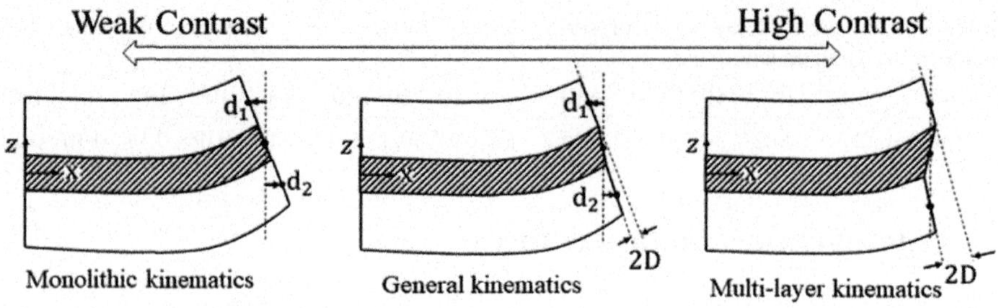

Fig. 1 Different kinematic regimes for laminate plate with different levels of contrast. Left: monolithic kinematics with D = 0 for extremely low contrast. Center: general kinematics for moderate contrast. d = 0 for extremely high contrast

3 Synthesis of Computational Schemes

3.1 Nonlinear Eigenvalue Problem in Wave Finite Element Scheme

The original system matrices $\tilde{\boldsymbol{M}}$ and $\tilde{\boldsymbol{K}}$ are obtained using an in-house FE package to handle the rheological properties. The equation of motion of the UC reads

$$\left[\tilde{\boldsymbol{K}} - \tilde{\boldsymbol{M}}\right]\boldsymbol{q} = \boldsymbol{F} \tag{2}$$

For a plane wave of 2D structures [8].

$$u(x, y, z, t) = U(z)\boldsymbol{e}^{i(-\kappa_x x - \kappa_y y + \omega t)} \tag{3}$$

where $U(z)$ is the wavemode, ω is the angular frequency, $\kappa_x = \boldsymbol{\kappa cos}\theta$ and $\kappa_y = \boldsymbol{\kappa sin}\theta$ are the components of wavevector $\boldsymbol{\kappa}$ in X and Y directions, respectively. The nodal displacements are related to periodicity conditions [1].

$$\begin{aligned} &\boldsymbol{q}_R = \lambda_x \boldsymbol{q}_L \\ &\boldsymbol{q}_T = \lambda_y \boldsymbol{q}_B \\ &\boldsymbol{q}_2 = \lambda_x \boldsymbol{q}_1, \boldsymbol{q}_3 = \lambda_y \boldsymbol{q}_{1,} \boldsymbol{q}_4 = \lambda_x \lambda_y \boldsymbol{q}_1 \end{aligned} \tag{4}$$

while equilibrium of the UC implies

$$\begin{aligned} &\boldsymbol{f}_1 + \lambda_x^{-1} \boldsymbol{f}_2 + \lambda_y^{-1} \boldsymbol{f}_3 + \lambda_x^{-1} \lambda_y^{-1} \boldsymbol{f}_4 = 0 \\ &\boldsymbol{f}_L + \lambda_x^{-1} \boldsymbol{f}_R = 0 \\ &\boldsymbol{f}_B + \lambda_y^{-1} \boldsymbol{f}_T = 0 \end{aligned} \tag{5}$$

Equation (4) enables the full set of displacements to be projected onto the reduced set, $\boldsymbol{q}_{red} = [\boldsymbol{q}_1, \boldsymbol{q}_L, \boldsymbol{q}_B, \boldsymbol{q}_I]^T$ using (4) and (5), providing a reduced EVP relating to propagating constants λ_x, λ_y and frequency ω,

$$\boldsymbol{\Lambda}_L\left[\widetilde{\boldsymbol{K}} - \widetilde{\boldsymbol{M}}\right]\boldsymbol{\Lambda}_R\boldsymbol{q}_{red} = [\boldsymbol{K} - \boldsymbol{M}]\boldsymbol{q}_{red} = 0 \tag{6}$$

The eigenvalues are the roots of $D = det(\boldsymbol{K} - \boldsymbol{M}) = 0$, the jth eigenvectors of $\boldsymbol{\phi}_{j,red}$ is the reduced set of wavemodes, the full wavemodes $\boldsymbol{\phi}_j$ writes

$$\boldsymbol{\phi}_j = \boldsymbol{\Lambda}_R\boldsymbol{\phi}_{j,red} \tag{7}$$

Equation (6) can be regarded as a NEP of wavevector $\boldsymbol{\kappa}$. The non-dimensional wavenumber μ and the scaled heading angle α as

$$\begin{gathered}\mu = \kappa\left[(\boldsymbol{L_x cos}\alpha)^2 + \left(\boldsymbol{L_y sin}\alpha\right)^2\right]^{0.5} \\ tan\alpha = \frac{\boldsymbol{L_y}}{\boldsymbol{L_x}}tan\theta\end{gathered} \tag{8}$$

so that the propagation constants can be written as

$$\begin{aligned}\lambda_x &= e^{-i\mu cos\alpha} \\ \lambda_y &= e^{-i\mu sin\alpha} \\ \frac{\mu_y}{\mu_x} &= \frac{\boldsymbol{L_y}}{\boldsymbol{L_x}}tan\theta\end{aligned} \tag{9}$$

Consequently, (6) can be formally written in the compact form:

$$\widehat{\boldsymbol{D}}(\mu)\boldsymbol{q}_{red} = 0 \tag{10}$$

where $\widehat{\boldsymbol{D}}(\mu) \in \mathbb{C}^{n,n}$.

3.2 *Contour Integral Method*

The NEP is solved using the Contour Integration method which is based on the calculation of the moments associated with the matrix function $\widehat{\boldsymbol{D}}(\mu)$, defined as

$$\boldsymbol{A}_p = \frac{1}{2\pi i}\int_{\Gamma}\mu^p\widehat{\boldsymbol{D}}(\mu)^{-1}\widehat{\boldsymbol{V}}d\mu \tag{11}$$

with $p = 0, 1, \ldots \overline{p}$ and $\overline{p} \in \mathbb{N}$, $\widehat{\boldsymbol{V}}(\mu) \in \mathbb{C}^{n,l}$ is a probe matrix in practice chosen at random to probe the matrix decomposition (11).

Moments $\boldsymbol{A}_p$ contain information about the spectrum of $\widehat{\boldsymbol{D}}(\mu)$ inside the contour Γ. To extract eigenvalues and eigenvectors, $2\overline{p}$ moments are used to construct $\overline{p}n \times \overline{p}l$ block Hankel matrices $\boldsymbol{B}_0$ and $\boldsymbol{B}_1 = \in \mathbb{C}^{\overline{p}n, \overline{p}l}$ as follows:

$$\boldsymbol{B}_0 = \begin{bmatrix} \boldsymbol{A}_0 & \dots & \boldsymbol{A}_{\overline{p}-1} \\ \vdots & \ddots & \vdots \\ \boldsymbol{A}_{\overline{p}-1} & \dots & \boldsymbol{A}_{2\overline{p}-2} \end{bmatrix} and \boldsymbol{B}_1 = \begin{bmatrix} \boldsymbol{A}_1 & \dots & \boldsymbol{A}_{\overline{p}} \\ \vdots & \ddots & \vdots \\ \boldsymbol{A}_{\overline{p}} & \dots & \boldsymbol{A}_{2\overline{p}-1} \end{bmatrix} \tag{12}$$

thus, all moments of order $p = 0, 1, \dots 2\overline{p} - 1$ must be computed using (11). The Singular Value Decomposition (SVD) of $\boldsymbol{B}_0 = \boldsymbol{V}\Sigma\boldsymbol{W}^{\boldsymbol{H}}$ is performed to obtain the Low Rank Approximation (LRA) $\boldsymbol{B}_0 \approx \boldsymbol{V}_0\Sigma_0\boldsymbol{W}_0^{\boldsymbol{H}}$. A tolerance ε_{SVD} is introduced to select the first $\overline{m}$ leading singular values (and corresponding columns of $\boldsymbol{V}$ and $\boldsymbol{W}$) according to $\sigma_1 > \dots \sigma_m \geq \varepsilon_{SVD} \geq \sigma_{\overline{m}+1} \approx \dots \approx 0$. After some algebraic manipulations, matrix $\widehat{\boldsymbol{B}} = \boldsymbol{V}_0^{\boldsymbol{H}}\boldsymbol{B}_1\boldsymbol{W}_0\Sigma^{-1}$ has the same eigenvalues of the original NEP inside the contour Γ. Also, providing $\boldsymbol{s}_i$ the eigenvectors of $\widehat{\boldsymbol{B}}$, the original eigenvectors can be retrieved from the first n rows of $\boldsymbol{V}_0\boldsymbol{s}_i$. Therefore, the NEP in (10) can be reformulated in terms of a reduced-sized linear eigenvalue problem with the same eigenvalues inside Γ.

Assume n_{sol} are the number of eigenvalues within Γ, it's expected that $\overline{m} > n_{sol}$ after the LRA. So, the number of moments and columns of the probe matrix $\widehat{\boldsymbol{V}}$ must satisfy $\overline{p}l \geq n_{sol}$. Note that $\overline{p} > 1$ is necessary only if solutions with algebraic multiplicity higher than 1 are expected. As a rule, the maximum order of moments $2\overline{p} - 1$ should be kept small [14, 15].

A numerical approximation of the integral in (12) is also necessary. Let a parametrization of the contour be $\Gamma = \gamma + \rho(cost + i\beta sint)$, with $t \in [0, 2\pi]$. Under these circumstances, the integral is approximated by an N-point trapezoidal rule, leading to the following discrete approximation of (12)

$$\boldsymbol{A} \simeq \frac{\rho}{N} \sum_{j=1}^{N} \mu_i^p \varpi_j \boldsymbol{X}_j \tag{13}$$

where μ_j are the integration points along Γ, $\varpi_j = \left(\beta cost_j + isint_j\right)$ are the weights and $\widehat{\boldsymbol{D}}\boldsymbol{X}_j = \widehat{\boldsymbol{V}}$ are the N linear systems to be solved. Note that, to ensure numerical stability, a shift and scale transformation is applied to the spectrum, so that the monomial basis $((\mu - \gamma)\rho)^2$ is used in (10) and (13) in place of μ^p.

3.3 Wave Identification Techniques Based on the Energetic Calculation

The eigenvalues and eigenvectors of HCS no longer take the correct order. An identification technique aims at mode pairing between two consecutive frequency steps is proposed, the Wave Assurance Criterion (WAC) proposed by [19] is defined as

$$WAC_{jk} = max\left(\frac{(\boldsymbol{\phi}_j^H \boldsymbol{\phi}_k)^2}{(\boldsymbol{\phi}_j^H \boldsymbol{\phi}_j)(\boldsymbol{\phi}_k^H \boldsymbol{\phi}_k)}\right) \tag{14}$$

where $\boldsymbol{\phi}_j$ and $\boldsymbol{\phi}_k$ are the jth and kth wavemodes, respectively. WAC_{jk} assigns the kth numerical mode to the jth mode.

To obtain better identification results, an extension of WAC is the Weighted Wave Assurance Criterion (WWAC) [19] which writes

$$WWAC_{jk} = max\left(\frac{(\boldsymbol{\phi}_j^H \mathbf{W} \boldsymbol{\phi}_k)^2}{(\boldsymbol{\phi}_j^H \mathbf{W} \boldsymbol{\phi}_j)(\boldsymbol{\phi}_k^H \mathbf{W} \boldsymbol{\phi}_k)}\right) \tag{15}$$

where $\boldsymbol{W}$ is the weighting matrix. If the mass matrix is applied, modal components related to the highest kinetic energy contributions are accentuated while the application of the stiffness matrix highlights the highest strain energy contributions.

3.4 Damping Loss Factor Using Modal Strain Energy Method

The DLF is defined in an energetic way to evaluate the energy dissipation in a quantitative way [20].

$$\eta = \frac{P_{diss}}{\omega(E_k + E_s)} \tag{16}$$

where P_{diss} denotes the total time-average power dissipated, E_k and E_s are the time-average kinetic energy and strain energy, respectively.

Considering the 2D structure, the time-average kinetic energy E_k writes:

$$E_k = \frac{\omega^2}{4} Re(\boldsymbol{\phi}^H \boldsymbol{M} \boldsymbol{\phi}) \tag{17}$$

where $\boldsymbol{M}$ is the mass matrix, superscript H denotes the conjugate transpose.

The time-average strain energy E_s of the laminate is obtained in a similar way:

$$E_s = \frac{1}{4} Re\left(\boldsymbol{\phi}^H \boldsymbol{K} \boldsymbol{\phi}\right) \tag{18}$$

where $\boldsymbol{K}$ is the stiffness matrix.

Provided that the DLF of the kth layer of the laminate is uniform through the thickness of the layer, the total dissipated power can be related to the strain energy of the individual layers by

$$P_{diss} = \frac{\omega}{2} Im\left(\boldsymbol{\phi}^H \boldsymbol{K} \boldsymbol{\phi}\right) \tag{19}$$

Substituting (17), (18), (19) into (16) gives

$$\eta_j = \frac{\frac{\omega}{2} Im\left(\boldsymbol{\phi}^H \boldsymbol{K} \boldsymbol{\phi}\right)}{\omega\left(\frac{\omega^2}{4} Re\left(\boldsymbol{\phi}^H \boldsymbol{M} \boldsymbol{\phi}\right) + \frac{1}{4} Re\left(\boldsymbol{\phi}^H \boldsymbol{K} \boldsymbol{\phi}\right)\right)} \tag{20}$$

Equation (20) provides the DLF for jth wave and is the main result of this section.

4 Numerical Calculation and Model Validation

4.1 Sandwich Panels

The thicknesses of the two skins are 0.1524 and 3.175 mm, respectively, the thickness of PVB core is 0.254 mm, the material properties of the panels are listed in Table 2.

The validation of the multi-scale behavior of the global laminate is depicted in Fig. 2, the 3 bending wavenumbers from WFE are validated with GLM [9] and HOHM [10], the wavenumbers validate the multi-scale behavior predicted in Fig. 1: the progressive wave at low frequency represents the bending dynamic of monolithic limit, the shear wave of the core at mid-frequency and finally jumps to the skins for the bi-layer limit at high frequency.

Table 2 Material properties of the sandwich plate

Properties (SI unit)	Aluminum	Polymer
Density (kg/m^3)	2070	950
Young's Moduli (Pa)	7.1E10	Rheological
Shear moduli (Pa)	2.6692E10	Rheological
Poisson ratio []	0.33	0.35
Damping ratio []	1E-3	Rheological

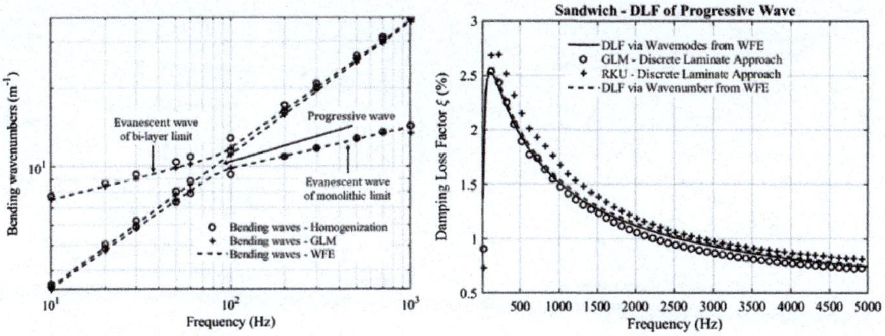

Fig. 2 WFE results for the sandwich plate. Left: Bending Waves. Right: Dominant DLF

The first propagating solution is retained because it corresponds to transversal displacements motion accompanied by in-plane shearing and transversal deformations, therefore, the DLF of this wave is named as "dominant DLF" and is exploited to examine the WFE method.

The dominant DLF results from WFE agree with GLM and RKU Method [11] as shown in Fig. 2.

4.2 *Laminated Glass*

The geometries and material properties are listed in Tables 3 and 4.

The high contrasts in the thicknesses and the stiffness of glass and PVB induce the significant multi-scale dynamic behavior in LG.

WFE results are compared with GLM [9] and HOHM [4]. Figure 3 shows an excellent agreement among all the three models. Moreover, the multi-scale behavior of the LG is correctly captured:

Table 3 Geometry of laminated glass

	Glass layer thickness (mm)	PVB layer thickness (mm)
LGa	10	0.76
LGb	5	0.38

Table 4 Material properties of the Glass and PVB

	Density (kg/m^3)	Poisson ratio []	Young's moduli (Pa)
Glass	2500	0.499	Rheological
PVB	1070	0.2	Rheological

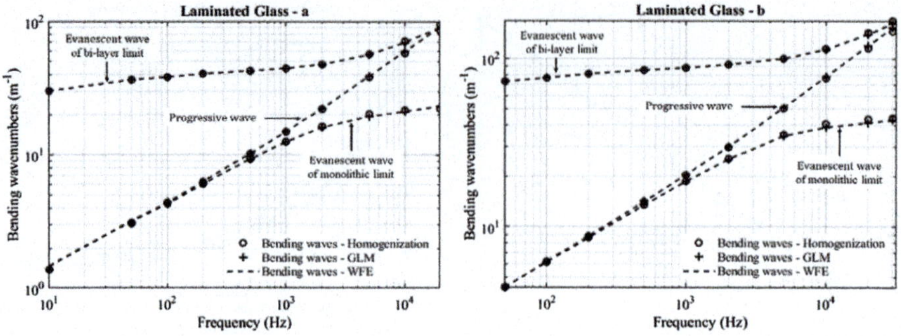

Fig. 3 Bending waves of LG for Ga (left) and Gb (right)

(a) At low-frequency range, the progressive wave is close to the evanescent wave of monolithic limit, the overall LG behaves as a monolithic plate.
(b) At high-frequency range, the progressive wave is close to the evanescent wave of the bi-layer limit, and the LG behaves like a bi-layer plate.

A more detailed representation of multi-scale dynamics in LG is shown in Fig. 4 using the progressive wavemode calculated by WFE:

(a) At low frequency, weak contrast exists since the stiffness of PVB still has the same order of magnitude as glass, the progressive waves are close to the evanescent waves of monolithic limit, and the whole structure behaves as a monolithic plate, which means $D = 0$ as depicted in Fig. 2.
(b) At high frequency, high contrast starts to occur, the progressive waves are close to the evanescent waves of the bi-layer limit, the whole LG behaves as a bi-layer plate, and the differential in-plane motion between the middle planes of the stiff layers disappears, which means $d = 0$.

The dominant DLF are in excellent agreement as shown in Fig. 5.

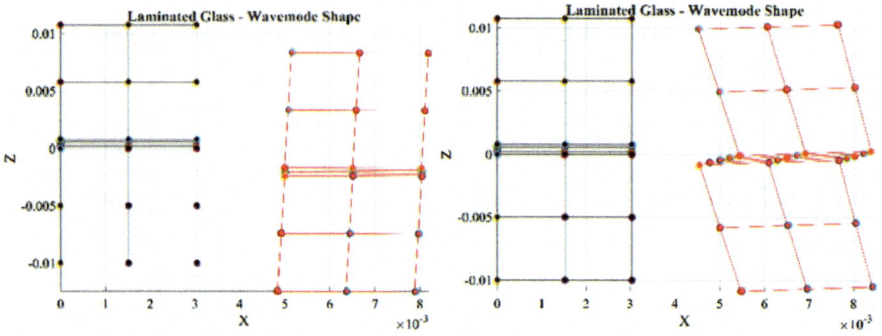

Fig. 4 Two limit dynamics of laminated glass with PVB core

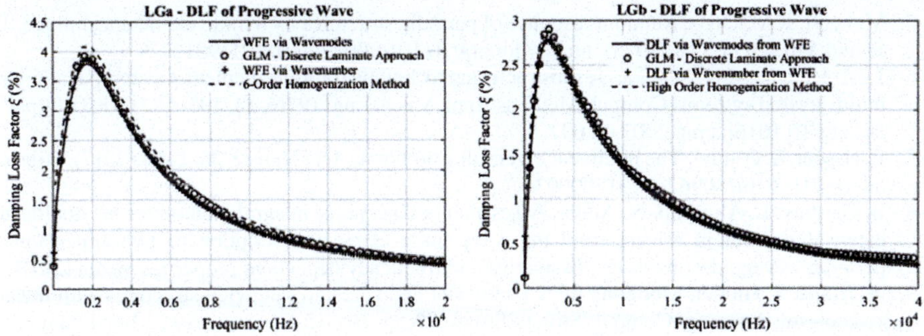

Fig. 5 Dominant damping loss factor for Ga (left) and Gb (right)

5 Conclusion

In this paper, we used WFE to study the multi-scale dynamics and energetic features of 2D metastructures. The main contributions are drawn as follows:

(a) The HCS and HDS of metastructures are clarified, and the different kinematic regimes are employed to distinguish the monolithic and bi-layer limits, thus inspiring a profound understanding of multi-scale dynamic behavior in laminated structures.
(b) The NEP induced by the changing propagating direction is proposed and tackled by the Contour Integral Method, such method is also beneficial for solving the NEP in Highly Dissipative Structures.
(c) With the energetic calculation extended from WFE, the wavenumbers are well identified by WWAC, and DLF is then calculated based on the MSEM.

Acknowledgements This work was supported by the Natural Sciences and Engineering Research Council of Canada-Discovery Grant Program. The authors also thank Dr Christophe Droz from Institut National de Recherche en Informatique et Automatique for his technical support.

References

1. L. Brillouin, *Wave Propagation in Periodic Structures*, 2nd edn. (McGraw-Hill Book Company, Canada, 1946)
2. S. Parthan, D.J. Mead, Free wave propagation in two-dimensional periodic plates. J. Sound Vib. **64**(3), 325–348 (1979). https://doi.org/10.1016/0022-460X(79)90581-9
3. A.L. Abrahamson, Flexural wave mechanics—an analytical approach to the vibration of periodic structures forced by convected pressure fields. J. Sound Vib. **28**(3), 247–258 (1973). https://doi.org/10.1016/S0022-460X(73)80105-1
4. M. Petyt, R.M. Orris, A finite element study of harmonic wave propagation in periodic structures. J. Sound Vib. **33**(2), 223–236 (1974). https://doi.org/10.1016/S0022-460X(74)80108-2

5. M. Petyt, R.M. Orris, Random response of periodic structures by a finite element technique. J. Sound Vib. **43**(1), 1–8 (1975). https://doi.org/10.1016/0022-460X(75)90199-6
6. J.-M. Mencik, M.N. Ichchou, A substructuring technique for finite element wavepropagation in multi-layered systems. Comput. Methods Appl. Mech. Eng. **197**(6–8), 505–523 (2008). https://doi.org/10.1016/j.cma.2007.08.002
7. F. Tisseur, S. G¨uttel, The nonlinear eigenvalue problem. Acta Numer. **26**, 1–94 (2017). https://doi.org/10.1017/S0962492917000034
8. B. Gardner, L. Alimonti, N. Atalla, Assessment of a contour integral method for the nonlinear eigenvalue problem arising from the wave finite element description of two-dimensional periodic waveguides involving dissipative and frequency dependent properties media (2018)
9. N. Atalla, S. Ghinet, Modeling thick composite laminate and sandwich structures with linear viscoelastic damping. Comput. Struct.. Struct. **89**(15–16), 1547–1561 (2011). https://doi.org/10.1016/j.compstruc.2010.09.008
10. S. Hans, C. Boutin, K. Viverge, Dynamics of contrasted stratified elastic and viscoelastic plates—application to laminated glass. Compos. B Eng. **212**(11), 108551 (2021). https://doi.org/10.1016/j.compositesb.2020.108551
11. E.E. Ungar, D. Ross, E.M. Kerwin Jr., Damping of plate flexural vibrations by means of viscoelastic laminae. Struct. Damping **6**(6), 49–97 (1959)
12. G. Savineau, P. D'Haene, Mechanical properties of laminated safety glass - fem study, in *Proceedings of GPD2007—Glass Performance Day* (Tampere, Finland, 2007)
13. K. Viverge, C. Boutin, Generalized plate model for highly contrasted laminates. Eur. J. Mech. A. Solids **55**, 149–166 (2016). https://doi.org/10.1016/j.euromechsol.2015.08.008
14. H. Tadano, T. Ikegami, T. Sakurai, J. Asakura, Error analysis for a matrix pencil of hankel matrices with perturbed complex moments. Jpn. Soc. Ind. Appl. Math. Lett. **1**(2), 76–79 (2009). https://doi.org/10.14495/jsiaml.1.76
15. P. Kravanja, M. Van Barel, Nonlinear eigenvalue problems and contour integrals. J. Comput. Appl. Math. **292**, 526–540 (2016). https://doi.org/10.1016/j.cam.2015.07.012
16. C. Kittel, *Introduction to Solid State Physics*, 8th edn. (Wiley, Canada, 2005)
17. E.L. Cardoso, R.O. Ruiz, L. Quinteros, V. Meruane, Phononic bandgap optimization in sandwich panels using cellular truss cores. Materials **14**(18), 5236 (2021). https://doi.org/10.3390/ma14185236
18. C. Bucher, M. Brehm, V. Zabel, An automatic mode pairing strategy using an enhanced modal assurance criterion based on modal strain energies. J. Sound Vib. **329**(25), 5375–5392 (2010). https://doi.org/10.1016/j.jsv.2010.07.006
19. D.L. Brown, R.J. Allemang, A correlation coefficient for modal vector analysis, in *Proceedings of the 1st International Modal Analysis Conference* (Orlando, 1982), pp. 110–116
20. L.L. Beranek, I.L. Ver, Noise and Vibration Control Engineering, 2nd edn. (Wiley, New York, United States, 1992)
21. B.R. Mace, E. Manconi, Estimation of the loss factor of viscoelastic laminated panels from finite element analysis. J. Sound Vib. **329**(19), 3928–3939 (2010). https://doi.org/10.1016/j.jsv.2010.04.014
22. N.K. Anifantis, F.D. Gounaris, Structural damping determination by finite element approach. Comput. Struct.. Struct. **73**(1), 445–452 (1999). https://doi.org/10.1016/S0045-7949(98)00257-0
23. I. Bartoli, A. Marzani, F.L. Di Scalea, E. Viola, Modeling wave propagation in damped waveguides of arbitrary cross-section. J. Sound Vib. **295**(3–5), 685–707 (2006). https://doi.org/10.1016/j.jsv.2006.01.021
24. I. Bartoli, F.L. di Scalea P.A. Rizzo, A. Marzani, E. Viola, Semi-analytical finite element formulation for modeling stress wave propagation in axisymmetric damped waveguides. J. Sound Vib. **318**(3), 488–505 (2008). https://doi.org/10.1016/j.jsv.2008.04.028
25. S.H. Crandall, The role of damping in vibration theory. J. Sound Vib. **11**(1), 3–IN1 (1970). https://doi.org/10.1016/S0022-460X(70)80105-5
26. S.H. Crandall, The hysteretic damping model in vibration theory. Proc. Inst. Mech. Eng., Part C Mech. Eng. Sci. **205**(1), 23–28 (1991). https://doi.org/10.1243/PIME_PROC_1991_205_086_02

Stabilization of a PV Fed Current Mode Controlled SEPIC with MPPT Using Piecewise Quadratic Slope Compensation

Kuntal Mandal and Abdelali El Aroudi

Abstract Instabilities in the form of subharmonic oscillations are addressed in this paper for a Single-Ended Primary Inductor Converter (SEPIC) used for Maximum Power Point Tracking (MPPT) in a Photovoltaic (PV) system under current mode control. It is widely known that this kind of system may exhibit undesired subharmonic oscillations when the duty ratio is larger than 50%. Piecewise quadratic (PWQ) slope compensation technique has been proven to be an efficient technique for eliminating this kind of undesired behavior in DC-DC switching converters. With this technique, the stability of the converter is guaranteed for all values of the duty ratio. In this work, the technique is applied on a SEPIC supplied from a photovoltaic energy source and linked at its output to a DC load. Under varying environmental conditions, the system is used for MPPT. A peak current mode controller is used for the current loop whose reference is provided by a Perturb and Observe (P&O) MPPT algorithm. The obtained results show that the system exhibits a fast MPPT of the PV generator under (a) sudden changes in the environmental conditions while the guaranteeing fast-timescale stability (b) change of inductance. The mathematical modeling, design procedures are validated by numerical simulations. The comparison with the conventional slope compensation technique is also given to show the suitability of the PWQ.

1 Introduction

In photovoltaic (PV) systems, a DC-DC converter is needed as an interface to extract the maximum power from the PV generator [14, 20, 23]. The power of a PV generator can vary widely depending on factors such as ambient temperature, shading, and the

K. Mandal (✉)
Universitat Rovira i Virgili, Tarragona, Spain
e-mail: dr.kuntal.mandal@gmail.com

A. El Aroudi
Universitat Rovira i Virgili, Tarragona, Spain
e-mail: abdelali.elaroudi@urv.cat

M. Belhaq (ed.), *Advances in Nonlinear Dynamics and Control of Mechanical and Physical Systems*, Springer Proceedings in Physics 301,
https://doi.org/10.1007/978-981-99-7958-5_12

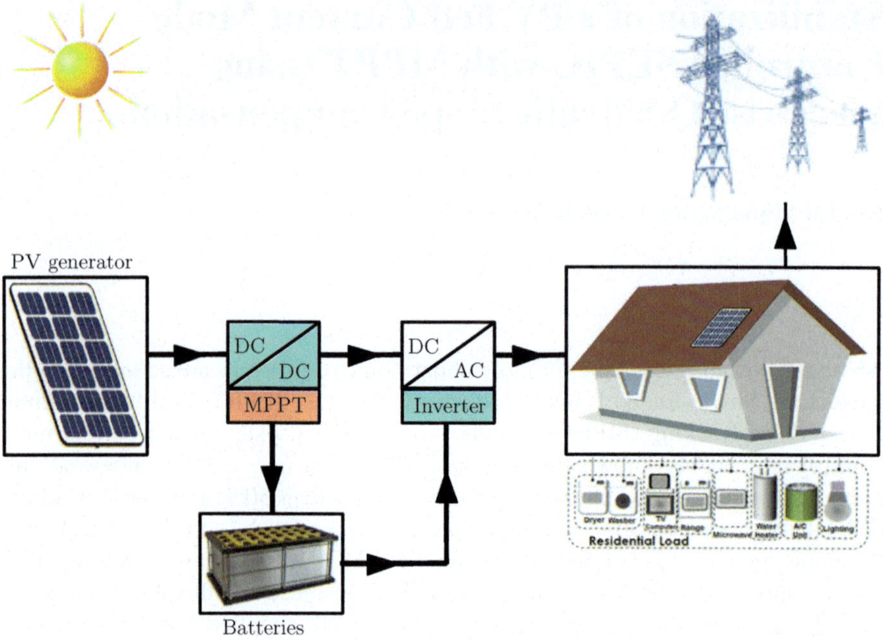

Fig. 1 A typical diagram of a PV system supplying load through power electronics interface and storage

angle of incidence of sunlight which affect the operating condition of the converter (Fig. 1).

A Single-Ended Primary Inductance Converter (SEPIC) is a type of DC-DC power converter used to step up or step down the DC voltage or current levels. It is a versatile and efficient converter that can both step down and step up the voltage levels, making it ideal for use in battery-powered systems or battery charging/discharging applications where the input/output voltages can vary widely. This makes it an excellent choice for use in PV applications. The SEPIC can be used in conjunction with a Maximum Power Point Tracking (MPPT) controller to extract the maximum available power from the PV generator [14, 20, 23]. The MPPT controller may adjust directly or indirectly the duty cycle of the SEPIC to match the output voltage of the PV generator to the load or battery bank, ensuring that the maximum power is transferred from its input to its output.

On the other hand, current mode control (CMC) is a predominant strategy in controlling DC-DC switched mode power electronic converters for different applications due to its many advantages such as fast system response, better system performances, and inherent over-current protection. However, it is well-known that power converters with CMC are prone to exhibit instabilities in the form of subharmonic oscillations [10, 12, Chap. 11] also named fast-timescale instability. There are several possible solutions to mitigate subharmonic oscillations in DC-DC converters. The slope compensation using an external sawtooth periodic signal with a constant slope is the conventional strategy used by power electronics engineers for eliminating subharmonic oscillations in switching converters under peak CMC [4, 12]. With the conventional version of this technique, a periodic sawtooth signal is subtracted from the reference signal such that by adequately adjusting its slope, the system is stabilized. The larger the slope of the ramp the wider is the stability range in terms of the duty cycle [5, 6]. In order to have a stable system for all values of duty cycles, the slope must be equal to one-half the absolute value of the falling slope of the sensed inductor current. However, the amplitude of the ramp guaranteeing stability must be known beforehand. To overcome this problem, recently, a Piecewise Quadratic (PWQ) slope compensation technique has been proposed. This technique can be used for stabilizing elementary DC-DC converters [8–11] and the amplitude of the compensating signal is self-adjusted to a suitable value guaranteeing the stability of the converter.

A PV system consisting of a PV generator interlinked to a DC-DC converter under CMC can exhibit the same advantages as in other applications [3, 15, 21, 22]. However, it may also exhibit a rich variety of nonlinear phenomena such as period doubling bifurcation causing subharmonic oscillations and chaotic behavior [24] with conventional linear ramp compensation/modulation. In this paper, we use the PWQ technique to stabilize a SEPIC under peak CMC when it is used for MPPT in a PV system. Simulation results are obtained from a circuit-level switched model of the converter to validate the theoretical predictions.

The rest of this paper is structured as follows. After this introduction, Sect. 2 presents the system description and its approximate mathematical modeling. In Sect. 3 numerical simulation of the system with PWQ and conventional compensation techniques are performed. Concluding remarks are given in the last section.

2 The SEPIC with the PWQ Compensation

2.1 System Description

The system under study (Fig. 2) is a PV-fed SEPIC under peak CMC to track the maximum power generated by a PV generator. The control approach is based on peak CMC of the inductor L_1 current i_{L_1} to generate the square driving signal u that drives the switch S of the converter. The reference current is generated by a P&O MPPT algorithm [14, 20, 23]. The switch S is closed during the time interval $[nT, (n + d[n])T] \in \mathbb{N}$ and it is open during the time interval $[(n + d[n])T, (n + 1)T] \in \mathbb{N}$, where $d[n]$ is the duty cycle of the driving signal u and T is the switching period. As shown in Fig. 2, the current loop comprises a sensing resistor R_s, a comparator, and an RS flip-flop. The switching frequency $f_s = 1/T$ is set by a T-periodic clock signal applied to the set entry of the RS flip-flop. The current of the first inductor i_{L_1}, sensed via the sensing resistor R_s and the voltage $R_s i_{L_1}$ is compared to the difference

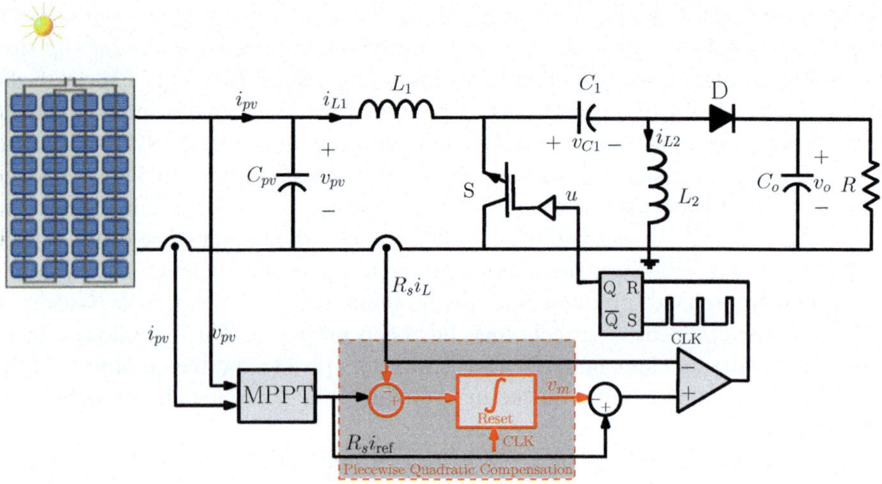

Fig. 2 The circuit diagram of a SEPIC with the PWQ compensation technique used in PV system for MPPT control

signal $R_s i_{\text{ref}} - v_m$. The comparator output signal is supplied to the reset entry of the flip-flop. During $[nT_s, (n + d[n])T]$, $n \in \mathbb{N}$, the switch S is closed and the current i_{L_1} increases, when the signal $R_s i_{L_1}$ attains the signal $R_s i_{\text{ref}} - v_m$, the output signal resets the control signal of the switch S which is then open and i_{L_1} decreases during $(n + d)T$, $(n + 1)T$, $n \in \mathbb{N}$ until the next rising edge of the clock signal. The state of the diode D is opposite to that of the switch S.

2.2 *Approximate Mathematical Modeling*

In order to simplify the analysis, the following assumptions are made in this paper:

1. The inductors (L_1 and L_2) and (C_{pv}, C_1 and C_o) capacitors are ideal with no equivalent series resistances.
2. The ripple voltages of the three capacitors and that of the inductor L_2 are much smaller than their average values within a switching cycle and can be neglected. If fast-timescale subharmonic oscillations are of concern, analysis of the capacitor voltages can be substituted by their steady-state values $V_{\text{pv}} = V_{C_1} = V_{\text{mpp}}$.
3. The switch (S) and diode (D) are ideal with no resistance and voltage drop.
4. Only the inductor current i_{L_1} dynamics is responsible for subharmonic oscillations.

Under these assumptions which are met for a practical design, the inductor current i_{L_1} is described by a piecewise linear state-space model that can be written as

$$\frac{\mathrm{d}i_{L_1}}{\mathrm{d}t} = \begin{cases} m_1 & \text{if } u=1 \\ m_0 & \text{if } u=0 \end{cases} \tag{1}$$

where $m_1 = V_{\mathrm{mpp}}/L_1$ and $m_0 = -DV_{\mathrm{mpp}}/(L_1(1-D))$ are the approximate rising and falling slopes of the inductor current i_{L_1} in steady-state, where D is the corresponding steady-state duty cycle which under peak CMC and PWQ. Under MPPT control, the average value of the inductor current is $I_{L1} = I_{\mathrm{mpp}}$. Since due to the power balance one has $I_{\mathrm{mpp}}V_{\mathrm{mpp}} = V_o^2/R$, where V_o is the steady-state average value of the output voltage v_o which for the SEPIC, is related to the input voltage V_{mpp} by the equation $V_o = DV_{\mathrm{mpp}}/(1-D)$. By substituting V_o by the previous expression, the power balance equation becomes

$$(RI_{\mathrm{mpp}} - V_{\mathrm{mpp}})D^2 - 2DRI_{\mathrm{mpp}} + RI_{\mathrm{mpp}} = 0 \tag{2}$$

The previous equation has two solutions. The steady-state duty cycle D is the physically valid one, i.e., the one fulfilling $0 < D < 1$ and this can be expressed as follows

$$D = \frac{RI_{\mathrm{mpp}} - \sqrt{RI_{\mathrm{mpp}}V_{\mathrm{mpp}}}}{RI_{\mathrm{mpp}} - V_{\mathrm{mpp}}} \tag{3}$$

The V_{mpp} is the MPP voltage that will be imposed by the MPPT controller through dictating the current reference i_{ref}. It is well-known that in order to guarantee the stability of a switching converter under Peak CMC for all duty cycle values in a switching converter with a conventional linear slope compensation scheme, the slope m_a of the external ramp signal has to be one-half the absolute value $-R_s m_0/2$ of the falling slope of the sensed inductor current $R_s i_{L_1}$ [12]. In the SEPIC, the slope of the ramp signal must be equal to $DV_{\mathrm{mpp}}/(L_1(1-D))$. Note that in order to tune the amplitude of the compensating ramp signal, the steady-state duty cycle D, the MPP voltage and the inductance value L_1 must be known beforehand.

For accurate stability analysis of the full-order state-space switched model with known switching conditions, one can use Floquet theory combined with Filippov method [7, 16]. This analysis can be carried out by using the monodromy matrix which is an effective tool to accurately analyze the stability of DC-DC switching converters [2, 18, 19]. The different kinds of instabilities can be identified by monitoring the eigenvalues of this matrix. Subharmonic oscillation is exhibited when one of the eigenvalues crosses the unit disk from the point $(-1, 0)$ in the complex plane. In case of PWQ compensation technique, the procedures for calculating the monodromy matrix in a DC-DC switching converter are detailed in [10].

3 Numerical Simulations with Switched Model

Numerical simulations are performed using PSIM© software to validate the theoretical results related to the PWQ slope compensation scheme. All parameter values used for the SEPIC and its control circuit are listed in Table 1. The design procedure of the passive components is given in [1, 13, 17]. Those used for the PV generator for Standard Test Conditions (STC, i.e., irradiance $S = 1000$ W/m^2, Air Mass (AM) –1.5, and Temperature $\Theta = 25\,°$C) are given in Table 2.

The i–v and i–p characteristic curves are obtained from the datasheet for several irradiance values as shown in Fig. 3.

3.1 System Performance Under Conventional Slope Compensation

To show the system performance under conventional slope compensation, ramp amplitude has been set to two different values $V_M = 0.5$ V and $V_M = 1.0$ V respectively and gradual change has been given to the irradiance level as shown in Figs. 4 and 5.

The MPPT algorithm gives optimized satisfactory performance when it has three steps which repeats its pattern with the selected perturbation in voltage (ΔV) and sampling period T_{mppt}.

It is interesting to note that when the ramp amplitude $V_M = 0.5$ V, the system exhibits subharmonic oscillations at the irradiance level $S = 1000$ W/m^2 (Fig. 6a) which is not desirable in practice. When $V_M = 1$ V, the subharmonic oscillation has disappeared for all irradiance levels but the MPPT performance is not optimized at $S = 200$ W/m^2 (Fig. 6b) because it gives four steps within an MPPT period.

Table 1 The power stage fixed parameter values of SEPIC

L_1	L_2	C_{pv}	C_1	C_o	f_s	R	R_s
580 µH	580 µH	25 µF	1 µF	47 µF	100 kHz	80 Ω	1 Ω

Table 2 The 'Sunmodule Plus SW 180 mono' PV module datasheet at STC

Parameter	Value
Number of series-connected cells	72
Maximum power P_{mpp}	180 Wp
Open-circuit voltage V_{oc}	44.6 V
Short-circuit current I_{sc}	5.40 A
Maximum power point voltage V_{mpp}	36.0 V
Maximum power point current I_{mpp}	5.01 A

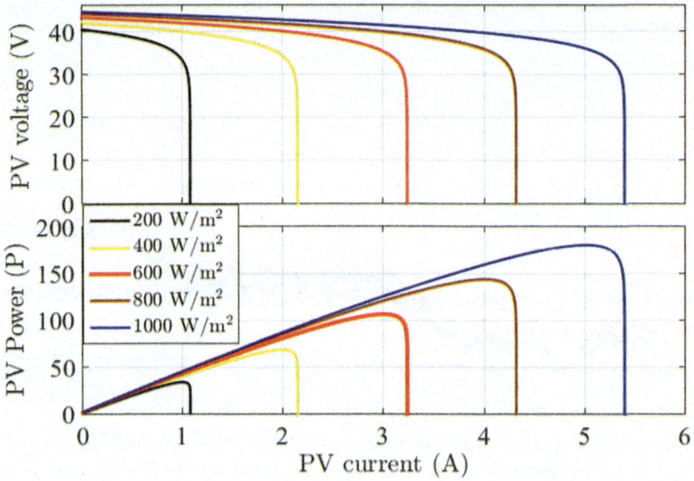

Fig. 3 The i–v and i–p curves of the 'Sunmodule Plus SW 180 mono' PV panel for different irradiance levels for fixed temperature 25 °C

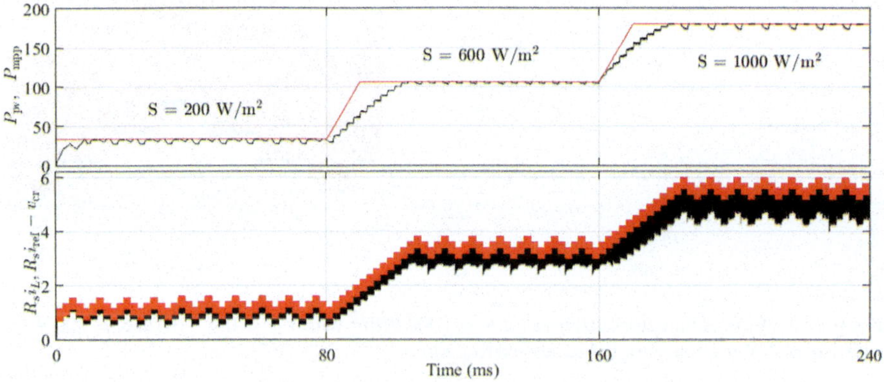

Fig. 4 The system response under different irradiance levels with conventional ramp compensation with $V_M = 0.5$ V for temperature 25 °C. The sampling time for the MPPT controller was $T_{mppt} = 2$ ms and the voltage perturbation amplitude is $\Delta V = 0.2$ V

3.2 System Performance Under PWQ Slope Compensation

Now with different irradiance levels, the system response has been captured with PWQ slope compensation technique as shown in Fig. 7. Due to the adaptive nature of the PWQ compensation technique, it is working satisfactorily with all irradiance levels as shown in Fig. 8. It is interesting to note that PWQ is working better than the conventional ramp even in transient and the MPPT performance is the optimal one.

Further, the simulation is extended to another situation where the value of the inductance ($L_1 = 580$ μH) is changed to lower value ($L_1 = 200$ μH). This change

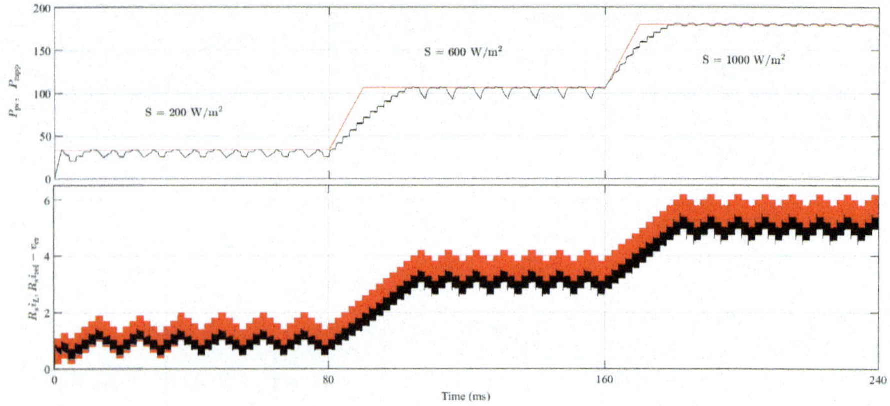

Fig. 5 The system response under different irradiance levels with conventional ramp compensation with $V_M = 1.0$ V for temperature 25 °C. The sampling time for the MPPT controller was $T_{\mathrm{mppt}} =$ 2 ms and the voltage perturbation amplitude is $\Delta V = 0.2$ V

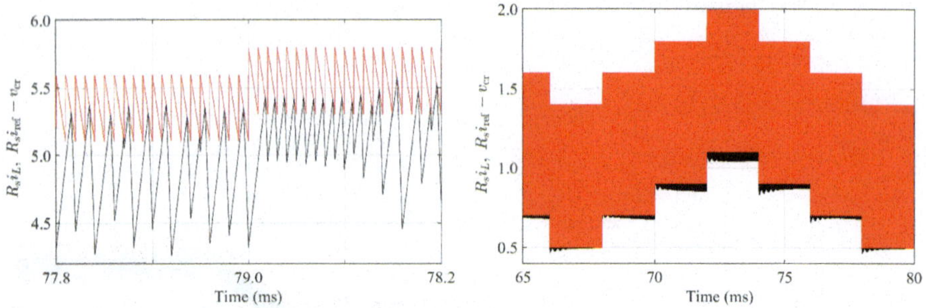

(a) (b)

Fig. 6 Waveform of inductor current (**a**) at $S = 1000$ W/m^2 with $V_M = 0.5$ V (**b**) at $S = 200$ W/m^2 with $V_M = 1$ V for conventional compensating ramp

could be attributed in practice to aging, different current rating, saturation nonlinearity, etc. Now keeping the environment conditions at temperature 25 °C and irradiance 1000 W/m^2, with reduced inductance value, the system is run by conventional compensating ramp and the results are shown in Fig. 9.

Initially with the nominal value ($L_1 = 580$ μH), the system response starts tracking the maximum power around $t = 0.11$ s. But at $t = 0.15$ s the change in inductance value ($L_1 = 200$ μH) is applied and the response shows chaotic behavior of the inductor current waveforms afterward. Then, the conventional compensating ramp is replaced by the PWQ compensation technique under the same environmental conditions. As shown in Fig. 10, in this case, the system response starts tracking the maximum power around $t = 0.11$ s and after the sudden change in the inductance value ($L_1 = 200$ μH) takes place at $t = 0.15$ s, the system response again starts tracking the maximum power at $t = 0.19$ s after a short initial transient dies out. Therefore,

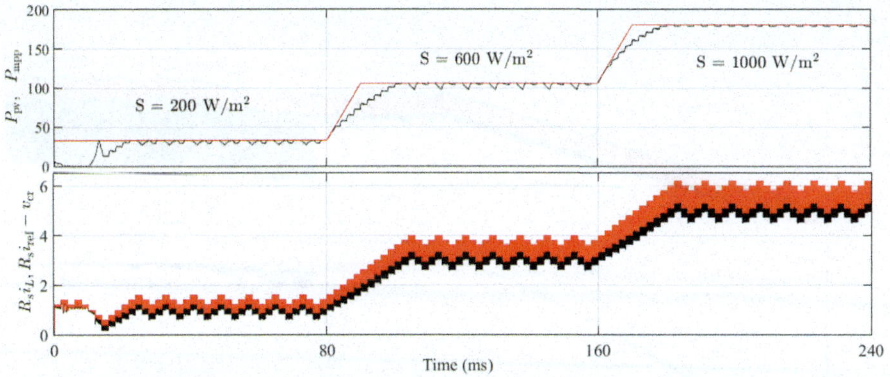

Fig. 7 The system response under different irradiance levels with conventional ramp compensation for temperature 25 °C. The sampling time for the MPPT controller was $T_{mppt} = 2$ ms and the voltage perturbation amplitude is $\Delta V = 0.2$ V

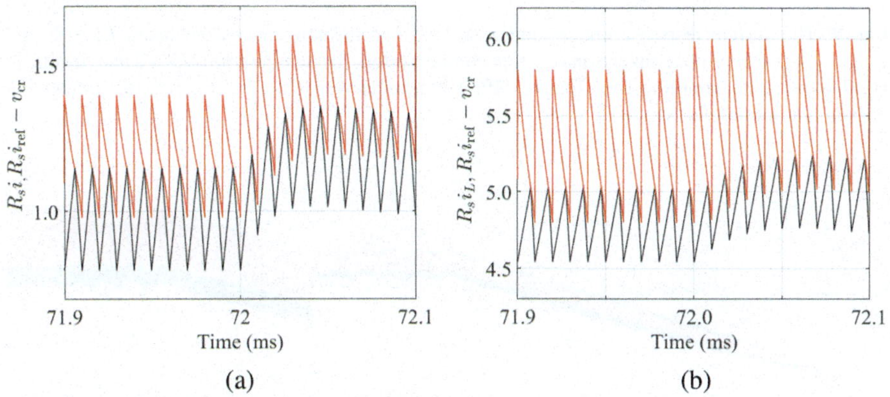

Fig. 8 In PWQ compensation, the Waveform of inductor current at **a** $S = 200$ W/m^2 **b** $S = 1000$ W/m^2. The sampling time for the MPPT controller was $T_{mppt} = 2$ ms and the voltage perturbation amplitude is $\Delta V = 0.2$ V

with PWQ compensation technique, the system shows stable behavior after a sudden change in the inductor value and excellent tracking performances.

4 Conclusions

This paper presented a piecewise quadratic slope compensation technique for eliminating subharmonic and chaotic oscillations in DC-DC switching converters under peak CMC. The amplitude of the modulating signal is self-adapted in terms of the inductance such that any drift due to aging or nonlinearity of the inductance is auto-

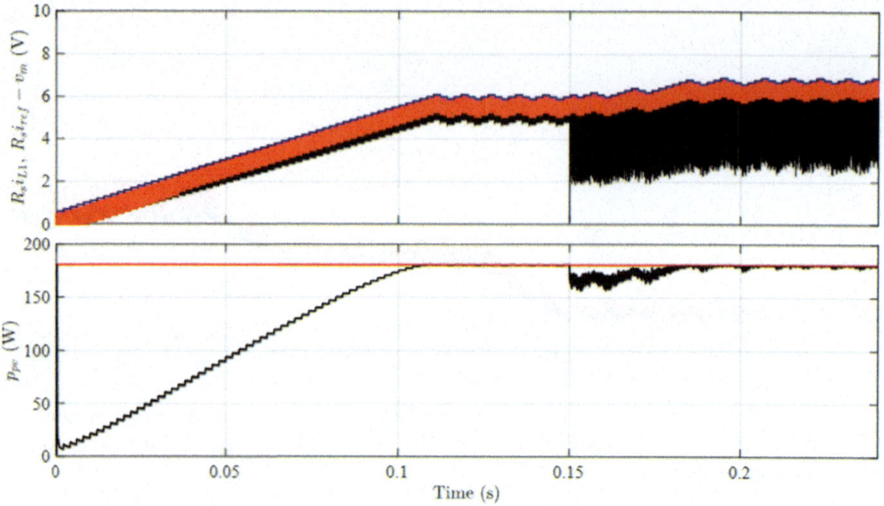

Fig. 9 The system response under sudden step change in inductance value from $L_1 = 580$ μH to $L_1 = 200$ μH with conventional ramp compensation for temperature 25 °C and irradiance 1000 W/m^2. The sampling time for the MPPT controller was $T_{\text{mppt}} = 2$ ms and the voltage perturbation amplitude is $\Delta V = 0.2$ V

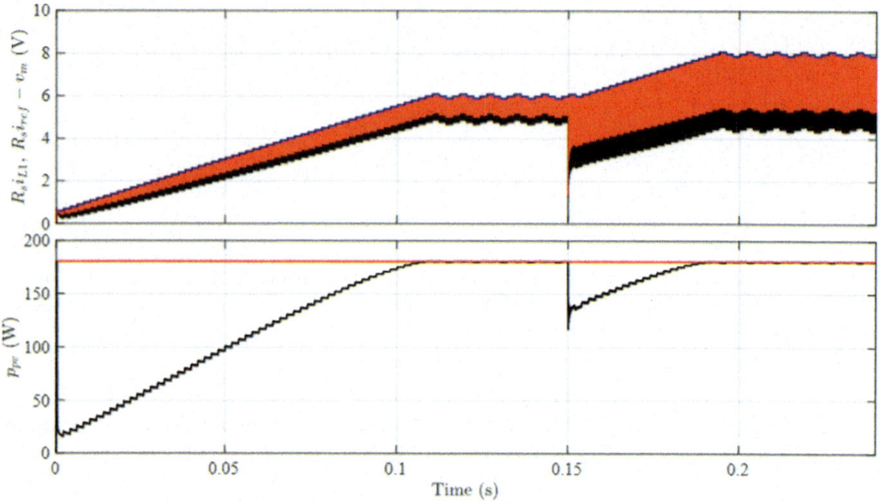

Fig. 10 The system response under sudden step change in inductance level with PWQ compensation for temperature 25 °C and irradiance 1000 W/m^2. The sampling time for the MPPT controller was $T_{\text{mppt}} = 2$ ms and the voltage perturbation amplitude is $\Delta V = 0.2$ V

matically taken into account. Therefore, the operation of the converter under the PWQ compensation scheme could allow a significant reduction of the size of the inductor which could result in an improvement of the power density. Moreover, since the stabilizing signal is generated from its own state variables, frequency mismatch, and phase shift problems, arising in practical situations when an external signal is injected for stabilization, are naturally avoided with the PWQ technique. The theoretical analysis shows that the technique can stabilize the system for all values of the duty cycles. Compared with the conventional slope compensation, the PWQ strategy does not require an external signal generation and extracts a compensation signal directly from the error between the inductor current and its reference. The PWQ strategy has the same stabilizing effect as the conventional slope compensation technique with ramp slope guaranteeing stability for all duty cycle values. This is achieved without the need for knowledge of the inductance value and sensing the input and output voltages. Future works will deal with the effect of the external voltage loop as well as that of the limited bandwidth of the current sensor on the results presented in this study. The extension of the technique to other control methods, such as average current mode control and voltage mode control, could also be a topic of further study.

Acknowledgements This work is supported and funded by the Agency for Management of University and Research Grants (AGAUR) under the Ministry of Research and Universities of the Government of Catalonia (Grant Reference No. 2020 BP 00260). A. El Aroudi acknowledges the support and funding of the Spanish Ministerio de Ciencia e Innovación under Grant PID2020-120151RB-I00.

References

1. Sepic converter analysis and design. Tech. Rep. AND90136/D, Onsemi, Arizona, USA (2021)
2. A. Abusorrah, K. Mandal, D. Giaouris, A. El Aroudi, M.M. Al-Hindawi, Y. Al-Turki, S. Banerjee, Avoiding instabilities in power electronic systems: toward an on-chip implementation. IET Power Electron. **10**(13), 1778–1787 (2017)
3. E. Bianconi, J. Calvente, R. Giral, E. Mamarelis, G. Petrone, C.A. Ramos-Paja, G. Spagnuolo, M. Vitelli, A fast current-based mppt technique employing sliding mode control. IEEE Trans. Ind. Electron. **60**(3), 1168–1178 (2013)
4. B. Bryant, M.K. Kazimierczuk, Modeling the closed-current loop of pwm boost dc-dc converters operating in ccm with peak current-mode control. IEEE Trans. Circuits Syst. I: Regul. Pap. **52**(11), 2404–2412 (2005)
5. L. Cheng, W.H. Ki, F. Yang, P.K.T. Mok, X. Jing, Predicting subharmonic oscillation of voltage-mode switching converters using a circuit-oriented geometrical approach. IEEE Trans. Circuits Syst. I: Regul. Pap. **64**(3) (2017)
6. A. El Aroudi, A new approach for accurate prediction of subharmonic oscillation in switching regulators-part i: Mathematical derivations. IEEE Trans. Power Electron. **32**(7), 5651–5651 (2017)
7. A. El Aroudi, D. Giaouris, H.H.C. Iu, I. Hiskens, A review on stability analysis methods for switching mode power converters. IEEE J. Emerg. Top. Circuits Syst. **5**(3), 302–315 (2015)
8. A. El Aroudi, R. Haroun, K. Mandal, M. Al-Numay, Auto-tuned quadratic slope compensation for current mode controlled dc-dc converters, in *2020 IEEE International Symposium on Circuits and Systems (ISCAS)* (Seville, Spain, 2020)

9. A. El Aroudi, K. Mandal, A. Abusorrah, M.M. Al-Hindawi, Y. Al-Turki, D. Giaouris, S. Banerjee, A novel nonlinear modulation technique for stabilizing dc-dc switching converters, in *2017 IEEE International Symposium on Circuits and Systems (ISCAS)* (Baltimore, MD, USA, 2002)
10. A. El Aroudi, K. Mandal, M.S. Al-Numay, D. Giaouris, S. Banerjee, Piecewise quadratic slope compensation technique for dc-dc switching converters. IEEE Trans. Circuits Syst. I: Regul. Pap. **67**(12), 5574–5585 (2020)
11. A. El Aroudi, K. Mandal, D. Giaouris, S. Banerjee, Slope compensation of dc-dc converters under peak current mode control. Electron. Lett. **53**(5), 345–346 (2017)
12. R.W. Erickson, D. Maksimovic, *Fundamentals of Power Electronics* (Kluwer, Springer, 2001)
13. J. Falin, Designing dc/dc converters based on sepic topology. Technical Report. Application Note 1484 (Texas Instruments, Dallas, Texas, 2008)
14. N. Femia, G. Petrone, G. Spagnuolo, M. Vitelli, Optimization of perturb and observe maximum power point tracking method. IEEE Trans. Power Electron. **20**(4), 963–973 (2005)
15. G. Garcerá, R. González-Medina, E. Figueres, Dynamic modeling of dc-dc converters with peak current control in double-stage photovoltaic grid-connected inverters. Int. J. Circuit Theory Appl. **40**(8), 793–813 (2012)
16. D. Giaouris, S. Banerjee, B. Zahawi, V. Pickert, Stability analysis of the continuous conduction mode buck converter via Filippov's method. IEEE Trans. Circuits Syst. I. **55**(4), 1084–1096 (2008)
17. W. Gu, Designing a sepic converter. Technical Report. Application Note 1484 (National Semiconductor, CA, USA, 2007)
18. K. Mandal, A. Abusorrah, M.M Al-Hindawi, Y. Al-Turki, S. Banerjee, A new software for stability and bifurcation analysis of switched dynamical systems, in *2014 International Symposium on Non Linear Theory and its Applications (NOLTA2014)* (Luzern, Switzerland, 2014)
19. K. Mandal, C. Chakraborty, A. Abusorrah, M.M. Al-Hindawi, Y. Al-Turki, S. Banerjee, An automated algorithm for stability analysis of hybrid dynamical systems. Eur. Phys. J. Spec. Top. **222**(3–4), 757–768 (2013)
20. N. Femia, G. Petrone, G. Spagnuolo, M. Vitelli, A technique for improving p&o mppt performances of double-stage grid-connected photovoltaic systems. IEEE Trans. Ind. Electron. **56**(11), 4473–4482 (2009)
21. S.M. Sousa, L.S. Gusman, T.A.S. Lopes, H.A. Pereira, J.M.S. Callegari, MPPT algorithm in single loop current-mode control applied to dc-dc converters with input current source characteristics. Int. J. Electr. Power Energy Syst. **138**, 1168–1178 (2022)
22. A. Urtasun, J. Samanes, E.L. Barrios, P. Sanchis, L. Marroyo, Control of a photovoltaic array interfacing current-mode-controlled boost converter based on virtual impedance emulation. IEEE Trans. Ind. Electron. **66**(5), 3496–3506 (2019)
23. W. Xiao, W.G. Dunford, P.R. Palmer, A. Capel, Regulation of photovoltaic voltage. IEEE Trans. Ind. Electron. **54**(3), 1365–1374 (2007)
24. M. Zhioua, A. El Aroudi, S. Belghith, J.M. Bosque-Moncusí, R. Giral, K.A. Hosani, M. Al-Numay, Modeling, dynamics, bifurcation behavior and stability analysis of a dc–dc boost converter in photovoltaic systems. Int. J. Bifurc. Chaos **26**(10) (2016)

Metamodeling and Sensitivity Analysis of a Piezoelectric Energy Harvester Based on Polynomial Chaos Expansions

Rabie Aloui, Walid Larbi, and Mnaouar Chouchane

Abstract Over the last decade, energy harvesting from ambient vibration sources has received substantial attention. Piezoelectric materials are widely used to convert ambient vibrations into electrical energy due to the efficiency of their electromechanical transduction. The finite element method is classically used to model this kind of system. Among its advantages, it makes it possible to capture the influence and effect of the different geometrical, mechanical, and electronic parameters of the problem on its electromechanical response and to study in particular the dependence between the eigenmodes of the system and its frequency response. The amount of electrical energy recovered by the system is largely influenced by the uncertainty of its geometric and physical (electrical and mechanical properties of the piezoelectric layers) parameters. The Monte Carlo method is widely used to assess this influence. Nevertheless, its prohibitive calculation time, in the case of dynamic modeling by finite element method, limits its use. Metamodels are proposed in the literature as a surrogate model allowing a robust modeling of these systems and a considerable reduction of the calculation time for analysis of uncertainty and sensitivity. In this context, this paper focuses on the metamodeling of a piezoelectric energy harvester based on polynomial chaos expansions (PCE) to predict the electrical output with low computational cost. The metamodel is built using a database created from finite element simulations previously fixed by the design of the experiment. The metamodeling is based on a regression technique using six degrees polynomial. The surrogate model is then used for a sensitivity analysis which consists of calculating Sobol indices with negligible computational cost. The main objective is to sort the uncertain parameters in a hierarchical order according to their influence on the harvester's responses. Simulation results using the metamodel show that the length of the har-

R. Aloui
R&D Department, Design Office, Safran Seats, Safran Group, Tunisia

W. Larbi (✉)
Structural Mechanics and Coupled Systems Laboratory, CNAM, Paris, France
e-mail: walid.larbi@lecnam.net

M. Chouchane
Mechanical Engineering Laboratory, LA-MA-05 ENIM Monastir, Tunisia
e-mail: mnaouar.chouchane@enim.rnu.tn

M. Belhaq (ed.), *Advances in Nonlinear Dynamics and Control of Mechanical and Physical Systems*, Springer Proceedings in Physics 301,
https://doi.org/10.1007/978-981-99-7958-5_13

vester and the thickness of the piezoelectric layer are the most influential parameters on the electrical output, while the width of the harvester and the thickness of the piezoelectric layer are the least influential on the harvester output. These results are shown to agree with those obtained by Monte Carlo simulations using the finite element model.

1 Introduction

Piezoelectric materials are widely used for electrical energy harvesting from strain mechanical energy because of their high conversion efficiency. Piezoelectric materials have found extensive applications in vast engineering areas including sensors and actuators. Various approaches have been proposed in the literature for modeling a typical composite cantilevered beam comprising piezoelectric layers; in particular, the finite element method used extensively [1–3]. This modeling approach is commonly used to solve the most complex problems, but it usually incurs a high computational cost.

Metamodeling using polynomial chaos expansions is used generally to quickly and accurately assess the output response of a system under different inputs and is particularly useful in cases where direct simulation is computationally expensive or not feasible. This approach involves creating a mathematical representation of a complex system by using polynomials to approximate its behavior [4]. The polynomial chaos expansion allows for the computation of statistical moments of the response, such as the mean, the variance, and higher order moments. This information can be used for various studies such as uncertainty quantification, model reduction, and design optimization. Baklouti et al. [5] used a Kriging metamodel to evaluate the performance of vibroacoustic models in the presence of uncertainties in the physical parameters using Monte Carlo simulations. The simulated results show the effectiveness of the Kriging metamodel in quantifying uncertainty in acoustic structural systems. Garcia-Macias et al. [6] presented metamodel-based approaches to study the individual and interactive effects of the mechanical properties of the carbon nanotube composite and grading profiles. Two surrogate models are used, Kriging [7] and high dimensional model representations (HDMR) metamodels, instead of the finite element model. In addition, uncertainty propagation is performed to determine the effect of stochastic uncertainties on the overall properties. Sensitivity analysis based on Sobol indices is a method for evaluating the importance of different input parameters on the output response of a model. This method provides a measure of the contribution of each input parameter to the total variance of the model's response [8, 9]. By quantifying the contribution of each parameter, it can identify the most important inputs, and provide insight into the relationship between inputs and outputs.

This paper focuses on the metamodeling of a piezoelectric energy harvester based on a finite element model to predict the electrical response with a low computational cost. The main goal is to study the sensitivity of the electrical output to the uncertain geometrical parameters of the harvester. Global sensitivity analysis is then performed using Sobol indices to identify the most influential parameters on the voltage output response. Both Monte Carlo simulations and polynomial chaos expansions are used to calculate Sobol indices and then both results are compared.

2 Finite Element Modeling of a Piezoelectric Energy Harvester

The finite element model of a piezoelectric energy harvester presented in Aloui et al. [10] is used in this research work. This model uses a standard discretization of the mechanical degrees of freedom and provides less restrictive assumptions for the electrical state which is fully described by few global discrete unknowns: *(i)* the electric charge contained in the electrodes and *(ii)* the voltage between the electrodes. The model is well adapted to practical applications since realistic electrical boundary conditions, such as equipotentiality on the electrodes and prescribed global electric charge, naturally appear. The harvester experiences bending vibrations as a result of the vertical movement of the base $w_b(t)$, as shown in Fig. 1. Considering low amplitude base vibration, the entire structure is assumed to be linearly elastic. In addition, shear deformations and rotary inertia effects are neglected according to the Euler-Bernoulli beam theory.

2.1 Matrix Formulation of the Electromechanical Problem

The governing damped electromechanical problem of a piezoelastic structure is defined by the following equations [3]:

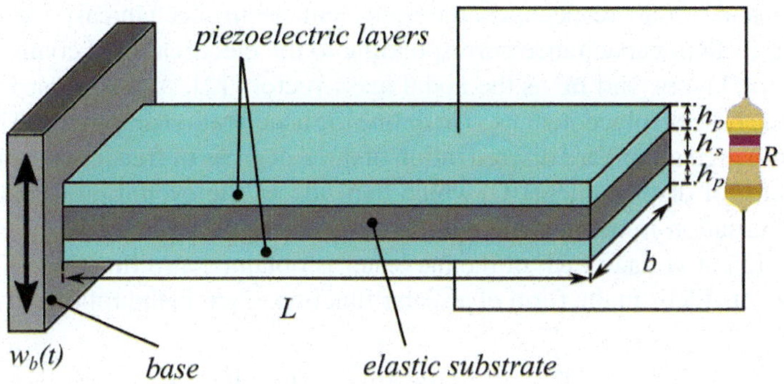

Fig. 1 A bimorph piezoelectric energy harvester under base excitation with a resistive load mounted in series

$$\mathbf{M}_m\ddot{\mathbf{U}}(t) + \mathbf{C}_m\dot{\mathbf{U}}(t) + \mathbf{K}_m\mathbf{U}(t) + \mathbf{K}_c\mathbf{V}(t) = \mathbf{F}(t) \tag{1}$$

$$\mathbf{K}_e\mathbf{V}(t) - \mathbf{K}_c^T\mathbf{U}(t) = \mathbf{Q}(t) \tag{2}$$

where $\mathbf{M}_m$ is the mass matrix, $\mathbf{K}_m$ is the stiffness matrix and $\mathbf{K}_c$ is the electromechanical coupling matrix, $\mathbf{K}_e$ is the diagonal capacitance matrix, $\mathbf{F}(t)$ is the vector of mechanical forces, $\mathbf{Q}(t)$ is the vector of electric charge outputs, $\mathbf{U}(t)$ is the vector of mechanical coordinates and $\mathbf{V}(t)$ is the vector of output voltages across the piezoelectric elements. Additionally, $\mathbf{C}_m$ is the mechanical damping matrix that deals with Rayleigh damping using the coefficients of proportionality α and β which are determined from physical experiments.

The electrical variables are also adapted to include any external circuit in the electromechanical problem. For the purpose of simulation, the structure is dissipated through a resistive load R. Ohm's law relates the voltage $\mathbf{V}(t)$ to the current $\dot{\mathbf{Q}}(t)$ as follows:

$$\mathbf{V}(t) = R\dot{\mathbf{Q}}(t) \tag{3}$$

2.2 *Frequency Response Functions*

Here, the frequency response functions (FRFs) are defined as the output responses of the harvester per base acceleration by transforming (1) and (2) to the frequency domain. The equivalent expression for nodal displacements FRFs relative to the different electrical configurations is given in the following equation [11]:

$$\frac{\mathbf{U}}{-\omega^2 w_b} = \left[-\omega^2\mathbf{M}_m + j\omega\mathbf{C}_m + \mathbf{K}_m + \frac{j\omega\,\mathbf{K}_c^{eq}\,\mathbf{K}_c^{eqT}}{\left(\frac{1}{R} + j\omega C^{eq}\right)}\right]^{-1}\mathbf{m}^* \tag{4}$$

where $\mathbf{K}_c^{eq}$ and C^{eq} are respectively the equivalent electromechanical coupling vector and the equivalent capacitance corresponding to the electrical load connection with piezoelectric layers, and $\mathbf{m}^*$ is the nodal mass vector [12]. As mentioned above, w_b presents the base displacement used to define the acceleration amplitude. This matrix formulation used a standard discretization of three degrees of freedom per node given in the vector of displacements $\mathbf{U}$, which depends on the excitation frequency; i.e., for each excitation frequency, there are associated displacements.

The resultant voltage FRF of the harvester is obtained as a function of the nodal displacements FRFs in the form of a scalar function given in the following equation:

$$\frac{V}{-\omega^2 w_b} = \frac{j\omega\,\mathbf{K}_c^{eqT}}{\left(\frac{1}{R} + j\omega C^{eq}\right)}\left(\frac{\mathbf{U}}{-\omega^2 w_b}\right) \tag{5}$$

3 Metamodeling Using Polynomial Chaos Expansions

The FRFs, given in (4) and (5), establish the dependence of the output responses on a set of uncertain inputs representing the physical parameters of the harvester. These parameters are given in the vector $\boldsymbol{\theta}$ assumed to contain M-independent random parameters. The output Y is determined using a deterministic model $\mathcal{M}$ which depends on the components of the vector $\boldsymbol{\theta}$.

Consider a random vector with independent components $\boldsymbol{\theta}$ with support $\mathcal{D}_{\theta}$ described by the joint probability density function f_{θ}. In addition, a finite variance of the model output $Y = \mathcal{M}(\boldsymbol{\theta})$ such that [8, 13]:

$$\mathbb{E}\left[Y^2\right] = \int_{\mathcal{D}_{\theta}} \mathcal{M}^2(\boldsymbol{\theta})\, f_{\theta} d\boldsymbol{\theta} \tag{6}$$

The PCE approximates the computational model $\mathcal{M}(\boldsymbol{\theta})$ by a sum of orthonormal polynomials [13]:

$$Y = \mathcal{M}^{\mathrm{PCE}}(\boldsymbol{\theta}) = \sum_{\alpha} y_{\alpha} \psi_{\alpha}(\boldsymbol{\theta}) \tag{7}$$

where ψ_{α} are multivariate orthonormal polynomials with respect to the input distributions f_{θ}, $\boldsymbol{\alpha}$ are multi-indices, and y_{α} are the deterministic corresponding coefficients.

In a PCE, the stochastic model (i.e., model uncertainty) is characterized in a decoupled form by the random basis functions, and the deterministic coefficients are given by the expansion in the (7). A square integrable random variable can be written in a mean square convergent series using random orthonormal polynomial bases called PCE for Hermite bases and generalized PCE for other bases. The random variables should be normally distributed for Hermite bases, while non-normal random variables should be transformed into standard normal random space using transformation schemes. For practical purposes, the expansion of Y in (7) should be truncated to include a limited number of polynomial terms.

Extending (7) to M random variables, the complete N-order PCE is defined as the set of all multidimensional polynomials ψ whose degree does not exceed N [14]:

$$Y(\theta_1, \theta_2 \ldots \theta_M) = \sum_{\alpha=0}^{P-1} y_{\alpha} \psi_{\alpha}(\boldsymbol{\theta}) \tag{8}$$

where P is the number of terms whose degree $\leq N$ such that:

$$P = \sum_{k=0}^{N} C_k^{M+k+1} = \frac{(M+N)!}{M!N!} \tag{9}$$

The main point of the characterization of uncertainty by PCE is represented in the determination of its coefficients y_{α}, which can be obtained by techniques such as

Galerkin's method by spectral projection, e.g., Monte Carlo sampling ... , or by collocation. Recent studies on PCE have shown that non-intrusive methods are computationally efficient for determining coefficients [15].

The PCE coefficient y_α is generally calculated using two approaches: (i) intrusive approaches, e.g., Galerkin scheme, and (ii) non-intrusive approaches, e.g., projection, least-squares regression. Here, the emphasis is on least-squares methods, under the name of statistical regression. The term non-intrusive term indicates that the chaos coefficients are evaluated over a set of input realizations $\boldsymbol{\theta} = (\theta_1, \theta_2 \ldots \theta_M)$, called the design of experiment. By considering $Y = (\mathcal{M}(\theta_1), \mathcal{M}(\theta_2) \ldots \mathcal{M}(\theta_M))$ as the set of outputs of the calculation model for each sampling point, the set of coefficients y_α is then calculated by minimizing the least-squares residual of the polynomial approximation of the true model:

$$\hat{y} = \arg\min_{y_\alpha} \frac{1}{N} \sum_{i=1}^{N} \left(\mathcal{M}\left(\boldsymbol{\theta}^{(i)}\right) - \sum_{\alpha=0}^{P-1} y_\alpha \psi_\alpha \left(\boldsymbol{\theta}^{(i)}\right) \right)^2 \tag{10}$$

where $\boldsymbol{\theta}^{(i)}$, $i = 1 \ldots n$, is a set of points building the design experience. Once the PCE metamodel has been built, it is necessary to assess its accuracy with respect to the real computational model. Several statistical error estimators exist, among which an effective measure is given by the mean squared error:

$$\mathrm{MSE} = \mathbb{E}\left[\mathcal{M}(\theta) - \mathcal{M}^{\mathrm{PCE}}(\theta)\right]^2 \tag{11}$$

A lower value of MSE corresponds to a more accurate metamodel. A variant estimator of MSE is the relative mean square error given by the following equation:

$$\mathrm{RMSE} = \frac{\mathrm{MSE}}{\sigma_Y^2} \tag{12}$$

with σ_Y^2 being the variance of the computational model response.

4 Variance-Based Sensitivity Analysis

Among the large number of variance-based methods proposed for sensitivity analyses, the method based on Sobol' indices are commonly used and widely applied to characterize parameter sensitivities. This method decomposes the variance of the model output into fractions which can be attributed to inputs or sets of inputs. The output response Y depending on a set of parameters θ_i regrouped into a vector $\boldsymbol{\theta}$ can be orthogonally decomposed as follows [16–18]:

$$Y(\boldsymbol{\theta}) = \mathcal{M}_0 + \sum_{i=1}^{k} \mathcal{M}_i(\theta_i) + \sum_{i<j}^{k} \mathcal{M}_{ij}(\theta_i, \theta_j) + \cdots + \mathcal{M}_{12\cdots k}(\boldsymbol{\theta}) \tag{13}$$

where the constant $\mathcal{M}_0$ and conditional expectations $\mathcal{M}_i, \mathcal{M}_{ij}, \cdots, \mathcal{M}_{12...k}$ are expressed as follows:

$$\mathcal{M}_0 = \mathbb{E}(Y) \tag{14a}$$

$$\mathcal{M}_i(\theta_i) = \mathbb{E}(Y|\theta_i) - \mathcal{M}_0 \tag{14b}$$

$$\mathcal{M}_{ij}(\theta_i, \theta_j) = \mathbb{E}(Y|\theta_i, \theta_j) - \mathcal{M}_0 - \mathcal{M}_i - \mathcal{M}_j \tag{14c}$$

...etc

This expansion is unique under the following conditions [16]:

$$\int_0^1 \mathcal{M}_{i_1 \ldots i_s}(\theta_{i_1}, \ldots, \theta_{i_s}) d\theta_{i_t} = 0,\ 1 \le t \le s,\ \{i_1, \ldots, i_s\} \subseteq \{1, \ldots, k\} \tag{15}$$

The variance of the output Y is decomposed as follows:

$$\text{Var}(Y) = \sum_{i=1}^{k} D_i(Y) + \sum_{i<j}^{k} D_{ij}(Y) + \cdots + D_{12\cdots k}(Y) \tag{16}$$

where $D_i(Y) = \text{Var}\left[\mathbb{E}(Y|\theta_i)\right]$, $D_{ij}(Y) = \text{Var}\left[\mathbb{E}(Y|\theta_i, \theta_j)\right] - D_i(Y) - D_j(Y)$ and so on for higher order interactions. The *Sobol' indices* are obtained as follows:

$$S_i = \frac{D_i(Y)}{\text{Var}(Y)},\ S_{ij} = \frac{D_{ij}(Y)}{\text{Var}(Y)},\ \ldots \tag{17}$$

The most important indices in practice are the main effect S_i and the total effect S_{Ti} which is defined below. These indices provide sufficient information for the sensitivity of the output system to its individual input variables. The simplest approach is to estimate the first order indices (main effects) $\{S_1, \ldots, S_k\}$ and to sort these indices in descending order [18]. The indices are all positive and varying between 0 and 1. The most important parameter that influences the output is the one with the highest index value (the closest to 1).

The total sensitivity index (total effect), denoted as S_{Ti}, includes all the effects associated with the parameter θ_i and defined as the sum of all the indices associated with θ_i which corresponds to the variance of Y induced by θ_i for fixed values of other variables [17].

$$S_{Ti} = 1 - \frac{\text{Var}\left[\mathbb{E}(Y|\theta_{\sim i})\right]}{\text{Var}(Y)} \tag{18}$$

where $\theta_{\sim i}$ denotes all factors except θ_i. The condition $S_{Ti} \approx 0$ is necessary and sufficient for θ_i to be a non-influential factor. If $S_{Ti} = 0$, then θ_i can be fixed at any value within its range of uncertainty without appreciably affecting the value of the output variance $\mathrm{Var}(Y)$. Thus, the total indices can be used in setting certain parameters at fixed value.

The difference between the two Sobol' indices S_{Ti} and S_i corresponds to the interactions effect of θ_i with the other parameters which is included into S_{Ti} but not in S_i. This information can be used to guide the design of experiments, prioritize inputs for further analysis, and support decision-making in complex systems. The Sobol indices can be calculated using Monte Carlo simulations or PCEs.

5 Case Study

The harvester consists of an aluminum substrate sandwiched between two piezoceramic PZT-5A layers as shown in Fig. 1. The electrical load R is connected in series to the piezoelectric layers. For simulation purposes, we used modal mechanical damping ratios $\xi_1 = 0.01$ and $\xi_2 = 0.012$ for the first two modes [19]. The geometrical dimensions and material properties of the harvester are given in Table 1.

The uncertain parameters selected are the length L and the width b of the harvester and the thickness h_s of the substrate and h_p of the piezoelectric layers. A tolerance of $\pm 10\%$ is allocated for each parameter, which is provided by the manufacturers with reference to the nominal values. A uniform distribution is assumed for these uncertain harvester parameters. In this study, a stochastic model with 1000 samples (specifying the size of the experimental design) is used to create the PCE metamodel by computing its coefficients using the regression technique. The maximum polynomial degree of the constructed metamodel Y^{PCE} is equal to 6.

Table 1 The nominal geometrical and material properties of the piezoelectric harvester given in the Fig. 1

Parameters	Descriptions	Units	PZT-5A	Aluminum
L	Beam length	mm	30	30
b	Beam width	mm	5	5
h_p , h_s	Layers thickness	mm	0.15	0.05
E_p , E_s	Young's modulus	GPa	61	70
ρ_p , ρ_s	Mass density	kg/m^3	7750	2700
e_{31}	Piezoelectric constant	c/m^2	–10.4	–
ϵ_{33}	Permittivity constant	nF/m	13.3	–

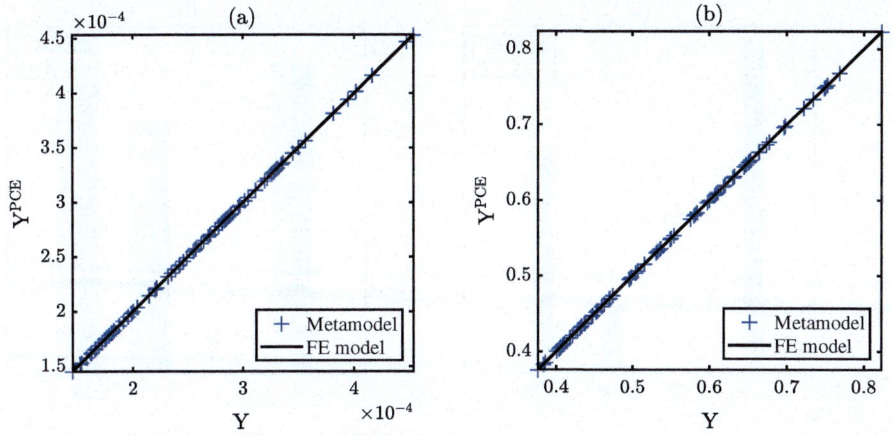

Fig. 2 The modulus of voltage FRF (given in [V/g]) at an excitation frequency of 100 Hz for: **a** a near Short circuit condition simulated by an electrical load $R = 100\,\Omega$, **b** a near open-circuit condition simulated by an electrical load $R = 10^7\,\Omega$

Figure 2 shows the modulus of the FRF voltage calculated by the PCE metamodel (denoted Y^{PCE}) compared to that of the direct simulation of the finite element model (denoted Y) using 80 samples. An excitation frequency of 100 Hz was considered for the evolution of the models, which is in the vicinity of the first vibrational mode. The scalar output Y varies according to the stochastic model of the geometric parameter samples. The simulations are given for two extreme values of load resistance that represent respectively short-circuit and open-circuit conditions as shown in Fig. 2a, b. For both conditions, the evolutions of the metamodel show output values that ideally overlap with the direct simulation using the finite element model. This shows the efficiency of the metamodel to approximate the system response with high accuracy. Furthermore, the ratio of metamodel computation time to finite element model evaluation time for the 80 samples is about 0.12.

Sobol indices are a measure of the contribution of each input to the total variance of the model's output. By quantifying the contribution of each input, the Sobol indices provide insight into the relationship between the inputs and the outputs and can be used to identify the most important inputs in a model. Figure 3 shows the Sobol indices calculated for the 4 considered uncertain parameters. The sample size chosen for the Monte Carlo simulations is equal to 10^5, so that the total cost (model evaluations) is about 6×10^5 evaluations of the full computational model. However, the PCE metamodel does not require to be specified with high degree of details as the last constructed model will be used in the analysis. The sensitivity analysis is based on the PCE model and calculates Sobol' indices accordingly by post-processing the coefficients of a PCE model without sampling. Sobol' indices computed using the two techniques are concordant, since they give index values that are very close to each other with a high precision as shown in Table 2. It is clearly seen that the length of the harvester L and the thickness of the piezoelectric layers h_p are the most influential

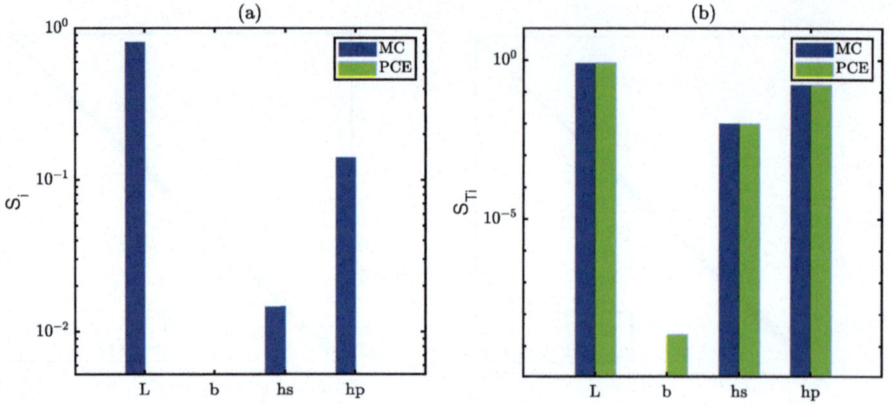

Fig. 3 Sobol indices for both Monte Carlo simulations (MC) and polynomial chaos expansions (PCE)corresponding to the four considered geometrical parameters: **a** the first order index (main effect), and **b** the total order index (total effect)

Table 2 Sobol indices for both Monte Carlo simulations (MC) and polynomial chaos expansions (PCE)

Parameters	First order indices (Main effects)			Total Sobol' indices (Total effects)		
	MC	PCE	Error (%)	MC	PCE	Error (%)
L	0.8195	0.8180	0.183	0.8352	0.8452	1.197
b	0.0052	0.0000	–	0.0000	0.0000	–
h_s	0.0147	0.0082	44.2	0.0101	00.0102	0.99
h_p	0.1414	0.1462	3.394	0.1710	0.1723	0.76

parameters on the output voltage of the harvester, whereas the width b has an almost negligible effect.

6 Conclusion

In this study, PCEs are used to represent the mathematical relation between the harvester responses and the input variables using polynomial approximations. The polynomials can be constructed in a way that allows for the computation of statistical moments of the responses as well as the Sobol indices of sensitivity at a low computational cost. Sobol indices can be calculated from the PCE by evaluating the variance of the polynomial terms that correspond to each input parameter. As can be seen in the presented results of this research, the PCE metamodel gives values of Sobol indices very close to those of Monte Carlo simulations with a very low computational cost. In addition, the sensitivity results show that the length L and the thickness of the piezoelectric layers h_p are the most significant parameters that affect the output voltage of the harvester. The width b of the harvester has a negligible effect on the model output. In conclusion, the PCE provides a powerful tool for approximating the behavior of a system and computing sensitivity indices such as the Sobol indices. By combining the two techniques, it is possible to gain a better understanding of the relationship between inputs and outputs of a complex system.

References

1. O. Thomas, J.-F. Deü, J. Ducarne, Vibrations of an elastic structure with shunted piezoelectric patches: efficient finite element formulation and electromechanical coupling coefficients. Int. J. Numer. Methods Eng. **80**(2), 235–268 (2009). https://onlinelibrary.wiley.com/doi/pdf/10.1002/nme.2632, https://doi.org/10.1002/nme.2632
2. Y. Amini, H. Emdad, M. Farid, Finite element modeling of functionally graded piezoelectric harvesters. Compos. Struct. **129**, 165–176 (2015)
3. W. Larbi, J.-F. Deü, Reduced order finite element formulations for vibration reduction using piezoelectric shunt damping. Appl. Acoust. (2018). https://doi.org/10.1016/j.apacoust.2018.04.016
4. A. Forrester, A. Sobester, A. Keane, *Engineering Design via Surrogate Modelling: A Practical Guide* (Wiley, New York, 2008)
5. A. Baklouti, K. Dammak, A. El Hami, Uncertainty analysis based on kriging meta-model for acoustic-structural problems. Appl. Sci. **12**(3), 1503 (2022). https://doi.org/10.3390/app12031503
6. E. García-Macías, R. Castro-Triguero, M.I. Friswell, S. Adhikari, A. Sáez, Metamodel-based approach for stochastic free vibration analysis of functionally graded carbon nanotube reinforced plates. Compos. Struct. **152**, 183–198 (2016). https://doi.org/10.1016/j.compstruct.2016.05.019
7. E. Denimal, J..-J.. Sinou,Advanced meta-modelling techniques and sensitivity analysis for rotordynamics in an uncertain context, in *Model Validation and Uncertainty Quantification, Volume 3, Conference Proceedings of the Society for Experimental Mechanics Series*. ed. by Z. Mao (Springer International Publishing, Cham, 2023), pp. 51–61. https://doi.org/10.1007/978-3-031-04090-0_6
8. L.L. Gratiet, S. Marelli, B. Sudret, Metamodel-based sensitivity analysis: polynomial chaos expansions and gaussian processes, in *Handbook of Uncertainty Quantification*. ed. by R. Ghanem, D. Higdon, H. Owhadi (Springer International Publishing, Cham, 2015), pp.1–37
9. R. Aloui, W. Larbi, M. Chouchane, Uncertainty quantification and global sensitivity analysis of piezoelectric energy harvesting using macro fiber composites. Smart Mater. Struct. **29**(9), 095014 (2020). IOP Publishing. https://doi.org/10.1088/1361-665X/ab9f12
10. R. Aloui, W. Larbi, M. Chouchane, Global sensitivity analysis of piezoelectric energy harvesters. Compos. Struct. **228**, 111317 (2019). https://doi.org/10.1016/j.compstruct.2019.111317
11. S.R. Anton, A. Erturk, D.J. Inman, Multifunctional unmanned aerial vehicle wing spar for low-power generation and storage. J. Aircraft **49**(1), 292–301 (2012). https://doi.org/10.2514/1.C031542
12. R. Aloui, W. Larbi, M. Chouchane, Sensitivity analysis of frequency response functions for load resistance of piezoelectric energy harvesters, in *Advances in Acoustics and Vibration II*. ed. by T. Fakhfakh, C. Karra, S. Bouaziz, F. Chaari, M. Haddar (Springer International Publishing, Cham, 2019), pp.136–148

13. S. Dutta, A.H. Gandomi, Chapter 15—Design of experiments for uncertainty quantification based on polynomial chaos expansion metamodels, in *Handbook of Probabilistic Models, Butterworth-Heinemann.* ed. by P. Samui, D. Tien Bui, S. Chakraborty, R.C. Deo (2020), pp.369–381
14. R.G. Ghanem, P.D. Spanos, *Stochastic Finite Elements: A Spectral Approach* (Springer, New York, NY, 1991). https://doi.org/10.1007/978-1-4612-3094-6
15. G. Blatman, B. Sudret, Adaptive sparse polynomial chaos expansion based on least angle regression. J. Comput. Phys. **230**(6), 2345–2367 (2011). https://doi.org/10.1016/j.jcp.2010.12.021
16. I. Sobol', Global sensitivity indices for nonlinear mathematical models and their Monte Carlo estimates. Math. Comput. Simul. **55**(1–3), 271–280 (2001). https://doi.org/10.1016/S0378-4754(00)00270-6
17. A. Saltelli (ed.), *Global Sensitivity Analysis: The Primer* (Wiley, Chichester, 2008). oCLC: 254420563
18. A. Saltelli, P. Annoni, I. Azzini, F. Campolongo, M. Ratto, S. Tarantola, Variance based sensitivity analysis of model output. Design and estimator for the total sensitivity index. Comput. Phys. Commun. **181**(2), 259–270 (2020). https://doi.org/10.1016/j.cpc.2009.09.018
19. A. Erturk, D.J. Inman, *Piezoelectric Energy Harvesting* (Wiley, Chichester, 2011). oCLC: ocn687714431

Small-Signal Dynamics of Interconnected Subsystems in Electric Vehicle

Kuntal Mandal and Abdelali El Aroudi

Abstract A general electric vehicle (EV) consists of many subsystems such as battery, DC-DC converter, DC-AC inverter, motor, and vehicle body along with their controllers. Therefore, the modeling, simulation, and validation are time-consuming and critical tasks due to its multiple time scale nature of the different subsystems and multi-domain modeling of different subsystems, i.e., battery model using physics, converter as well as inverter model using electrical, and motor-vehicle body model using mechanical engineering concept. Moreover, the selection of the type of modeling has a direct impact on control decisions of high-performance vehicle designs and their analysis. In this paper, small-signal transfer function analysis has been used to capture the overall dynamics of the EV along with controller design and stability analysis. To do this, a constant battery voltage DC-DC converter with equivalent inverter-fed motor-vehicle load in constant torque mode and constant power mode is considered. This assumption is valid because the time constant of the DC-DC converter is low compared to the other subsystems in the EV, therefore the overall dynamics is governed by the closed-loop control of the DC-DC converter.

1 Introduction

There are many dynamical systems encountered in nature and in engineering, that have a large number of subsystems connected together. In these systems, several time scales appear for modeling different subsystems. For instance, in Electric Vehicle (EV) and microgrids, there are multiple subsystems connected together and the time constants of different subsystems are different by multiple order. For example, in EV, the time constant of the battery is in hour range, the switching period of the DC-DC

K. Mandal (✉) · A. El Aroudi
Universitat Rovira i Virgili, Tarragona, Spain
e-mail: dr.kuntal.mandal@gmail.com

A. El Aroudi
e-mail: abdelali.elaroudi@urv.cat

M. Belhaq (ed.), *Advances in Nonlinear Dynamics and Control of Mechanical and Physical Systems*, Springer Proceedings in Physics 301,
https://doi.org/10.1007/978-981-99-7958-5_14

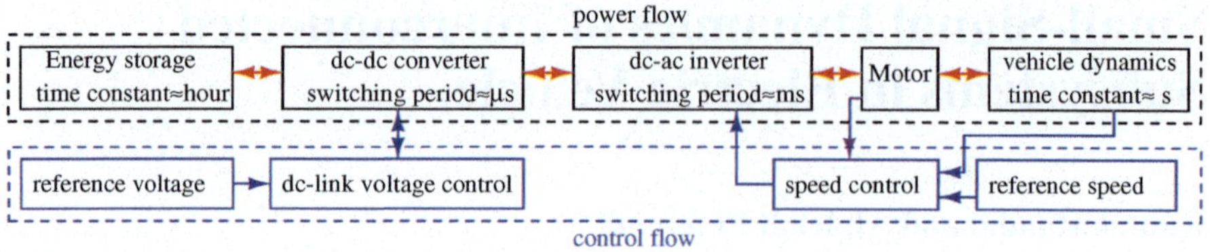

Fig. 1 Interconnection of different subsystems in EV with different time constants through block diagram representation. Reference speed can be denoted by desired drive cycle

converter is in microsecond range, the switching period of the DC-AC motor is in millisecond range, and vehicle body is in the second range as shown in Fig. 1.

Therefore, challenges arise when the subsystems on the different timescales cannot be treated separately and emergent behavior is displayed, which is only observable in the overall interconnected coupled model. This means that the analysis of the dynamics on each timescale does not lead to a reliable description of the full system. To address such challenges such as multi-scale interactions, the overall simulation step size of the mathematical model is in the nanosecond range for the overall system [2, 3].

In the past, stability analysis has been carried out for specific subsystem with the variation of the critical parameters of the EV to show different types of undesirable behaviors [7, 13, 15, 16, 18]. In this paper, the overall EV is modeled by taking the fastest dynamics of the DC-DC converter among the dynamics of the other subsystems with input source constant and output load as a coupled inverter-fed motor-vehicle system. Through small-signal analysis, the stability analysis with the variation of the parameters is shown in low-frequency scale.

The rest of this paper is organized as follows. After this introduction, Sect. 2 describes the studied system. In Sect. 3 small-signal analysis of the system with the variation of the parameters is presented. Concluding remarks are given in the last section.

2 System Description

A battery electric vehicle consists of many interconnected subsystems such as the battery, bidirectional boost/buck converter, three-phase voltage source inverter, motor, and vehicle body along with overall control including thermal management, battery management system, auxiliary circuit, and other protection circuits. Depending on the applications and cost requirement, different motors (Permanent Magnet Brushless DC Motor, Induction Motor, Permanent Magnet Synchronous Motor, Switched Reluctance Motor are in use for different vehicle bodies (two-wheeler, three-wheeler, four-wheeler, bus, or truck) [8, 10, 14].

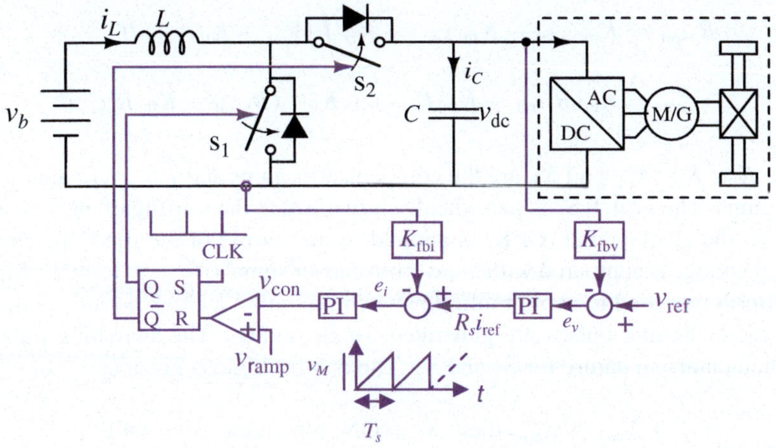

Fig. 2 Two-loop control for DC-link voltage of bidirectional DC-DC converter. The load of the converter is inverter-fed motor-vehicle coupled system which can operate as constant torque or constant power

The bidirectional DC-DC converter is useful for the selection of battery and motor ratings. During regenerative braking, the power flows in the reverse direction from the motor to battery for charging the battery. It is a process of recovering the kinetic energy of the EV wheel at the time of braking with increased efficiency and range per charge of the EV [4].

Based on the time constants of the different subsystems, it has been observed that the DC-DC converter is the lowest, i.e., the dynamics is the fastest. Irrespective of the type of the vehicle, the operation of the motor-vehicle coupled system runs either in constant torque or constant power mode [6]. On the road, the reference speed of the EV is dictated by the drive cycle which is varying in nature. A typical drive cycle consists of four-speed profiles such as idling, acceleration, deceleration, and constant speed. Though the motor exceptionally or rarely follows the torque-speed curve during operation, these curves can be viewed as covering all load points.

Therefore, to study the dynamics of the whole system, we can consider the dynamics of the closed-loop DC-DC converter with constant voltage input and constant power or constant current load as shown in Fig. 2.

A two-loop control is selected for regulating the output DC-link voltage of the bidirectional DC-DC converter, because this voltage is the input of the motor through VSI. In the outer voltage control loop, the sensed DC-link voltage (v_{dc}) is subtracted from the reference voltage (v_{ref}) and the error signal (e_v) passed through proportional-integral (PI) compensator to generate the reference signal ($R_s i_{ref}$) for the inner loop. Then the sensed inductor current signal ($K_{fbi} i_L$) is subtracted from the reference signal ($R_s i_{ref}$) where R_s is the unity sensor gain. The error (e_i) is sent through another proportional-integral (PI) compensator to generate the control voltage (v_{con}). The reference signal and control voltage are given as

$$R_s i_{\text{ref}} = K_{\text{pv}}(v_{\text{ref}} - K_{\text{fbv}} v_{\text{dc}}) + K_{\text{iv}} \int (V_{\text{ref}} - K_{\text{fbv}} v_{\text{dc}}) dt \tag{1}$$

$$v_{\text{con}} = K_{\text{pi}}(R_s i_{\text{ref}} - K_{\text{fbi}} R_s i_L) + K_{\text{ii}} \int (R_s i_{\text{ref}} - K_{\text{fbi}} R_s i_L) dt \tag{2}$$

where K_{pv}, K_{iv}, K_{pi}, and K_{ii} are the compensators gains and K_{fbv}, K_{fbi} are the feedback gains. The switch S_1 is periodically turned ON at the starting of each switching cycle as the clock signal (CLK) is applied to the set pin of the flip/flop. Then, the control voltage is compared with a sawtooth carrier signal (v_{ramp}) to reset the flip/flop. The flip/flop is used to avoid multiple switching within a clock period. In this process, the switching pulses are generated for S_1 and S_2. The switching pulses are complementary in nature for S_1 and S_2. The control logic is given as

$$\text{if } \; v_{\text{con}} \geq v_{\text{ramp}} \text{ then } \; S_1 = \text{ON, otherwise } \; S_1 = \text{OFF} \tag{3}$$

where $v_{\text{ramp}} = V_M(t \bmod T_s)$, V_M is the amplitude. T_s denotes the switching period.

3 Small-Signal Analysis

The small-signal frequency-domain analysis is desired for the design of controllers for a system as well as analyzing stability. This study is carried out for the tuning of the controller parameters with the goal of achieving desired performance [17]. The small-signal transfer functions have been obtained for the two-loop controlled bidirectional DC-DC converter. The corresponding block diagram is shown in Fig. 3. This diagram is obtained based on the closed-loop circuit diagram given in Fig. 2.

The loop gain of the two-loop controlled boost converter is calculated as follows

$$T_{\text{LG}}(s) = G_{\text{cv}}(s) \left[\frac{G_{\text{ci}}(s) G_{\text{ramp}}(s) G_{\text{id}}(s)}{1 + G_{\text{ci}}(s) G_{\text{ramp}}(s) G_{\text{id}}(s) H_i(s)} \right] G_{\text{vi}}(s) H_v(s) \tag{4}$$

where $G_{\text{vi}}(s) = \left(K_{\text{pv}} + \frac{K_{\text{iv}}}{s}\right)$, $G_{\text{ci}}(s) = \left(K_{\text{pi}} + \frac{K_{\text{ii}}}{s}\right)$, $G_{\text{ramp}}(s) = \frac{1}{V_M}$, $H_v(s) = K_{\text{fbv}}$, $H_i(s) = K_{\text{fbi}}$.

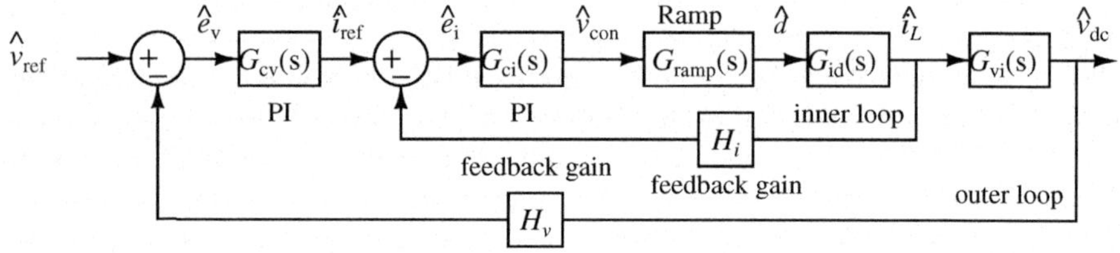

Fig. 3 Transfer function block diagram of two-loop controlled bidirectional DC-DC converter in boost mode

Table 1 Bidirectional DC-DC converter specifications and parameters

Specification	Value	Parameter	Value
Battery voltage	48 V	Inductance (L)	200 μH
Current ripple	10%	Capacitance (C)	447 μF
Voltage ripple	5%	Switching frequency ($f_s = 1/T_s$)	20 kHz

Now control-to-inductor current transfer function $G_{\mathrm{id}}(s)$ and inductor current to output voltage transfer function $G_{\mathrm{vi}}(s)$ are related to the DC-DC converter and the expressions are different for constant torque (i.e., current) and constant power case.

In case of constant torque load to the DC-DC converter, the expressions are given as follows [11].

$$G_{\mathrm{id}}(s) = \frac{\hat{i}_L(s)}{\hat{d}(s)} = \frac{\frac{I_L}{1-D} + \frac{sCV_{\mathrm{dc}}}{(1-D)^2}}{1 + \frac{s^2LC}{(1-D)^2}} \tag{5}$$

$$G_{\mathrm{vi}}(s) = \frac{\hat{v}_{\mathrm{dc}}(s)}{\hat{i}_L(s)} = \frac{\frac{V_{\mathrm{dc}}}{L} - \frac{sL}{1-D}}{1 + \frac{sCV_{\mathrm{dc}}}{I_L(1-D)}} \tag{6}$$

where I_L and D are the steady-state inductor current and duty ratio, respectively.

In case of constant power load to the DC-DC converter, these expressions are given as follows [9].

$$G_{\mathrm{id}}(s) = \frac{\hat{i}_L(s)}{\hat{d}(s)} = \frac{V_{\mathrm{dc}}}{(1-D)^2} \frac{sC}{1 - \frac{sL}{(1-D)^2}\frac{P}{V_{\mathrm{dc}}^2} + \frac{s^2LC}{(1-D)^2}}, \tag{7}$$

$$G_{\mathrm{vi}}(s) = \frac{\hat{v}_{\mathrm{dc}}(s)}{\hat{i}_L(s)} = \frac{1 - \frac{sL}{(1-D)^2}\frac{P}{V_{\mathrm{dc}}^2}}{\frac{sC}{1-D}} \tag{8}$$

It is to be noted that control-to-inductor current transfer function $G_{\mathrm{id}}(s)$ is stable in the case of constant torque. Only the performance is the concern for controller design. But it is unstable in case of constant power load. Therefore, stability and performance both must be achieved by two-loop control.

From Fig. 3, the closed-loop transfer function is given by

$$\frac{H_v.\hat{v}_{\mathrm{ref}}}{\hat{v}_{\mathrm{dc}}} = \frac{T_{\mathrm{LG}}(s)}{1 + T_{\mathrm{LG}}(s)} \tag{9}$$

First, the constant torque case is considered. Therefore, putting the (5) and (6) in (4) and using the power-stage and controller stage parameters from Tables 1 and 2, the Bode plot of the loop gain is obtained as shown in Figs. 4 and 5 by varying the PI controller gains K_{pv} and K_{iv} of the outer loop respectively.

Table 2 DC-link voltage controller parameters

Parameter	Value
DC-link voltage reference (v_{ref})	3 V
Voltage feedback gain (K_{fbv})	1/32
Current feedback gain (K_{fbi})	1/60
Voltage loop proportional gain (K_{pv})	1
Voltage loop integral gain (K_{iv})	200 s^{-1}
Current loop proportional gain (K_{pi})	0.5
Current loop integral gain (K_{ii})	50 s^{-1}
Ramp amplitude (V_M)	1 V
Sensed resistance (R_s)	1 Ω

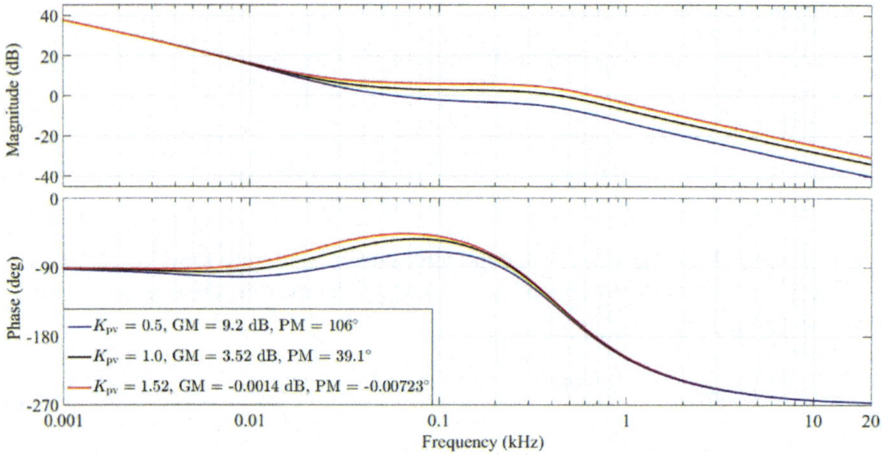

Fig. 4 Bode plot of loop gain of the two-loop controlled bidirectional DC-DC converter at full load $I = 36$ A with varying outer loop proportional gain K_{pv}. Other parameters are given in Table 2

It is observed that by tuning or adjusting the values of the PI controller gains, the desired gain and phase margins can be achieved. The gains $K_{pv} = 1$ and $K_{iv} = 200\ s^{-1}$ are selected for the outer voltage control loop for regulating the DC-link voltage in the presence of disturbances.

Now the Bode plot of the loop gain (4) is obtained as shown in Fig. 6 with the variation of DC-link capacitance. It is observed from the diagram that below a critical value ($C = 296$ μF) the system is unstable as indicated by the negative Gain Margin (GM) and Phase Margin (PM). With the increasing value of the capacitance, the stability of the system improves, but due to the better transient response to the disturbance $C = 447$ μF is selected.

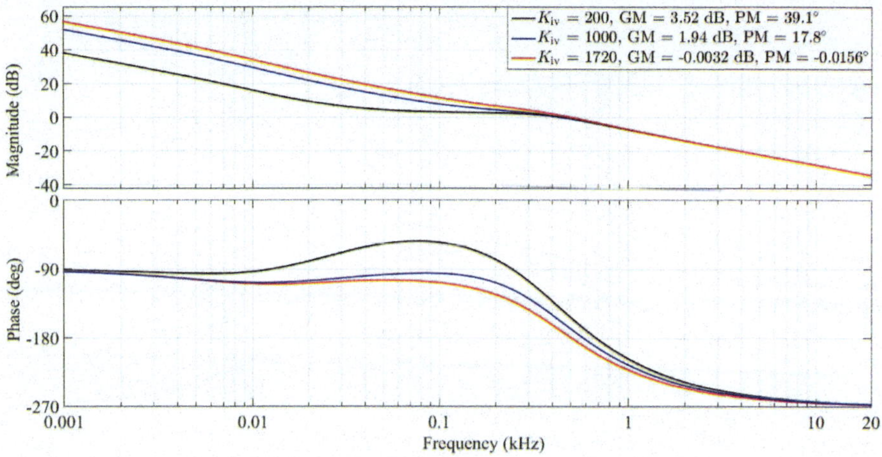

Fig. 5 Bode plot of the loop gain of the two-loop controlled linearized bidirectional DC-DC converter at full load $I_o = 36$ A with varying outer loop proportional gain K_{iv}. Other parameters are given in Table 2

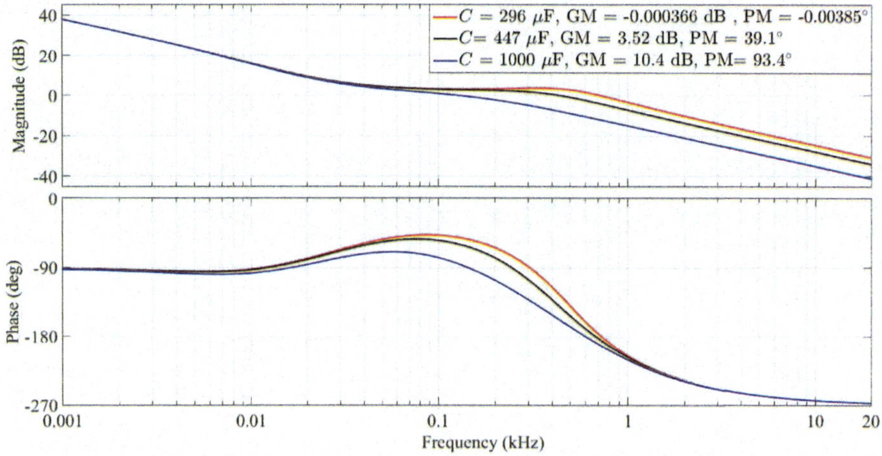

Fig. 6 Bode plot of two-loop controlled linearized bidirectional DC-DC converter at full load $I_o = 36$ A with varying C. Other parameters are given in Table 2.

Now keeping $C = 447$ μF, the Bode plots have been drawn for different values of the load current (I_o) starting from zero, full load, etc. as shown in Fig. 7. From the diagram, we can see that at $I_o = 55$ A the system is unstable.

Now this study has been extended to the constant power load case also. Putting the (7) and (8) in (4) and using the power stage and controller stage parameters from the power-stage and controller stage parameters from Tables 1 and 2, the Bode plot of the loop gain is obtained as shown in Fig. 8 with the variation of load power P.

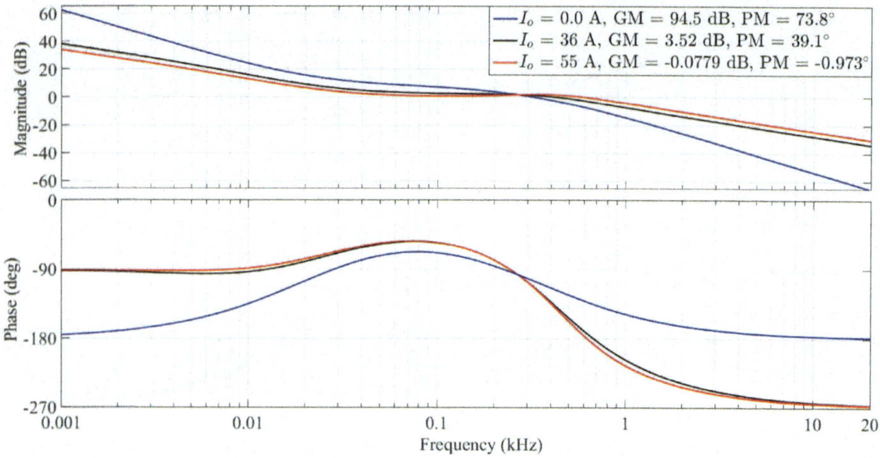

Fig. 7 Bode plot of two-loop controlled bidirectional DC-DC converter with varying load current I_o. Other parameters are given in Table 2

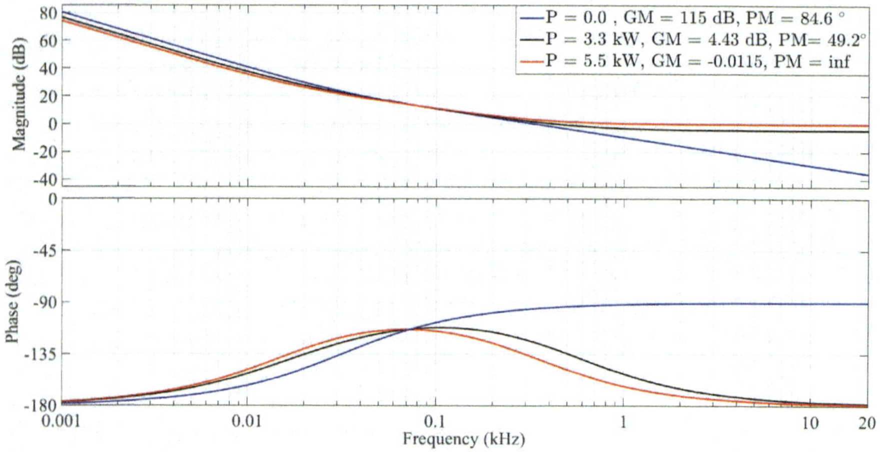

Fig. 8 Bode plot of loop gain of the two-loop controlled bidirectional DC-DC converter with varying load power P starting from zero. Other parameters are given in Table 2

Starting from zero load power, the system is unstable at $P = 5.5$ kW as the GM and PM are negative.

Similarly, this analysis can be extended to other subsystems of the EV for tuning the controller gains [5]. As the small-signal analysis is based on the state-space averaged model which ignores switching dynamics, it is valid for low-frequency range only. But there are sophisticated mathematical tool for small-signal analysis [1, 12] of the switched nonlinear system which considers switching dynamics. These methods are not discussed and used in this paper.

4 Conclusion

In this paper, the small-signal transfer function analysis of the DC-DC converter with constant input battery voltage and constant current or constant power load has been carried out. A step-by-step procedure for the selection of the controller gains is given along with its tuning. The effect of the stability margins with the variation of the parameters is also shown for the reliable and desired performance of the system. As the time constants of the different subsystems vary from microseconds to hours in EV, the study has been confined to the dominant dynamics of closed-loop DC-DC converter. In the future, this analysis can be extended to the exact EV model considering all the interconnected subsystems with switching dynamics. Such analysis provides more useful guidelines for the design and control of the interconnected subsystems under parameter variations. The analysis of detailed interactions among the subsystems is of further research direction.

Acknowledgements This work is supported and funded by the Agency for Management of University and Research Grants (AGAUR) under the Ministry of Research and Universities of the Government of Catalonia (Grant Reference No. 2020 BP 00260) and by the Spanish Ministerio de Ciencia e Innovación under Grants PID2020-120151RB-I00.

References

1. Y. Al-Turki1, A. El Aroudi, K. Mandal, D. Giaouris, A. Abusorrah, M.A. Hindawi, S. Banerjee, Non-averaged control-oriented modeling and relative stability analysis of dc-dc switching converters. Int. J. Circuit Theory Appl. **46**(3), 565–580 (2018)
2. M. Amrhein, P.T. Krein, Dynamic simulation for analysis of hybrid electric vehicle system and subsystem interactions, including power electronics. IEEE Trans. Veh. Technol. **54**(3), 825–836 (2005)
3. Y. Cao, R.C. Kroeze, P.T. Krein, Multi-timescale parametric electrical battery model for use in dynamic electric vehicle simulations. IEEE Trans. Transp. Electrification **2**(4), 432–442 (2016)
4. E.O.H. Catata, P.J. dos Santos Neto, T.A.S. Barros, E.R. Filho, Modeling, control and regenerative braking of bldc machines in electric bicycles, in Congresso Brasileiro de Automática-CBA, vol. 2 (2020)
5. A. Chauhan, K. Mandal, A. Panda, Performance improvement of electric vehicle using reset switch and bias. IFAC-PapersOnLine **55**(1), 144–149 (2022)
6. H.E. Hadraoui, M. Zegrari, A. Chebak, O. Laayati, N. Guennouni, A multi-criteria analysis and trends of electric motors for electric vehicles. World Electr. Veh. J. **13**(65) (2022)
7. Q. Han, S. Niu, L. He, Bifurcation and stability of resonance of electric vehicle powertrain: Focus on the influence of electromagnetic parameters. Proc. Inst. Mech. Eng. Part D J. Automob. Eng. **236**(47) (2021)
8. A. Karki, S. Phuyal, D. Tuladhar, S. Basnet, B.P. Shrestha, Status of pure electric vehicle power train technology and future prospects. Appl. Syst. Innov. **3**(35), 5228–5249 (2020)
9. Y. Li, K.R. Vannorsdel, A.J. Zirger, M. Norrish, D. Maksimovic, Current mode control for boost converters with constant power loads. IEEE Trans. Circuits Syst.-I **59**(1), 198–206 (2012)
10. M.S.H. Lipu, M. Faisal, S. Ansari, M.A. Hannan, T.F. Karim, A. Ayob, A. Hussain, M.S. Miah, M.H.M. Saad, Review of electric vehicle converter configurations, control schemes and optimizations: challenges and suggestions. Energies **10**(477), 1–37 (2021)

11. D. Maksimovic, *Power electronics for electric drive vehicles*. Technical report (Power Electronics Seminar at IIT Kharagpur, 2016)
12. K. Mandal, S. Banerjee, C. Chakraborty, A new algorithm for small-signal analysis of dc-dc converters. IEEE Trans. Ind. Inform. **10**(1), 628–636 (2014)
13. M.I.T. Meli, D. Yemélé, G.D. Leutcho, Dynamical analysis of series hybrid electric vehicle powertrain with torsional vibration: Antimonotonicity and coexisting attractors. Chaos, Solitons Fractals **150** (2021)
14. T. Mo, Y. Li, K. Tak Lau, C.K. Poon, Y. Wu, Y. Luo, Trends and emerging technologies for the development of electric vehicles. Energies **15**(17), 1–34 (2022)
15. H. Nademi, K. Sedghisigarchi, L. Vanfretti, Coordinated stability assessment of power converter in electric vehicle charging station using predictive control reconfiguration, in *2019 IEEE 28th International Symposium on Industrial Electronics (ISIE)* (2019)
16. J.F. Padial, A. Casal, Bifurcation in car-following models with time delays and driver and mechanic sensitivities. Revista de la Real Academia de Ciencias Exactas, Físicas y Naturales. Serie A. Matemáticas **116**(189) (2022)
17. C. Raga, A. Lázaro, A. Barrado, A. Martín-Lozano, I. Quesada, Step-by-step small-signal modeling and control of a light hybrid electric vehicle propulsion system. Energies **12**(21), 4082 (2019)
18. F.D. Rossaa, M. Gobbib, G. Mastinub, C. Piccardia, G. Previati, Bifurcation analysis of a car and driver model. Veh. Syst. Dyn. **52**, 142–156 (2014)

Nonlinear Piezoelectric Vibration Energy Harvesting of a Cantilever Beam Using Homotopy Analysis Method

Airton Nabarrete, Jorge Luis Nabarrete, and J. M. Balthazar

Abstract The vibration energy collectors based on piezoelectric resonators are promising elements for energizing remotely located systems. However, differences between the resonant frequency of these traditional harvesters and the vibration frequency can drastically decrease the collected energy and make them ineffective. Appropriate mathematical models, different analyses, and optimization techniques to tune the resonant frequency of piezoelectric collectors have been researched. In this study, the model of an inverted vertical cantilever beam with a piezoelectric patch and a tip mass is used for energy harvesting. The beam is subjected to base excitations that can induce large lateral displacements of the tip, and consequently large deformation for the piezoelectric patch. Applying the homotopy analysis method (HAM) to the coupled electromechanical governing equations of motion, novel analytical solutions of the transverse displacement of the cantilever beam, its amplitude and phase as well as the output voltage obtained from the piezoelectric patch are derived. The analytical solutions are derived for the transversal displacements of the beam, even if it presents a varying cross-sectional area. The analytical solution considers the nonlinear behavior characteristics emphasizing the capabilities of a first-order approximation of HAM to present highly accurate closed-form solutions. The accuracy of this approximation of HAM is confirmed by comparison to numerical integration methods.

A. Nabarrete (✉)
Aeronautics Institute of Technology - ITA, São José Dos Campos, Brazil
e-mail: nabarrete@ita.br

J. L. Nabarrete
Engineering consultant, São Bernardo do Campo, Brazil

J. M. Balthazar
Federal University of Technology - Paraná - UTFPR, Ponta Grossa, Brazil

College of Egineering - Unesp, Bauru, Brazil

M. Belhaq (ed.), *Advances in Nonlinear Dynamics and Control of Mechanical and Physical Systems*, Springer Proceedings in Physics 301,
https://doi.org/10.1007/978-981-99-7958-5_15

1 Introduction

Widely investigated, energy harvesting exploiting ambient vibration energy has generated very interesting results for powering small electronic components. Among the possible transduction mechanisms, piezoelectric transduction has received the most attention for converting ambient vibrations to useful electrical energy [1]. The design of an energy harvesting device must be tailored to the ambient energy available. Since many times external excitations are unknown the parameters of an energy harvester need to be changed to be able to operate in these situations. Adaptive response systems or the use of a number of cantilever beams tuned to different frequencies would be a possible yet impractical and difficult solution [2]. Aiming to harvest a reasonable amount of energy when the excitation frequency is low can be a difficult undertaking for applications in which the physical dimensions of host structures are small.

The nonlinear transverse vibration of a cantilever beam carrying a lumped mass has been studied by many authors [3–5]. The nonlinear displacement–curvature relation of the beam due to the large transverse displacement has been analyzed uniquely by some researchers when designing a slender structure to have the first torsion resonance frequency much higher than the excitation frequency [5]. With the concentrated mass kept symmetrical with respect to the beam center line, the vibration can be considered purely planar, neglecting the torsion modes of the beam in the analysis.

The energy harvesting device used in this study is composed of an inverted vertical cantilever beam with a piezoelectric patch subjected to base excitations. The beam has a tip mass adjusted to obtain a structural behavior near buckling, having the same geometric configuration used by Friswell et al. [6] to achieve a low effective resonance frequency. Due to large deformations, this geometrically nonlinear device exhibits highly nonlinear behavior characteristics such as multiple solutions, jump phenomena, and bifurcations. Friswell et al. [6] investigated the bimorph cantilever configurations with series and parallel connections of piezoceramic layers employing Rayleigh–Ritz-type discrete formulation. Another one focused on numerical solutions in comparison with an experimental analysis of the energy harvesting device based on an inverted vertical cantilever beam [6]. Looking for analytical investigations for the beam model one will find simplified single-degree-of-freedom solutions.

Applying the homotopy analysis method (HAM), introduced by Liao [7, 8], to the coupled electromechanical governing equations of motion, novel analytical solutions of the transverse displacement of the cantilever beam, of its amplitude and phase as well as the output voltage are derived. This method boasts some significant distinguishable characteristics that are desirable in nonlinear analysis theory. Primarily, its wide field of application is due to the fact that it is not dependent on small or large parameters and therefore weak as well as strong nonlinear problems can be dealt with. Moreover, for the approximation of a nonlinear problem and its desired solution great freedom is provided for the choice of different base functions [9]. Furthermore, it offers an unusually convenient way to adjust and control the convergence region of the solution series [10]. The aim is to address the subject of piezoelectric energy harvesting from a purely analytical perspective emphasizing the capabilities of a first-order approximation of HAM to present highly accurate closed-form solutions.

2 Nonlinear Model of the Cantilever Beam with Lumped Mass

Figure 1 shows a slender cantilever beam with length L. The beam carries a lumped tip mass m_t, with moment of inertia I_t, at a position L_t from the base of the beam. The horizontal and vertical displacements of any point of the beam are v and u, respectively. The mechanical properties of the beam are assumed to be uniform along its length. These properties include the cross-sectional area A, mass density ρ, equivalent Young's modulus E and second moment of area I. The base excitation is denoted by the function $v_b(t) = Y_0 \cos(\Omega t)$. An arbitrary point on the beam P is considered at a distance s from the base and along the neutral axis of the beam. This point undergoes a rigid body translation due to the base excitation and a further displacement due to the elastic beam deformation. Thus the horizontal and vertical displacements of point P are given by $(v_b(t) + v(s, t), -u(s, t))$. The rotation of the beam at s is given by $\theta(s, t)$.

The model parameters used by Friswell et al. [6] for the energy harvester as well as for the piezoelectric patch are specified in Table 1. The piezoelectric constant is d_{31} and s represents the distance along the neutral axis of the beam. The length, thickness, and width of the piezoelectric patch are represented by L_C, h_C, and b_C, respectively.

The coupled electromechanical governing equations for the piezoelectric patches attached to the cantilever beam with a tip mass are given by

$$
\begin{aligned}
&\left[I_t N_5^2 + m_t + \rho A N_1 + \left(\rho A N_3 + m_t N_4^2 + I_t N_5^4\right) v^2\right] \ddot{v} \\
&+ \left(\rho A N_3 + m_t N_4^2 + I_t N_5^4\right) v \dot{v}^2 + \left(E I N_6 - \rho A g N_9 - m_t g N_4 + 2 E I N_7 v^2\right) v \\
&- \Theta_1 V - \Theta_2 v^2 V = -(\rho A N_2 + m_t) \ddot{v}_b
\end{aligned}
\tag{1}
$$

and

$$
C_p \dot{V} + \frac{1}{R_l} V + \Theta_1 \dot{v} + \Theta_2 v^2 \dot{v} = 0 \tag{2}
$$

subject to the initial conditions

$$
v(0) = A, \quad \dot{v}(0) = 0 \quad \text{and} \quad V(0) = 0 \tag{3}
$$

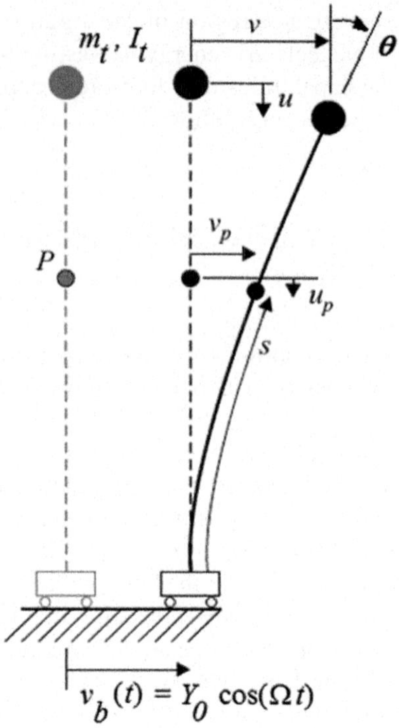

Fig. 1 Inverted cantilever beam with tip mass subject to base excitation (adapted from Friswell et al. [6])

Table 1 Summary of model parameters for the energy harvester

$\lambda_t = \frac{1}{\left(1-\cos\left(\frac{\pi L_t}{2L}\right)\right)}$	$N_8 = \int_0^L (\psi'(s))^4 (\psi''(s))^2 ds = \lambda_t^6 \left(\frac{\pi^8}{4096}\right) \frac{1}{L^7}$
$\psi(s) = \lambda_t (1 - \cos(\frac{\pi s}{2L}))$	$N_9 = \int_0^L \left(\int_0^s (\psi'(\xi))^2 d\xi\right) ds = \lambda_t^2 \left(\frac{\pi^2-4}{16}\right)$
$\gamma_c = d_{31} b_c E (h + h_c)$	$C_1 = I_t N_5^2 + m_t + \rho A N_1$
$N_1 = \int_0^L (\psi(s))^2 ds = \lambda_t^2 (\frac{3\pi-8}{2\pi}) L$	$C_2 = \rho A N_3 + m_t N_4^2 + I_t N_5^4$
$N_2 = \int_0^L \psi(s) ds = \lambda_t (\frac{\pi-2}{\pi}) L$	$C_3 = \rho A N_3 + m_t N_4^2 + I_t N_5^4$
$N_3 = \int_0^L \left(\int_0^s (\psi'(\xi))^2 d\xi\right)^2 ds = \lambda_t^4 \left(\frac{\pi^2(2\pi^2-9)}{384}\right) \frac{1}{L}$	$C_4 = EIN_6 - \rho A g N_9 - m_t g N_4$
$N_4 = \int_0^{L_t} (\psi'(s))^2 ds = \lambda_t^2 \left(\frac{\pi^2}{8}\right) \frac{1}{L}, L_t = L$	$C_5 = 2EIN_7$
$N_5 = \psi'(L_t) = \lambda_t (\frac{\pi}{2}) \frac{1}{L}, L_t = L$	$C_6 = \rho A N_2 + m_t$
$N_6 = \int_0^L (\psi''(s))^2 ds = \lambda_t^2 \left(\frac{\pi^4}{32}\right) \frac{1}{L^3}$	$\Theta_1 = \gamma_c \int_0^{L_c} \psi''(s) ds = \gamma_c \psi'(L_c)$
$N_7 = \int_0^L (\psi'(s)\psi''(s))^2 ds = \lambda_t^4 \left(\frac{\pi^6}{2^9}\right) \frac{1}{L^5}$	$\Theta_2 = \frac{3}{2}\gamma_c \int_0^{L_c} \psi''(s)(\psi'(s))^2 ds = \frac{1}{2}\gamma_c (\psi'(L_c))^3$

where v is the horizontal elastic displacement at the tip mass, V is the voltage output of the piezoelectric patch, Θ_1, Θ_2 are parameters of the charge produced by the piezoelectric patch and C_P is its capacitance, R_l is the load resistor and g is the gravitational constant. The base excitation is given by $v_b(t) = Y_0 \cos(\Omega\ t)$. The values of the parameters $N_i,\ i = 1, 2, \dots, 6$, Θ_1 and Θ_2 are given in Table 1.

First, we solve (2) to obtain a solution for V as a function of v and $\dot{v}$. Substituting the obtained solution expression of V into (1) we end up with a single differential equation to solve for v. Finally, we find the solution for V.

In the solution of (2) it is considered that for large values of t the exponential term is negligible, thus

$$V(v, \dot{v}, t) = -\ R_l\ \dot{v}(\Theta_1 + \Theta_2 v^2) \tag{4}$$

Substituting (4) into (1),

$$\begin{aligned}(C_1 + C_2 v^2)\ddot{v} + C_3\, v\, \dot{v}^2 + C_4\, v + C_5\, v^3 + \Theta_1^2 R_l\, \dot{v} + 2\Theta_1\Theta_2 R_l\, \dot{v}\, v^2 + \\ \Theta_2^2 R_l\, \dot{v}\, v^4 = C_6\, Y_0\, \Omega^2 \cos(\Omega t - \varphi)\end{aligned} \tag{5}$$

where the constants $C_i,\ i = 1, 2, \dots, 6$ are given in Table 1. Note that, for convenience and without loss of generality, we introduce the phase angle φ in the expression of the harmonic load as a quantity to be determined.

Defining the variables $\tau = \Omega t, \quad v(t) = Au(\tau)$ and inserting them into (5)

$$\begin{aligned}&(C_1 A\Omega^2 + C_2 A^3\Omega^2 u^2)\frac{\partial^2 u(\tau)}{\partial \tau^2} + C_3 A^3\Omega^2 u\left(\frac{\partial u(\tau)}{\partial \tau}\right)^2 + C_4 Au + C_5 A^3 u^3 + \\ &(\Theta_1^2 R_l A\, \Omega + 2\Theta_1\Theta_2 R_l A^3 \Omega\, u^2 + \Theta_2^2 R_l A^5 \Omega\, u^4)\frac{\partial u(\tau)}{\partial \tau} = F_1 \cos(\tau) + F_2 \sin(\tau)\end{aligned} \tag{6}$$

subject to the initial conditions

$$u(0) = 1, \quad \frac{\partial u(0)}{\partial \tau} = 0 \tag{7}$$

with the constants F_1 and F_2 satisfying

$$\left(F_1^2 + F_2^2\right)^{\frac{1}{2}} = C_6 Y_0 \Omega^2, \quad \varphi = \tan^{-1}\left(\frac{F_2}{F_1}\right) \tag{8}$$

3 Homotopy Analysis for Electromechanical Beam Model

The periodic solution to (6) can be expressed by a set of base functions in the form

$$u(\tau) = \sum_{k=1}^{\infty} [\alpha_k \sin(k\,\tau) + \beta_k \cos(k\,\tau)] \tag{9}$$

where the initial guess for $u(\tau)$ is $u_0(\tau) = \cos(\tau)$, which satisfies the initial conditions of (7). To ensure the *rule of solution expression* given by (9), the linear operator is

$$\mathscr{L}[\phi(\tau,q)] = \Omega^2\left[\frac{\partial^2\phi(\tau,q)}{\partial\tau^2} + \phi(\tau,q)\right] \tag{10}$$

with the property $\mathscr{L}[D_1 \sin(\tau) + D_2 \cos(\tau)] = 0$, with D_1 and D_2 as constants.

The nonlinear operator is defined in accordance with (6), considering $q \in [0, 1]$ as embedding parameter. The zero-order deformation equation is given by

$$(1-q)\mathscr{L}[\phi(\tau,q) - u_0(\tau)] = q\,c_0 H(\tau)\mathscr{N}[\phi(\tau,q), \Lambda(q)] \tag{11}$$

where $c_0 \neq 0$ is the convergence-control parameter and $H(\tau)$ a nonzero auxiliary function. Considering the values 0 and 1 for q, the function $\phi(\tau,q)$ varies from the initial guess $u_0(\tau)$ to the desired solution

$$\phi(\tau,0) = u_0(\tau), \quad \phi(\tau,1) = u(\tau), \quad \Lambda(0) = A_0, \quad \Lambda(1) = A \tag{12}$$

Differentiating the zero-order equation m times with respect to q, dividing it by $m!$ and setting $q = 0$, the mth order deformation equation is obtained. In order to satisfy the *rule of solution expression* and the *rule of coefficient ergodicity*, a nonzero auxiliary function is chosen to be $H(\tau) = \cos(2\kappa\tau)$ and it can be determined uniquely as $H(\tau) = 1$. According to the property of the linear operator, if the terms $\sin(\tau)$ and $\cos(\tau)$ exist, the secular terms $\tau\ \sin(\tau)$ and $\tau\ \cos(\tau)$ would appear in the final solution and some restrictions must be applied. One form is to enforce the constant term and the coefficients of $\sin(\tau)$ and $\cos(\tau)$ to be equal to zero, and use the conditions in (8). Finally, solving the first-order deformation equation considering the initial conditions it follows

$$
\begin{aligned}
u_1(\tau) = \frac{c_0}{\Omega^2}\Bigg\{ & \frac{A_0^3}{32}\left(C_2\,\Omega^2 + C_3\,\Omega^2 - C_5\right)[\cos(3\tau) - \cos(\tau)] + \\
& \frac{R_l}{16}\left(\Theta_1\Theta_2 A_0^3\,\Omega + \frac{3}{8}\Theta_2^2 A_0^5 \Omega\right)[\sin(3\tau) - \sin(\tau)] + \\
& \frac{R_l}{384}\Theta_2^2 A_0^5\,\Omega[\sin(5\tau) - \sin(\tau)]
\end{aligned} \tag{13}
$$

Thus, the first-order approximation of $v(t)$ becomes

$$
v(t) = A\,u(\tau) \cong A_0\,[u_0(\tau) + u_1(\tau)] \tag{14}
$$

To obtain the solution of voltage V, the (14) is inserted into (4) considering the steady-state solution for large values of time.

4 Discussion of Results

For the problem at hand, the usual application of HAM for two coupled differential equations does not achieve unique solutions for the amplitude and phase of the horizontal displacement of the beam since the number of zero terms to avoid secular terms in the final solution is greater than the number of unknown parameters when both differential equations are handled simultaneously. By this approach the horizontal displacement amplitude of the beam is derived to be a function of the external excitation frequency and at the same time a fixed value independent from the harmonic base excitation. This is due to the fact that both differential equations are solved independently. In this case, the free choice of initial guesses and base functions cannot avoid the problem at stake.

To overcome this ambiguity, the first-order differential equation is first solved as a function of v yielding $V(\dot{v}, v, t)$, which is inserted into the second-order differential equation to obtain a single nonlinear differential equation of the horizontal displacement for the tip mass that can be solved by HAM, as depicted in Fig. 2. Using the initial conditions and substituting the solution of v to obtain the output voltage V, the electromechanical problem is solved. It is to be remarked that the zero initial conditions for the displacement and velocity of the tip mass used for the numerical analysis by Friswell et al. [6] is not feasible when applying HAM, rather it is assumed that at $t = 0$, the vertical beam is displaced by an amplitude A_0 to be solved. For the solution approach it is noteworthy that the phase angle φ is introduced in the expression for the harmonic excitation rather than in the solution function to allow for a more suitable set of base functions. The dimensions and other parameters for the beam and energy harvester are the same used by Friswell et al. [6].

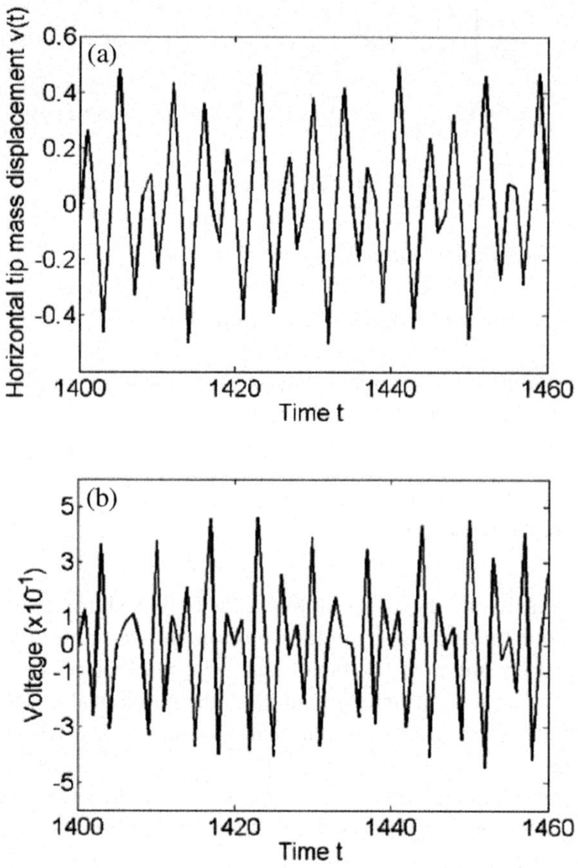

Fig. 2 Nonlinear displacement (**a**) and voltage (**b**) responses of the energy harvester

5 Conclusions

By means of the HAM, this research describes the study of analytical solutions to the forced nonlinear vibrations of a slender cantilever beam. The accuracy of a first-order approximation of HAM is compared to numerical solutions from numerical integration methods. The right choice of the auxiliary linear operator, nonzero auxiliary function, and initial guess allows for the right interplay of all components involved in arriving at accurate solutions. The interplay and correlation of the horizontal tip displacement and voltage output to maximize the energy harvested by this strongly nonlinear electromechanical system have been studied and amplitude frequency as well as response curves for both variables have been achieved. The results demonstrate a good qualitative comparison to the literature. In Fig. 3, the main achievement

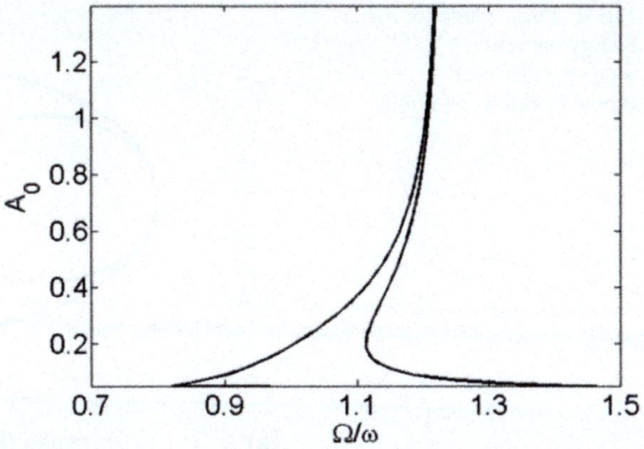

Fig. 3 Amplitude-frequency curve for parameter values

of the analysis is the frequency response curves that display jump phenomena of coexisting solutions that, from a numerical point of view, are extremely difficult to produce. Further investigation exhibits periodic as well as chaotic response for different values of the tip mass corresponding to pre- and post-buckled regimes of the response, as shown in Fig. 4. The effect of the tip mass is analyzed by means of phase planes of the energy harvester system.

Fig. 4 Phase plane for the energy harvester: **a** pre-buckled and **b** post-buckled conditions

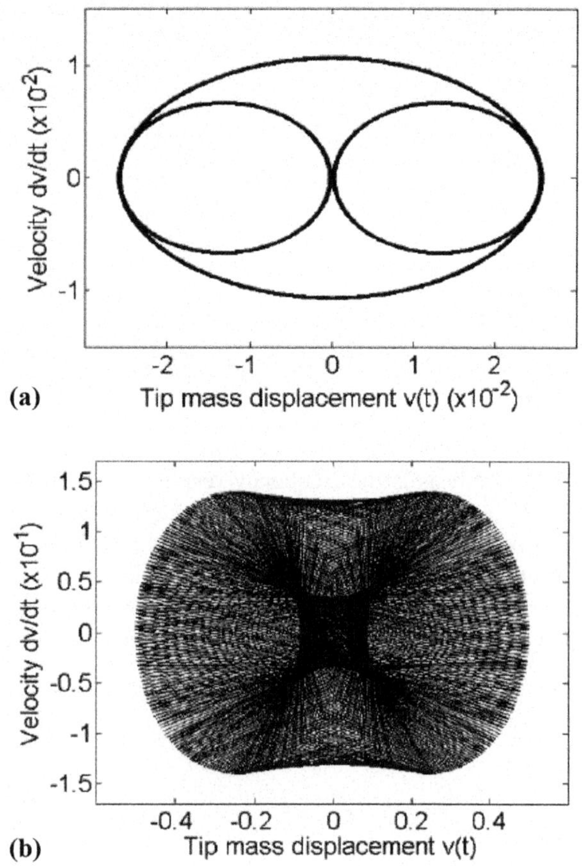

Acknowledgements The author Airton Nabarrete is grateful to the CNPq for funding this research, grant no. 317388/2021-5.

References

1. A. Erturk, D.J. Inman, A Distributed parameter electromechanical model for cantilevered piezoelectric energy harvesters. ASME. J. Vib. Acoust. **130**(4), 041002 (2008). https://doi.org/10.1115/1.2890402
2. M. Ferrari, V. Ferrari, M. Guizzetti, B. Andò, S. Baglio, C. Trigona, Improved energy harvesting from wideband vibrations by nonlinear piezoelectric converters. Sens. Actuat. A **162**(2), 425–431 (2010). https://doi.org/10.1016/j.sna.2010.05.022
3. M.P. Cartmell, The equations of motion for a parametrically excited cantilever beam. J. Sound Vib. **143**(3), 395–406 (1990). https://doi.org/10.1016/0022-460X(90)90731-E

4. E. Esmailzadeh, G. Nakhaie-Jazar, Periodic behaviour of a cantilever beam with end mass subjected to harmonic base excitation. Int. J. Non-Linear Mech. **33**(4), 567–577 (1998). https://doi.org/10.1016/S0020-7462(97)00038-3
5. A. Erturk, D.J. Inman, An experimentally validated bimorph cantilever model for piezoelectric energy harvesting from base excitations. Smart Mater. Struct. **18**(2), 1–18 (2009). https://doi.org/10.1088/0964-1726/18/2/025009
6. M.I. Friswell, S.F. Ali, O. Bilgen, S. Adhikari, A.W. Lees, G. Litak, Non-linear piezoelectric vibration energy harvesting from a vertical cantilever beam with a tip mass. J. Intell. Mater. Syst. Struct. **23**(13), 1505–1521 (2012). https://doi.org/10.1177/1045389X12455722
7. S.-J. Liao, A.T. Chwang, Application of homotopy analysis method in nonlinear oscillations. ASME J. Appl. Mech. **65**(4), 914–922 (1998). https://doi.org/10.1115/1.2791935
8. S.-J. Liao, On the homotopy analysis method for nonlinear problems. Appl. Math. Comput. **147**(2), 499–513 (2004). https://doi.org/10.1016/S0096-3003(02)00790-7
9. S. Shahlaei-Far, A. Nabarrete, J.M. Balthazar, Nonlinear vibrations of cantilever Timoshenko beams: a homotopy analysis. Lat. Am. J. Solids Struct. **13**(10), 1866–1877 (2016). https://doi.org/10.1590/1679-78252766
10. S. Shahlaei-Far, A. Nabarrete, J.M. Balthazar, Homotopy analysis of a forced nonlinear beam model with quadratic and cubic nonlinearities. J. Theor. Appl. Mech. **54**(4), 1219–1230 (2016). https://doi.org/10.15632/jtam-pl.54.4.1219

Dynamic Modelling and Fault Diagnosis of a High Contact Ratio Gear

Omar D. Mohammed

Abstract High contact ratio (HCR) gears are needed in different cases for having lower generated stresses and smoother operation as compared to the standard low contact ratio (LCR) gears. For the purpose of gear fault diagnosis and to avoid possible failure consequences due to tooth cracks, it is important to study the possibility of fault detection with crack propagation. A reduction in the gear mesh stiffness due to the existence of a crack can be used to detect and assess tooth damage. Time-varying gear mesh stiffness has an impact on the system's dynamic response, and then the gear diagnosis indicators applied to the response signal will be affected. The current paper studies the influence of increasing the tooth addendum for having a HCR on the gear diagnosis indicators, namely the RMS, kurtosis, peak value and crest factor. Twenty different crack cases are modelled by using both the LCR and HCR designs. For every crack case, the mesh stiffness is evaluated and input to a gear dynamic model that is built for obtaining the system dynamic response. The results of the applied gear diagnosis indicators show that in the HCR design case, the indicators change with the crack size will have a lower increase than in the LCR one. Therefore, the LCR design case can be detected earlier than the HCR case.

1 Introduction

Due to the increasing demands for higher loads and smoother operation requirements, design engineers are under pressure to deliver suitable design solutions. One solution which is very likely to be employed is to move from the standard full-depth tooth form to a long addendum form for increasing the gear contact ratio. High contact ratio (HCR) gears are highly applied and more research focus should get to facilitate tooth design and modelling. Highly modified carburized and ground gears were employed by [14] for better performance by using high contact ratio forms for industrial use

O. D. Mohammed (✉)
Mechanical Engineering Department, Prince Mohammad Bin Fahd University PMU, Khobar, Saudi Arabia
e-mail: osily@pmu.edu.sa

M. Belhaq (ed.), *Advances in Nonlinear Dynamics and Control of Mechanical and Physical Systems*, Springer Proceedings in Physics 301,
https://doi.org/10.1007/978-981-99-7958-5_16

at low cost. A nonlinear dynamic model of HCR gears was presented by [5] with multiple clearances by the lumped mass method. The system exhibits different types of nonlinear behaviours and dynamic evolution mechanisms with the changes in the studied parameters. A non-involute HCR spur gear design of a reduced sliding contact, enhanced fillet strength, better lubrication condition and no interference was presented by [6] as a comparison to the traditional HCR involute spur gear design. The proposed HCR spur gear can achieve better surface durability and fillet strength. The dynamic responses of high-contact spur gears were investigated by [3] at different operating speeds. The results showed that HCR gears are highly efficient in applications requiring high load-carrying capacity. The HCR gears have low operating vibration, noise, and dynamic transmission error as compared to the standard low contact ratio (LCR) gears. The contact strength of a gear drive was enhanced using a novel HCR spur gear design [13]. This design was developed using the novel hob cutter with variable tooth thickness to minimize contact stress. A parametric study was conducted for the gear parameters such as gear ratio, gear teeth, pressure angle, addendum factor and addendum correction factor to determine optimum contact stress. The dynamic response and tooth spalling fault detection gear-bearing system were studied by [4] using a spur gear design of a HCR. The results show that the dynamic response with ellipsoid tooth spalling defect exhibits rich dynamic behaviour using this design. Crack propagation of HCR gears with the existence of a tooth root crack was studied by [10]. A method is proposed for mesh stiffness calculation based on the predicted crack path for different backup ratios and pressure angles. Fault diagnosis of HCR gears as compared to those of LCR requires more attention.

In the current article, the time-varying mesh stiffness (TVMS), the dynamic response and the fault detection of an HCR design are studied using a gear dynamic model. The applied method is used for LCR gear design to compare the results and investigate the early fault detection possibility.

2 The Gear Model and Contact Ratio

Unlike standard LCR gears, which have a typical contact ratio of 1.2–1.6, HCR gears have a contact ratio greater than 2.0. HCR gears have two or three teeth in contact during the gear engagement cycle. HCR gears as compared to LCR ones are smoother and more durable. Because HCR gears have more teeth in contact they have a significantly lower applied load on each tooth as the load is spread across more teeth and then lower stresses are generated [11]. Increasing the tooth addendum is one design solution for having a higher contact ratio while keeping the same gear ratio. However, this option is limited by the tooth strength and the generated tip thickness. In the current research, both the LCR and HCR designs are studied. In the LCR case, the contact ratio is obtained as 1.63, and when the addendum is increased from 2.00 to 3.00 mm the contact ratio is increased to 2.30. The gear design parameters are shown in Table 1.

Table 1 Gear design parameters

	Pinion	Gear
No. of teeth, Z	25	30
Module, m (mm)	2	
Tooth width, w (mm)	20	
Pressure angle, α (deg.)	20°	
Young's modulus, E (N/ mm^2)	2×10^5	
Poisson's ratio	0.3	

3 Crack Model and Time-Varying Mesh Stiffness TVMS

Twenty different crack cases are modelled by using both the LCR and HCR designs. The crack model is shown in Fig. 1 and the studied crack cases are; 0 mm (the healthy case), 0.05 mm and then 0.1–1.8 mm with an increment of 0.1 mm, Table 2. A Matlab™ code was developed to calculate the TVMS analytically according to the method adopted in [1, 2, 7, 8] for the LCR design. The method is adjusted for calculating the TVMS for the HCR design, and the obtained TVMS for both the LCR and HCR cases are shown in Figs. 2 and 3. It is clear in Fig. 2 how the stiffness is varying between two certain levels representing the one and two teeth engagement intervals. In Fig. 3 the stiffness is varying between two other levels representing the two and three teeth engagement intervals.

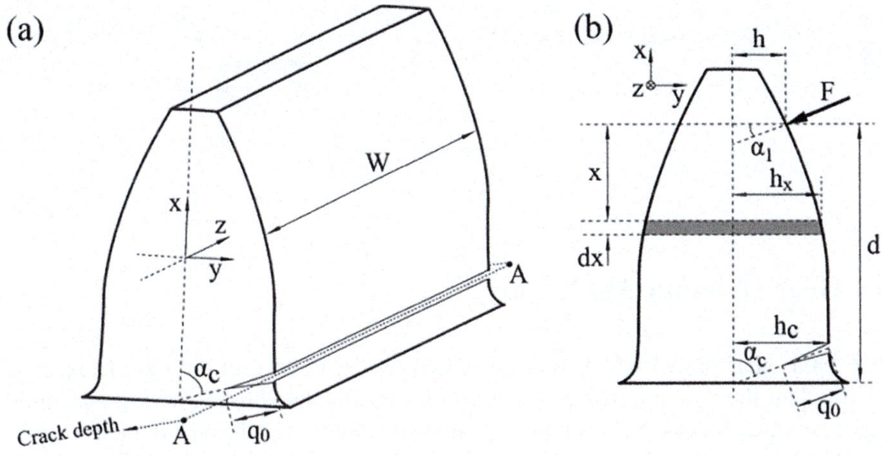

Fig. 1 Modelling of gear tooth crack. **a** modelling of a cracked tooth, **b** tooth notation [7]

Table 2 Crack propagation cases

Case	Crack depth (mm)	Case	Crack depth (mm)	
1	0	11	0.9	Crack path angle $\alpha_c = 70°$
2	0.05	12	1.0	
3	0.1	13	1.1	
4	0.2	14	1.2	
5	0.3	15	1.3	
6	0.4	16	1.4	
7	0.5	17	1.5	
8	0.6	18	1.6	
9	0.7	19	1.7	
10	0.8	20	1.8	

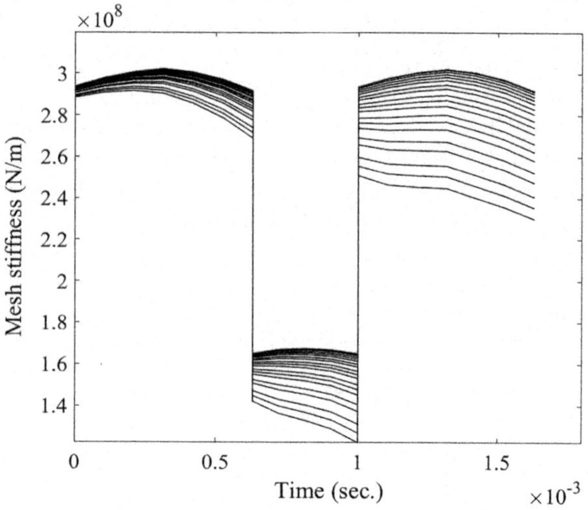

Fig. 2 Time-varying mesh stiffness for the LCR design

4 Gear Dynamic Modelling

A 6 DOF gear model which was presented by [9] is used in the current research to simulate the dynamic response based on the calculated time-varying gear mesh stiffness value for every crack case. A schematic diagram of the 6 DOF model, which has 3 DOF (one rotational and two translational) for each gear disc, is shown in Fig. 4. A MATLAB™ code using the ODE45 function was built for modelling the system equations of motion. The simulation is performed for the healthy case, after which is repeated for the crack cases. The equations of motion for this model can be explained as follows, where *p* refers to the pinion and *g* refers to the gear. The model parameters are stated in Table 3.

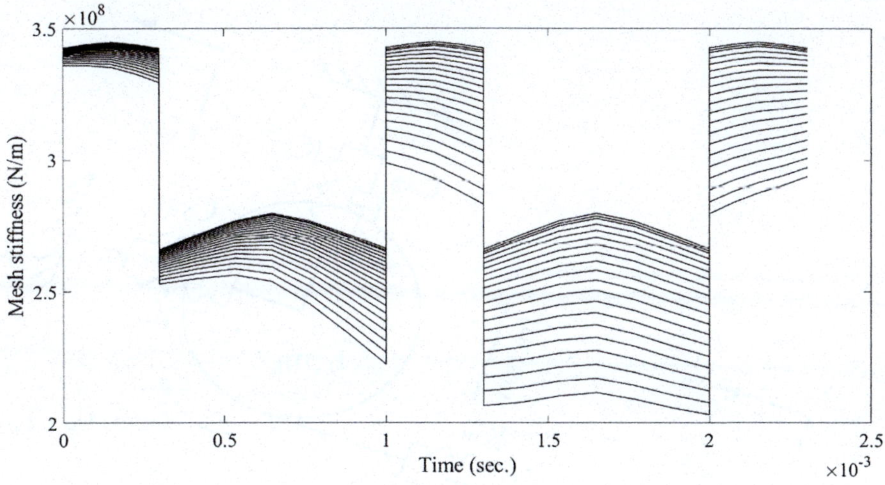

Fig. 3 Time-varying mesh stiffness for the HCR design

The equations of motion in the 'x' direction for the pinion and gear are

$$m_p \ddot{x}_p = -K_{xp} x_p - C_{xp} \dot{x}_p + F_p \tag{1}$$

$$m_g \ddot{x}_g = -K_{xg} x_g - C_{xg} \dot{x}_g + F_g \tag{2}$$

The equations of motion in the 'y' direction for the pinion and gear are

$$m_p \ddot{y}_p = -N - K_{yp} y_p - C_{yp} \dot{y}_p \tag{3}$$

$$m_g \ddot{y}_g = N - K_{yg} y_g - C_{yg} \dot{y}_g \tag{4}$$

The equations of motion in the 'θ' direction of the pinion and gear are

$$I_p \ddot{\theta}_p = r_p N + T_p + M_p \tag{5}$$

$$I_g \ddot{\theta}_g = -r_g N - T_g + M_g \tag{6}$$

In these equations,

$$N = K_m\left[\left(y_p - y_g\right) - \left(r_p\theta_p - r_g\theta_g\right)\right] + C_m\left[\left(\dot{y}_p - \dot{y}_g\right) - \left(r_p\dot{\theta}_p - r_g\dot{\theta}_g\right)\right]$$

$$N = N_1 + N_2 \ldots N_n \text{ (} n \text{ is the number of teeth in contact)}$$

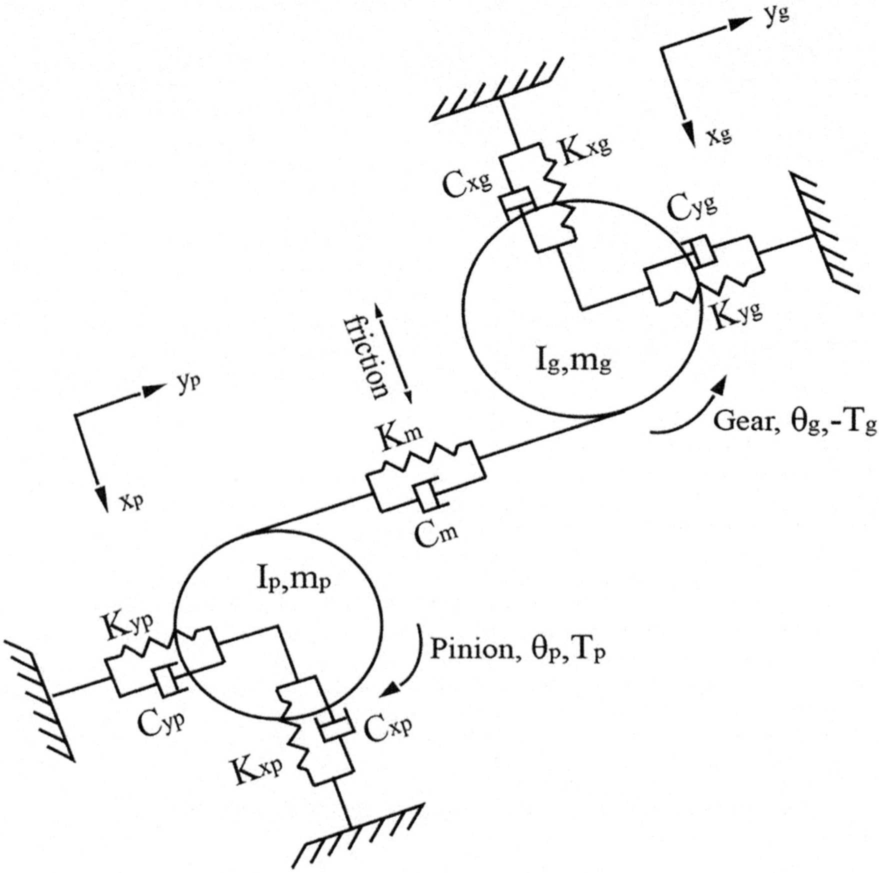

Fig. 4 Dynamic model of a one-stage gear system with 6 DOF [9]

Table 3 Dynamic model parameters

Parameter	Pinion	Gear
Mass, m_p and m_g (kg)	0.3083	0.4439
Mass moment of inertia, I_p and I_g (kg m^2)	0.96×10^{-4}	2×10^{-4}
Radial stiffness K of the bearing in the x, y direction	6.56×10^8	
Radial damping C of the bearing in the x, y direction	1.8×10^3	
Damping of tooth meshing, C_m (N s/m)	67	
Coefficient of friction, μ	0.06	

$$F_p = F_{p1} + F_{p2} \ldots F_n,\ F_{p1} = \mu N_1 \text{ and} F_{p2} = -\mu N_2$$

$$F_g = F_{g1} + F_{g2} \ldots F_n,\ F_{g1} = -\mu N_1 \text{ and } F_{g2} = \mu N_2$$

Here, the notation p is used for thc pinion, which is the driving smaller gear, and the notation g is used for the gear, which is a driven larger gear in this study.

5 Results and Discussion

The vibration signal is generated for every crack case and for both the LCR and HCR designs. The generated signal is named "the original signal" for every crack case. To enhance the extraction of the hidden crack impact the residual signal of a specific faulty case can be generated, which involves subtracting the whole original signal of the healthy case from the original signal of that faulty case, see Fig. 5.

A reduction in the mesh stiffness due to the existence of a crack is used to detect tooth damage. The stiffness reduction will be presented in the TVMS and then an impact on the system's dynamic response will be detected using the diagnosis indicators applied to the generated signals. Four different diagnosis indicators, namely the RMS, kurtosis, peak value and crest factor are calculated from the collected signals. Those statistical indicators are stated in (7))–(10) [12, 15]. Figure 6a–d shows the performance of the applied gear diagnosis indicators with increasing crack size. The results show that in the HCR design case, the performance of the indicators with

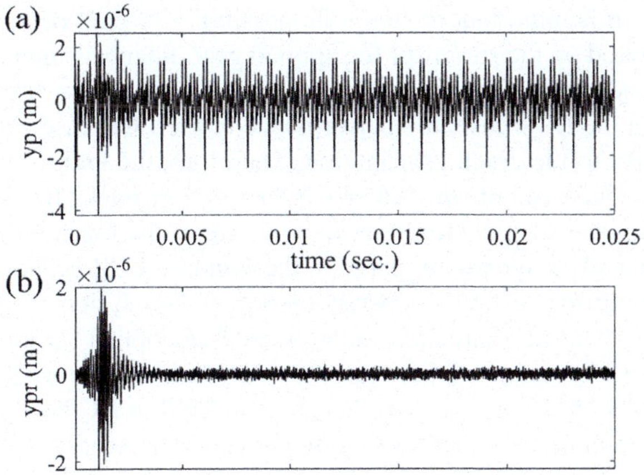

Fig. 5 The generated vibration signal of the HCR gear of 1.8 mm crack. **a** the original signal, **b** the residual signal

the crack size has a lower increase than in the LCR case. In the HCR case the load will be shared on more teeth and the stiffness reduction shows a lower impact. Thus, in the four studied indicators, the LCR design case shows higher values than the corresponding ones of the HCR case.

$$\text{RMS} = \sqrt{\frac{1}{N}\sum_{n=1}^{N}(x(n) - \bar{x})^2}, \text{ where } \bar{x} = \frac{1}{N}\sum_{n=1}^{N}x(n). \tag{7}$$

$$\text{Kurtosis} = \frac{\frac{1}{N}\sum_{n=1}^{N}(x(n) - \bar{x})^4}{[\frac{1}{N}\sum_{n=1}^{N}(x(n) - \bar{x})^2]^2} \tag{8}$$

$$\text{Peak value} = \max|x(n)| \tag{9}$$

$$\text{Crest Factor} = \frac{\max|x(n)|}{\text{RMS}} \tag{10}$$

6 Conclusions

Using high contact ratio HCR gears is important in different applications. The load will be shared on more teeth and this is essential for having lower generated stresses and smoother operation as compared to the standard low contact ratio LCR gears. For the purpose of gear tooth diagnosis and to avoid possible failure consequences due to tooth cracks, it is important to study the possibility of fault detection with crack propagation. Based on the results of the applied gear diagnosis indicators and as the change of the indicators with the crack size in the HCR design case shows a lower increase than in the LCR case, the performance of the LCR case can be concluded as more sensitive to the crack propagation. Therefore, the fault in the LCR design case can be detected earlier than in the HCR case. Accordingly, by using HCR gears of more teeth in contact it is important to be aware of the lower possibility of fault detection in the early stage as compared to the standard LCR gears.

The early detection of faults depends on the studied system, the quality of the collected data, the background noise and the method applied for fault detection. In this study, the performance of the main fault detection indicators was investigated to understand how changing the gear design from LCR to HCR will require further work on how to improve the fault detection in the early stage.

Fig. 6 The fault detection indicators obtained using the residual signals. **a** RMS, **b** Kurtosis, **c** Peak value, **d** Crest factor

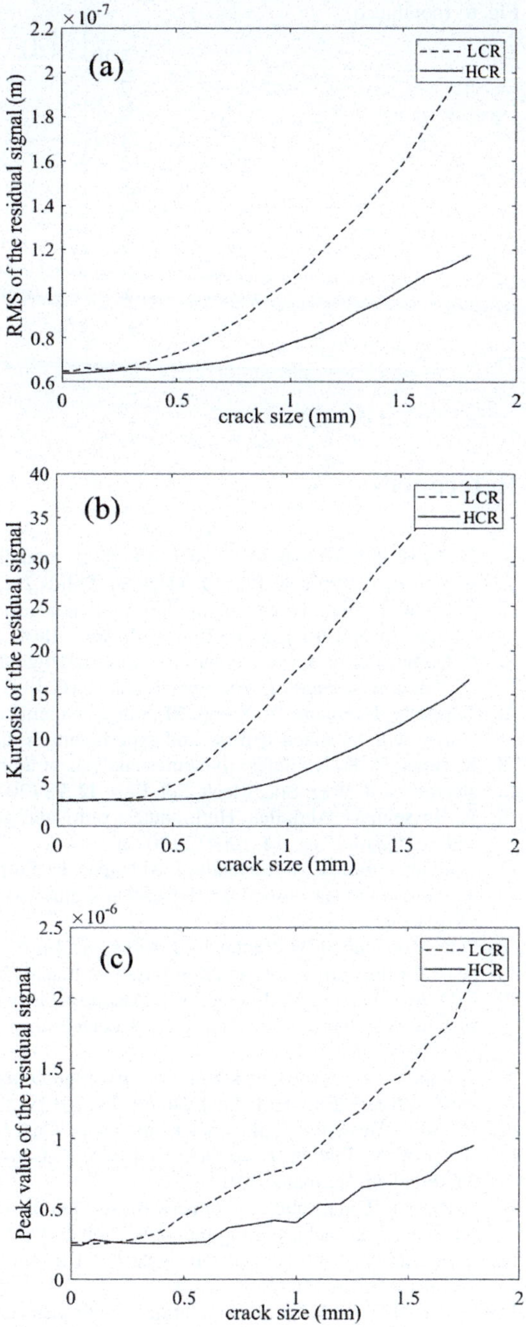

Fig. 6 (continued)

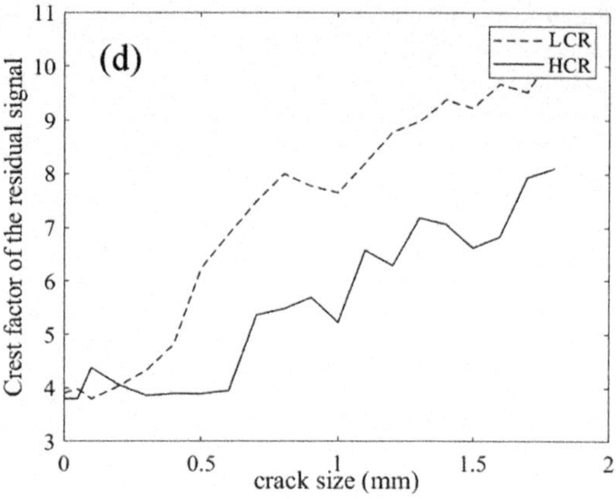

References

1. F. Chaari, T. Fakhfakh, M. Haddar, Analytical modelling of spur gear tooth crack and influence on gear mesh stiffness. Euro. J. Mech. A, Solids **28**(3), 461–468 (2009)
2. Z. Chen, Y. Shao, Dynamic simulation of spur gear with tooth root crack propagating along tooth width and crack depth. Eng. Fail. Anal. **18**(8), 2149–2164 (2011)
3. O. Dogan, Performance evaluation of low- and high-contact ratio spur gears: dynamic response and bending strength. J. Vib. Eng. Technol. **10**, 1337–1347 (2022)
4. Z. Cheng, K. Huang, Y. Xiong, M. Sang, Dynamic analysis of a high-contact-ratio spur gear system with localized spalling and experimental validation. Mechanics **10**(154), 1–18 (2022)
5. K. Huang et al., Nonlinear dynamics analysis of high contact ratio gears system with multiple clearances. J. Braz. Soc. Mech. Sci. Eng. **42**, 98 (2020)
6. A. Hussein, M. Abdullah, High-contact ratio spur gears with conformal contact and reduced sliding. Results Eng. **14**, 100412 (2022)
7. O.D. Mohammed, M. Rantatalo, J. Aidanpaa, U. Kumar, Vibration signal analysis for gear fault diagnosis with various crack progression scenarios. Mech. Syst. Signal Process. **41**, 176–195 (2013)
8. O.D. Mohammed, M. Rantatalo, J. Aidanpaa, Improving mesh stiffness calculation of cracked gears for the purpose of vibration-based fault analysis. Eng. Fail. Anal. **34**, 235–251 (2013)
9. O.D. Mohammed, M. Rantatalo, J. Aidanpaa, Dynamic modelling of a one-stage spur gear system and vibration-based tooth crack detection analysis. Mech. Syst. Signal Process. **54–55**, 293–305 (2015)
10. Y. Pandya, A. Parey, Crack behavior in a high contact ratio spur gear tooth and its effect on mesh stiffness. Eng. Fail. Anal. **34**, 69–78 (2013)
11. R. Miller, Designing high contact-ratio gears. Gear Solutions (2017)
12. A. Parey, N. Tandon, *Fault Detection of Spur Gears Using Vibration Monitoring* (Lambert, Saarbrucken, Germany, 2010)
13. S. Rajesh, P. Marimuthu, P. Dinesh Babu, Optimization of contact stress for the high contact ratio spur gears achieved through novel hob cutter. Forschung im Ingenieurwesen (1) (2022)
14. C.D. Schultz, High contact ratio gearing: a technology ready for implementation?. AGMA Technical paper, 14FTM06 (2014)
15. S. Wu, M. Zuo, A. Parey, Simulation of spur gear dynamics and estimation of fault growth. J. Sound Vib. **317**(3–5), 608–624 (2008)

Time Integration of Constrained Multi-catenary Systems

Damir Sedlar and Zeljan Lozina

Abstract The multi-catenary systems offer interesting possibilities in practical application especially in combination with slider. However, the motion of the catenaries involves geometric nonlinearity, and by adding sliders, the problem becomes very complex because a set of differential-algebraic equations is obtained. Choosing a time integration scheme and the right parameters is one of the key steps in solving this differential-algebraic system. In this paper, several time integration methods from the Newmark scheme family are analyzed. As an example, catenaries with a slider were chosen, which were modeled using an isogeometric approach and four time integration schemes were analyzed.

1 Introduction

The multi-catenary systems offer interesting possibilities for miscellaneous tension structures and shaping different three-dimensional forms such as cable-stayed bridges, roofs, power lines, zip-lines, pantograph-catenary interaction. This wide variety of application was achieved thanks to catenary flexibility. Because of catenary flexibility these structures are exposed to the large displacements which lead to the geometric nonlinearities which is the main difficulty encountered in dynamic analysis of multi-catenary systems. Typically, the problem of catenary dynamics is solved analytically in simpler cases [10], and in more complex cases by applying the finite element method [2]. The sliding joint is often the subject of numerous scientific papers because it frequently appears in applications. Modeling and simulating sliding joint dynamics using finite elements is very demanding because the problem of searching for point of contact between the slider and actual finite element arises [11].

D. Sedlar (✉) · Z. Lozina
University of Split, Faculty of Electrical Engineering, Mechanical Engineering and Naval Architecture, Rudjera Boskovica 32, 21000 Split, Croatia
e-mail: dsedlar@fesb.hr

Z. Lozina
e-mail: lozina@fesb.hr

M. Belhaq (ed.), *Advances in Nonlinear Dynamics and Control of Mechanical and Physical Systems*, Springer Proceedings in Physics 301,
https://doi.org/10.1007/978-981-99-7958-5_17

Isogeometric analysis (IGA) is a recently introduced method of computational analysis with the goal of merging design and analysis into one model by employing Non-Uniform Rational B-Splines (NURBS) [1]. Since its good results, the research in the field of IGA is in increasing path and has been applied to the various problems. Due to its complexity the constrained multi-catenary systems were only recently investigated using IGA, [5, 8, 9]. In this paper, the multi-catenary system is modeled and analyzed by employing IGA approach, too. The obtained equations of motion for a constrained multi-catenary system present the form of a mixed set of second-order differential and algebraic equations (DAE). The augmented Lagrangian method is used to solve holonomic constrained problem which consists of slider joint. In this work, different aspects of the implementation of Newmark family schemes for integration of DAE system are analyzed. The important role in solving such complex system has time step and numerical damping which need to be carefully selected depending on chosen Newmark scheme.

The rest of the paper is organized as follows: in the second section, the geometric modeling of the catenary and the governing equations using IGA are presented; the third section explains the time integration methods used in this paper; in the fourth section, an example is given on which different time integration schemes are analyzed; and the conclusion is given in the last section.

2 Isogeometric Analysis of the Catenary

In this paper, the NURBS and B-Spline basis functions are used to describe the geometry of catenary. Representation of catenary geometry descriptions using NURBS is shown in several recently published papers [5, 8], and the same representation is used here. The catenary is geometrically represented as a B-Spline curve which is obtained using B-Spline basic functions and control points. In the general case, the control points do not interpolate the B-Spline curve. There is a coordinate in parameter space along the B-Spline curve which is called the knot vector. The size of the knot vectors is equal to $n + p + 1$, where n is the number of basis functions and control points, and p is polynomial order. In the knot vector, there can be several coordinates with the same value, which is called repeated knots. What is named an element in the finite element approach is defined here as non-zero knot spans in the knot vector.

2.1 The Governing Equations and Discretization of the Catenary

The geometry of catenary is defined through the parameterization of the form:

$$\mathbf{x} = \mathbf{RP} \tag{1}$$

where **R** is matrix of basis functions and **P** is the coordinate vector of control points. Adopting an isogeometric approach, the same basis functions used to define geometry are also used to define displacement

$$\mathbf{u} = \mathbf{RQ} \tag{2}$$

where **R** is matrix of basis functions and **Q** is the displacement vector of control points. Augmenting Hamilton's principle and performing variation and integration over a catenary domain leads to the governing equations [9], in matrix form:

$$\mathbf{M}\ddot{\mathbf{q}} + \mathbf{F} + \mathbf{B}(\boldsymbol{\Lambda} + p\boldsymbol{\Phi}) - \mathbf{F}_e = 0 \tag{3}$$

$$\boldsymbol{\Phi} = 0 \tag{4}$$

where **M** is the mass matrix, **F** is the internal forces, **B** is the matrix of constraint gradients, $\boldsymbol{\Lambda}$ is Lagrange multipliers, p is penalty coefficient, $\boldsymbol{\Phi}$ is matrix of constraints, and $\mathbf{F}_e$ is externally applied forces. The integration of (3), which involves linearization and an incremental procedure, leads to the system equations:

$$\mathbf{M}\Delta\ddot{\mathbf{q}} + (\mathbf{K}_T + \mathbf{K}_C)\Delta\mathbf{q} + \mathbf{B}^{\mathrm{T}}\Delta\boldsymbol{\Lambda} = \mathbf{r} \tag{5}$$

$$\mathbf{B}\Delta\mathbf{q} = -\boldsymbol{\Phi} \tag{6}$$

where the increment of the variables is marked with $\boldsymbol{\Delta}$, $\mathbf{K}_T$ is tangent stiffness matrix, **r** is residual vector, and $\mathbf{K}_C$ is constraint stiffness matrix which is defined as follows:

$$\mathbf{K}_C = \frac{\partial}{\partial \mathbf{q}}(\mathbf{B}\boldsymbol{\Lambda} + p\boldsymbol{\Phi}) \tag{7}$$

3 Time Integration Schemes

Time integration methods, to solve time-dependent differential equations, are usually divided into explicit and implicit methods. Explicit methods [7] solve the equation at time $t + \Delta t$ (current state) from the previous state t. On the other side, implicit methods [6] solve the equation using the current state at time $t + \Delta t$ and the previous state t. Explicit methods require attention when choosing the time step and it becomes impractically small for certain problems. With implicit methods, it is possible to go with a larger step, but it is necessary to solve the system of equations in each step.

In this paper, the time integration of governing equation is performed with implicit step-by-step Newmark method [6] with kinematical hypotheses.

$$\mathbf{q}_{t+\Delta t} = \mathbf{q}_t + \Delta t\dot{\mathbf{q}}_t + (0.5 - \beta)\Delta t^2\ddot{\mathbf{q}}_t + \beta\Delta t^2\ddot{\mathbf{q}}_{t+\Delta t} \tag{8}$$

$$\dot{\mathbf{q}}_{t+\Delta t} = \dot{\mathbf{q}}_t + (1 - \gamma)\Delta t\ddot{\mathbf{q}}_t + \gamma\Delta t\ddot{\mathbf{q}}_{t+\Delta t} \tag{9}$$

where β and γ are free parameters expressed as

$$\beta = \frac{1}{4}(1 + \alpha_f - \alpha_m)^2 \tag{10}$$

$$\gamma = \frac{1}{2}(1 + 2\alpha_f - 2\alpha_m)^2 \tag{11}$$

Depending on the values of parameters α_f and α_m, the following methods can be esteemed [3]:

- $\alpha_f = 0$ and $\alpha_m = 0$—Newmark method without numerical dissipation,
- $\alpha_f - \alpha_m > 0$—Newmark method with numerical dissipation,
- $\alpha_f = (0, 0.3]$ and $\alpha_m = 0$—HHT method [4],
- $\alpha_m < \alpha_f \leq 0.5$—generalized α method.

Table 2 shows the selected values for the parameters α_f and α_m that were used in this work.

The system of Equations (5) and (6) together with integration equations (10) and (11) was solved iteratively using the standard Newton-Raphson method (Table 1).

4 Sliding Catenary Example

In this section, nonlinear problem of two catenary in two-dimensional space is analyzed, the example is the same as in [9]. The two catenaries are shown in the initial position in Fig. 1, catenary 1 from A to C and catenary 2 from B to D. The catenaries are in static equilibrium position and their initial shape was obtained by static analysis and applying self-weight. The catenary 1 is supported on two hinged supports. Point D of catenary 2 can move freely and point B slides along catenary 1.

In the initial position, there is a small gap between points A and B on catenaries 1 and 2, according to Fig. 2.

Table 1 Time integration schemes—parameters

Schemes	α_f	α_m
Newmark without numerical dissipation	0.00	0.00
Newmark with numerical dissipation	0.80	0.20
HHT	0.20	0.00
Generalized α	0.45	0.20

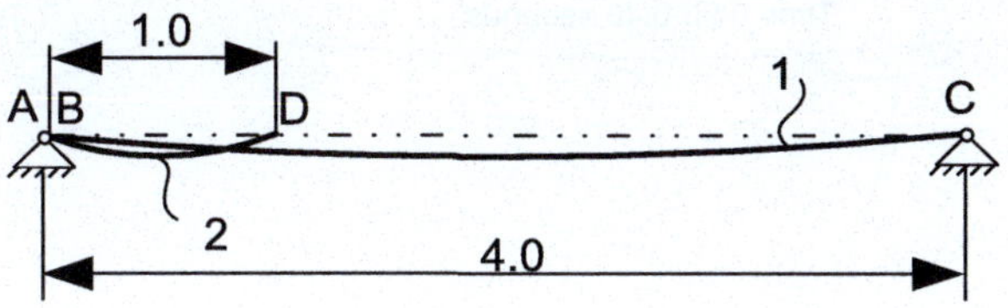

Fig. 1 Two catenary initial equilibriums

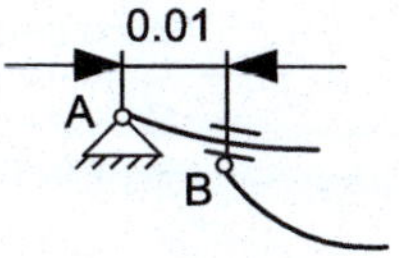

Fig. 2 Initial position of points A and B

The properties of two catenaries which are analyzed are given in Table 2. Catenaries have the same properties except for length.

Both catenaries are modeled with six control points, the used polynomial degree is 3 and knot vectors are evenly distributed where the first and last knots are repeated four times. In this way, it was achieved that the first and last control points are located on the curve itself, the catenary. The catenaries are released from their initial position, Fig. 1, to move under the influence of their own weight. The simulation was carried out for the first two and a half seconds, and Fig. 3 shows the position of the catenaries for particular time moments. From the figure, it can be seen how the catenary 2 continuously oscillates while its starting point B slides along the catenary 1.

The time step during numerical time integration was 0.001 s. Four different time integration schemes are made according to Table 1.

The movement of the end point D, of the second catenary, was specially observed and analyzed because it has the largest displacements and accelerations. Point D is the most sensitive to time integration, and all difficulties are best demonstrated by observing its movement. The horizontal and vertical displacements of point D are shown in Figs. 4 and 5, respectively. There is no displacement in these figures applying the Newmark without numerical dissipation time integration scheme because after 0.04 s the scheme did not converge anymore. The remaining three schemes converged properly in the observed time period. In these figures, it can be seen that HHT and Generalized-α scheme have the same behavior from the beginning to the

Table 2 The properties of catenaries

Property	Catenary 1	Catenary 2
Length (m)	4	1
Cross-sectional area (m^2)	$0.001^2\pi$	$0.001^2\pi$
Density (kg/m^3)	3.0E+03	3.0E+03
Modulus of elasticity (Pa)	1.0E+11	1.0E+11

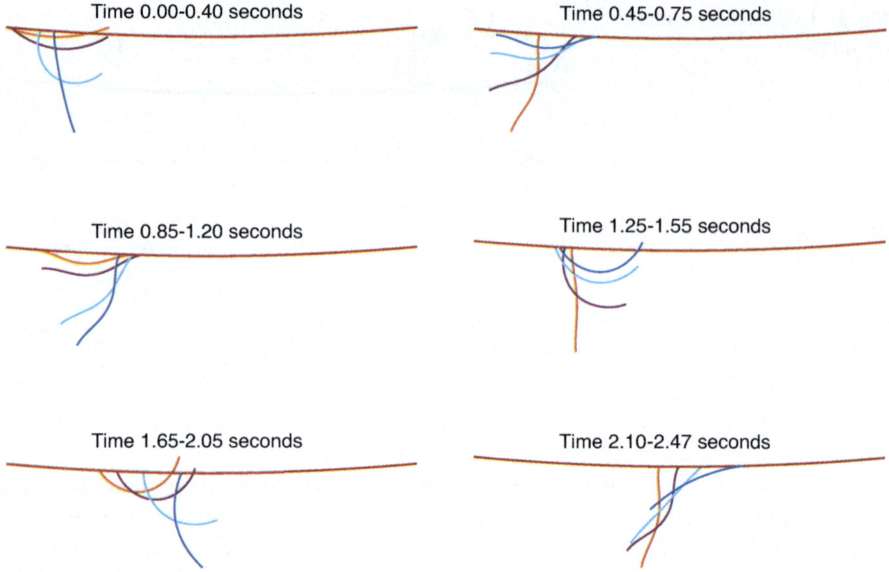

Fig. 3 Sliding and swinging of the pendulum

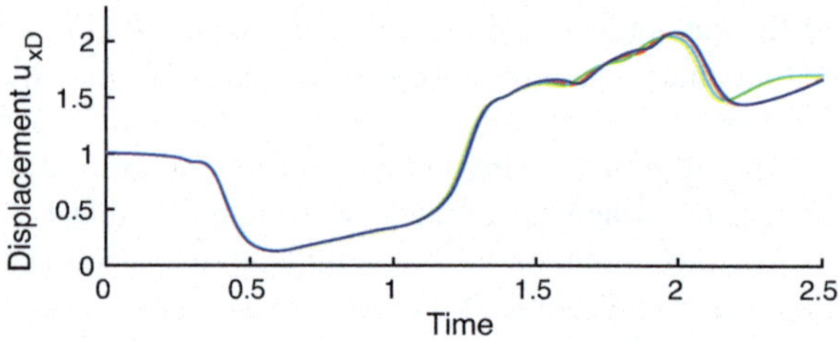

Fig. 4 Horizontal displacement of point D, green—Newmark with numerical dissipation, blue—HHT, red—Generalized α

end of the observed time period without significant differences between them. After 1.5 s, Newmark with numerical dissipation scheme shows a slightly different behavior than the other two schemes. From Fig. 5, which shows the vertical displacements of point D, it can be noted that at the very end of the observation interval, there was a significant decrease in the displacement amplitude. It is also visible that the slight decrease starts already at the time of second peak, a little before the first second, but it is most pronounced at the end of the time interval.

The horizontal and vertical velocities of point D are shown in Figs. 6 and 7. The velocity curves follow the same trend as the displacement curves, that is, the biggest difference is between Newmark with numerical dissipation and the remaining two schemes.

The horizontal and vertical accelerations of point D are shown in Figs. 8 and 9. Also, in these figures, the acceleration curve shows the same trend as in the displacement and velocity cases. However, what is clearly noticeable and different compared to

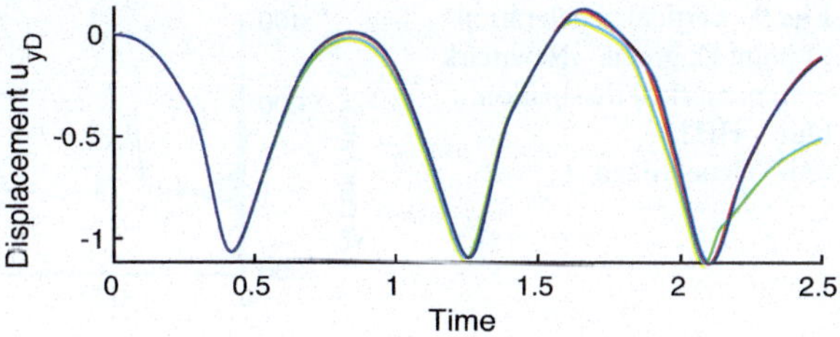

Fig. 5 Vertical displacement of point D, green—Newmark with numerical dissipation, blue—HHT, red—Generalized α

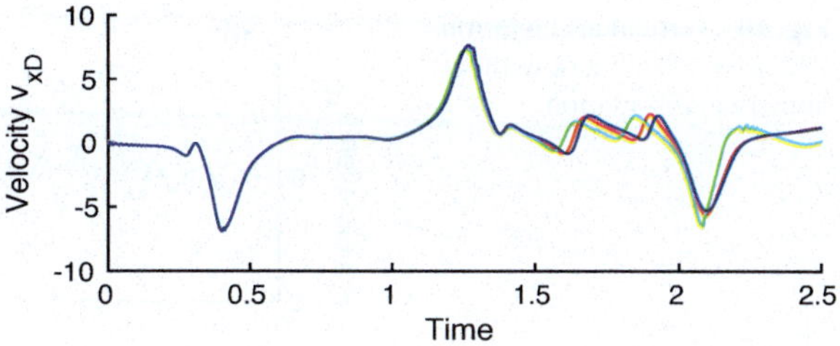

Fig. 6 Horizontal velocity of point D, green—Newmark with numerical dissipation, blue—HHT, red—Generalized α

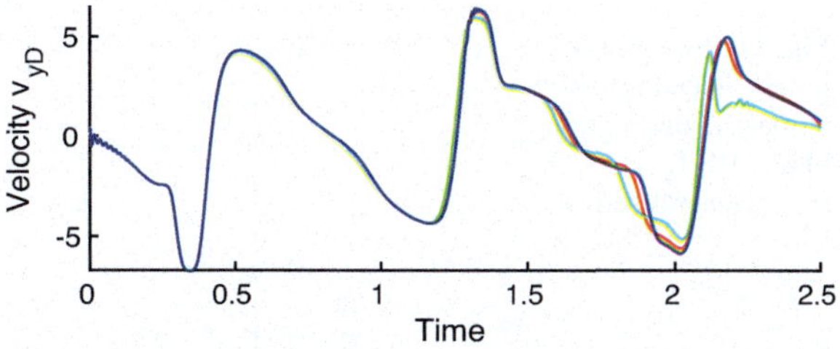

Fig. 7 Vertical velocity of point D, green—Newmark with numerical dissipation, blue—HHT, red—Generalized α

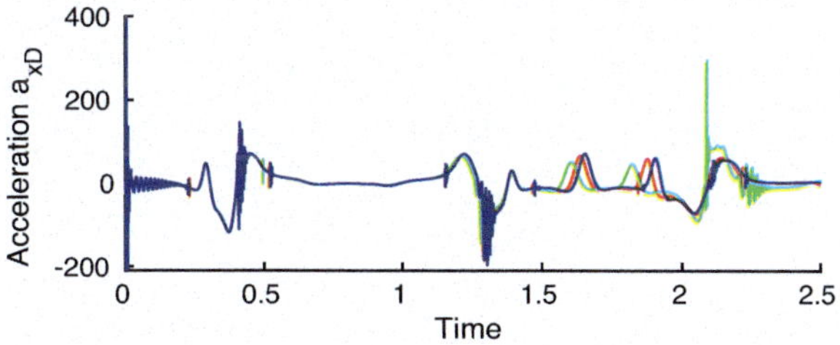

Fig. 8 Horizontal acceleration of point D, green—Newmark with numerical dissipation, blue—HHT, red—Generalized α

previous curves is the occasional occurrence of oscillations for all three time integration schemes. And in this case, Newmark with numerical dissipation scheme is the most different compared to HHT and Generalized α schemes. On the acceleration figures, certain oscillations can be noticed at the very beginning of the time interval, for a period below 0.25 s, which can be explained by the probably insufficiently well and accurately achieved initial, equilibrium position.

As already mentioned, the Newmark without numerical dissipation scheme did not show convergence. Figure 10 shows the acceleration of point D before the pro-

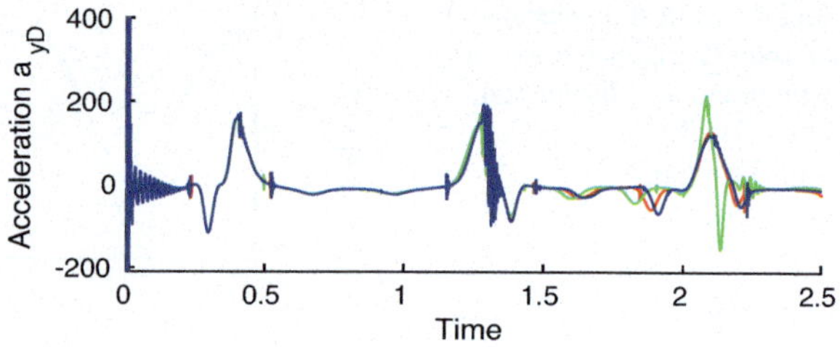

Fig. 9 Vertical acceleration of point D, green—Newmark with numerical dissipation, blue—HHT, red—Generalized α

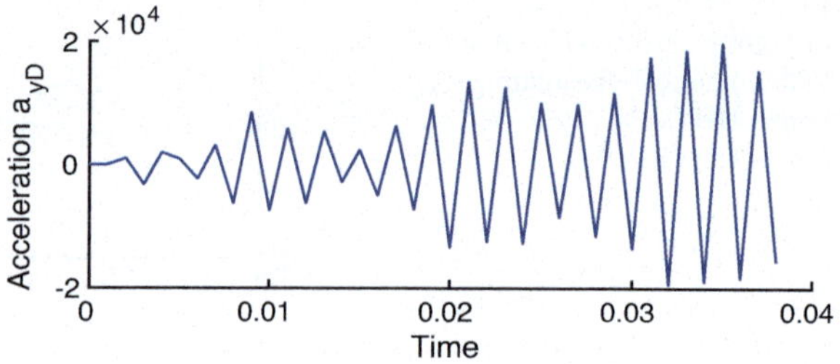

Fig. 10 Vertical acceleration of point D, Newmark without numerical dissipation

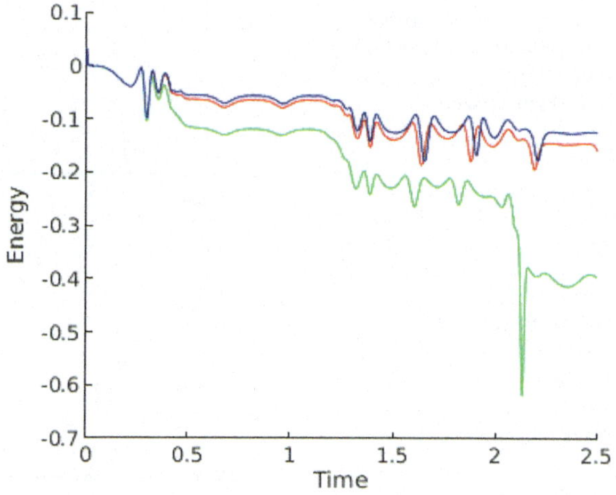

Fig. 11 Energy loss, green—Newmark with numerical dissipation, blue—HHT, red—Generalized α

gram broke, from which it is visible that the acceleration suddenly and continuously increases to large values, and is interrupted around 0.04 s.

All these schemes introduce some artificial energy loss to achieve convergence. Figure 11 shows the energy loss for the three applied time integration schemes and the investigated time period. What is observed in the displacement, velocity, and acceleration figures, Figs. 4, 5, 6, 7, 8, and 9, clearly shown why there was a deviation between Newmark with numerical dissipation scheme and the other two applied schemes.

In the Newmark with numerical dissipation scheme, there is a much faster loss of energy than in other schemes, which results in a smaller amplitude of the displacement shown in Fig. 5. A particularly significant loss of energy occurred at the end of the observed time interval.

5 Conclusion

The aim of this paper was to investigate time integration methods from the Newmark family schemes on the problem of multi-catenary system with a slider using an isogeometric approach. The introduction of numerical dissipation in time integration has already been shown to be useful in solving numerical stability, and in this study, it has been confirmed that these problems cannot be solved without a certain degree of numerical dissipation. Special attention should be paid to the degree of introduced dissipation, because as expected it has a negative effect if a long period of time is observed. Additionally, the IGA approach in combination with Newmark integration schemes proved to be efficient in solving this type of problem, and even a small degree of dissipation led to numerical stability.

Acknowledgements This paper is supported by the University of Split, Faculty of Electrical Engineering, Mechanical Engineering and Naval Architecture, Dynamic and Vibration group.

References

1. Y. Bazilevs, J. Cottrell, T.J. Hughes, *Isogeometric Analysis: Toward Integration of CAD and FEA* (Wiley, New York, 2009)
2. Z. Chen, Y. Wu, Y. Yin, C. Shan, Formulation and application of multi-node sliding cable element for the analysis of suspen-dome structures. Finite Elem. Anal. Des. **46**(9), 743–750 (2010). https://doi.org/10.1016/j.finel.2010.04.003
3. M. Geradin, D.R. Rixen, *Mechanical Vibrations—Theory and Application to Structural Dynamics* (Wiley, New York, 1997)
4. H.M. Hilber, T.J.R. Hughes, R.L. Taylor, Improved numerical dissipation for time integration algorithms in structural dynamics. Earthq. Eng. Struct. Dyn. **5**(3), 283–292 (1997)
5. Z. Lozina, D. Sedlar, I. Tomac, Isogeometric approach to the dynamics of the catenary exposed to large displacements. Acta Mechanica **230**(1), 351–366 (2019). https://doi.org/10.1007/s00707-018-2321-1
6. N. Newmark, A method of computation for structural dynamics. ASCE J. Eng. Mech. Div. **85**(3), 67–94 (1959)
7. G. Rio, A. Soive, V. Grolleau, Comparative study of numerical explicit time integration algorithms. Adv. Eng. Softw. **36**(4), 252–265 (2005)
8. D. Sedlar, Z. Lozina, A. Bartulovic, Nonlinear static isogeometric analysis of cable structures. Arch. Appl. Mech. **89**, 713–729 (2019). https://doi.org/10.1007/s00419-018-1489-0
9. D. Sedlar, Z. Lozina, A. Bartulovic, Multi-catenary isogeometric discretization with the sliding constraint. Acta Mechanica Sinica **37**, 80–91 (2021). https://doi.org/10.1007/s10409-020-01021-7

10. H. Thai, S. Kim, Nonlinear static and dynamic analysis of cable structures. Finite Elem. Anal. Des. **47**(3), 237–246 (2011). https://doi.org/10.1016/j.finel.2010.10.005
11. B. Zhou, M. Accorsi, J. Leonard, Finite element formulation for modeling sliding cable elements. Comput. Struct. **82**(2–3), 271–280 (2004). https://doi.org/10.1016/j.compstruc.2003.08.006

A Novel Pseudorandom Number Generator Based on the Conservative Chaotic System with Non-smooth Nonlinearities

Peter A. Meleshenko, Mikhail E. Semenov, and Olesya I. Kanishcheva

Abstract Conservative chaos plays a special role in the modern field of engineering sciences including digital communications and cryptography. In this paper, we propose a novel pseudorandom number generator (PRNG) which is based on a relatively simple oscillatory system consisting of non-smooth nonlinearities and exhibiting chaotic dynamics. Main properties of this PRNG are discussed and analyzed. Numerical results show that the proposed generator has a good performance and successfully passes the standard NIST test for randomness (within the NIST Statistical Test Suite). A comparison of the proposed PRNG to various chaotic-based PRNGs is also presented. The obtained results for the security claim and speed characteristics make it possible to consider this PRNG as a promising source of pseudorandom numbers in the field of chaotic-based digital communications and cryptography.

1 Introduction

Chaos and chaotic systems are the most inspiring and intensively developing areas in the field of modern nonlinear dynamics. Since the concept of deterministic chaos was introduced by Lorenz in 1963 [1], an interest in this phenomenon is not decreasing but intensifies due to extensive researches in the field of novel and perspective chaotic systems with unusual properties such as Rössler system [2, 3], Chua's system [4–6],

P. A. Meleshenko · M. E. Semenov (✉) · O. I. Kanishcheva
Digital Technologies Department, Voronezh State University, Universitetskaya sq. 1, 394018 Voronezh, Russia
e-mail: mkl150@mail.ru

P. A. Meleshenko · M. E. Semenov
Department of Geometry and Functional Theory, Voronezh State University, Universitetskaya sq. 1, Voronezh 394018, Russia

M. E. Semenov
Geophysical Survey of Russia Academy of Science, Lenina av. 189, 249035 Obninsk, Russia

Applied Mathematics and Mechanics Department, Voronezh State Technical University, XX-letiya Oktyabrya st. 84, 394006 Voronezh, Russia

M. Belhaq (ed.), *Advances in Nonlinear Dynamics and Control of Mechanical and Physical Systems*, Springer Proceedings in Physics 301,
https://doi.org/10.1007/978-981-99-7958-5_18

Arneodo system [7–11], Lorenz-84 system [12, 13], and Sprott system [14–16]. Of course, this list is not completely full and can be significantly extended, however we listed the most valuable systems that inspire scientists since they were proposed.

A concept of conservative chaos plays a special role in modern development of the chaos theory [14, 16–22]. The history of conservative chaos started in 1994 when J.C. Sprott proposed a simple nonlinear chaotic system which is volume-conservative but not energy-conservative and known now as the Sprott A system [14]. This system exhibits a set of complicated dynamical regimes including periodic motion, quasi-periodic motion (or invariant tori), and chaos. Recent results in this field [16, 19] show that the dynamics of the Sprott A system are not only controlled by the system parameters but also affected by initial conditions (in this sense the Sprott A system is unique comparing with the known dissipative systems). Additionally, the Sprott A system has similar (at least formally) properties to classical Hamiltonian systems, such as no dissipation, no attractors, time-reversal symmetry, symmetry of non-zero Lyapunov characteristic exponents with respect to zero, and integer dimension. These features make systems with conservative chaos as promising sources of chaotic signals in various engineering tasks, especially in the field of modern chaotic-based secure communications and encryption [19].

At the same time, all aforementioned systems are dealing with smooth nonlinearities. There are only small amount of publications dedicated to investigation of dynamical regimes in non-smooth or piece-wise smooth systems (see, e.g., [23–28] and related references). Non-smooth dynamical systems have received increased attention motivated particularly by engineering applications. In the field of smooth dynamics (where systems are described by complex nonlinear equations containing, nevertheless, smooth nonlinearities), many results were obtained under certain restrictions to smoothness of the nonlinear term. However, in modern applications, some phenomena cannot be explained within the smooth paradigm and the non-smooth dynamics becomes significant in the modeling of such systems. A significant growth in the field of non-smooth systems has been due to the fact that these systems exhibit a big amount of complex dynamical phenomena including chaos.

An important area in the field of the modern communication systems is related to the development of the high-efficiency and fast operating PRNGs [29–37]. Random and pseudorandom numbers are required for a wide applied areas such as cryptography, information security, numerical simulations, programming, decision support, etc. [38]. Random number generation is especially important in cryptography because random numbers are used at various stages of cryptographic algorithms to initialize symmetric encryption and remote access sessions, generate cryptographic keys, passwords, one-time access codes, etc. [39]. To achieve a high level of security, effective sources of random numbers are required; however, due to the complexity of implementing the source of entropy [40] many cryptographic systems use PRNGs.

At the same time, many of the known PRNGs do not provide sufficient randomness or have some features that reduce security. Commonly used PRNGs include, for example, the Mersenne twister [41], the Blum-Blum-Shub generator [42] and the linear congruential generator [43]. One of the high-speed methods for generating pseudorandom numbers is to use the Xorshift algorithm, which is based on the shift operation, and the exclusive OR operation [44, 45]. Nonlinear modifications of this algorithm (Xorshift+ and Xorshift*) are successfully passed statistical tests, but, at the same time, are not cryptographically strong. Thereby, most algorithms for pseudorandom numbers generation are either high speed or cryptographically strong. The second volume of the famous book by Donald Knuth "The Art of Computer Programming" [46] gives an overview of generators of uniformly distributed pseudorandom numbers and criteria for assessing their quality. The author puts forward the important idea that the complexity of the generation algorithm is not related to the quality of the generated numerical sequence. One of the ways to implement computationally efficient and cryptographically strong algorithms is to use chaotic mappings [29–31, 33, 34].

In this work, we propose a novel PRNG based on conservative chaotic system with non-smooth nonlinearities [21]. The proposed PRNG passes main NIST tests as well as establishes relatively high speed of generation comparing to other known PRNGs. Thus, the proposed PRNG with non-smooth nonlinearities can be considered as a promising source of pseudorandom numbers in the field of modern chaotic-based communication technologies and cryptography.

2 Preliminaries: Chaos in the System with Non-smooth Nonlinearities

2.1 Main Equations

Consider the dynamics of a particle with mass $m = 1$ moving in the non-smooth single-well potential

$$V(x) = |x|$$

under external excitation $f(t) = A\cos(\omega t + \xi)$, where ξ is an initial phase of the external force.

The Lagrange function in this case can be presented in the form[1]:

$$\mathcal{L}(x, \dot{x}, t) = \frac{\dot{x}^2}{2} - |x| + xf(t).$$

Using the standard form of the Lagrange equation

$$\frac{\mathrm{d}}{\mathrm{d}t}\frac{\partial \mathcal{L}}{\partial \dot{x}} - \frac{\partial \mathcal{L}}{\partial x} = 0,$$

an equation of motion can be obtained in the form:

[1] Here and below a dot over the variable means the derivative with respect to time.

$$\ddot{x} + \operatorname{sign} x = f(t), \quad x(0) = x_0, \quad \dot{x}(0) = \dot{x}_0, \tag{1}$$

where $\operatorname{sign} x$ is the standard signum function ($\operatorname{sign} x = \frac{\mathrm{d}|x|}{\mathrm{d}x}$). A solution to unperturbed (1) (when $f(t)$ is zero) can be obtained in explicit analytical form. Consider two cases: $x_0 > 0$ and $x_0 < 0$. In the first case, we have

$$\ddot{x}_+ + 1 = 0, \quad x_+(0) = x_0, \quad \dot{x}_+(0) = \dot{x}_0,$$

and the corresponding solution is

$$x_+(t) = -\frac{t^2}{2} + \dot{x}_0 t + x_0.$$

In the second case:

$$\ddot{x}_- - 1 = 0, \quad x_-(0) = x_0, \quad \dot{x}_-(0) = \dot{x}_0,$$

and the corresponding solution is

$$x_-(t) = \frac{t^2}{2} + \dot{x}_0 t + x_0.$$

Next, these solutions should be sewn at the points where the sign of the solution changes. Due to the fact that motion in this potential is finite, a solution to unperturbed (1) has an oscillatory nature [47]. However, this solution cannot be presented in terms of sin or cos and, roughly speaking, is not a periodic function in a common sense, but it periodically changes depending on the sign of x. The solution in this case is a set of quadratic parabolas with opposite extremes that follow one by one and must be sewn at time instants where the sign of x changes (at the same time these instants essentially depend on the initial conditions). The period of oscillations can be obtained using the general expression [47]:

$$T = 2\int_{x_1}^{x_2} \frac{\mathrm{d}x}{\sqrt{2\,[E - V(x)]}} = 2\int_{x_1}^{x_2} \frac{\mathrm{d}x}{\sqrt{2\,[E - |x|]}},$$

where $x_{1,2}$ are classical turning points determined by $E = V(x)$ (in this case $x_1 = -E$, $x_2 = E$). After simple integration, we have $T = 4\sqrt{2E}$. If initial deviation x_0 and the corresponding initial velocity $\dot{x}_0 = 0$, then the period of oscillations is determined by $T = 4\sqrt{2|x_0|}$ and the corresponding time instants at which the sign of solution changes are

$$t_\pm = T/4 + (T/2)n = (1 + 2n)\sqrt{2|x_0|}, \quad n = 0, 1, 2, \ldots \tag{2}$$

Thus, the sewing conditions read $x_-(t_\pm) = x_+(t_\pm)$ and $\dot{x}_-(t_\pm) = \dot{x}_+(t_\pm)$.

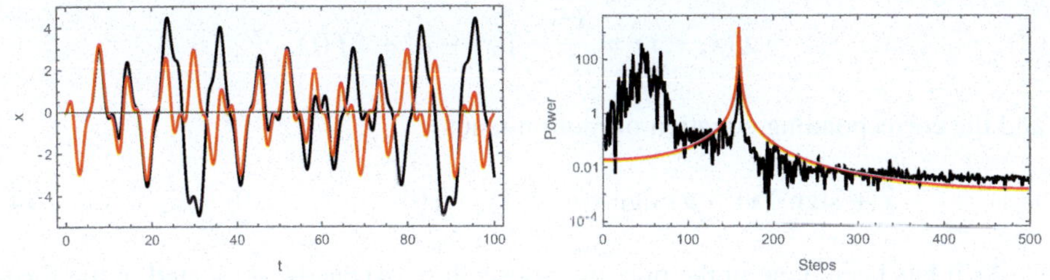

Fig. 1 Left panel: sensitivity of the solution to (1) to initial conditions. Parameters of the excitation force are $A = 3$, $\omega = 2$, $\xi = 0$. Initial conditions are $x_0 = 0$, $\dot{x}_0 = 0$ (black curve) and $x_0 = 0.01$, $\dot{x}_0 = 0$ (red curve). Right panel: log plot of the Fourier spectrum of the solution to (1) (black curve) comparing to the spectrum of the excitation force (red curve). Parameters and initial conditions are $A = 3$, $\omega = 2$, $\xi = 0$, $x_0 = 0$, $\dot{x}_0 = 0$

Equation (1) can also be presented in the form of equivalent system within the Hamiltonian approach. Introducing the following notations $x = q$, $\dot{x} = p$, and $z = t$, we have

$$\begin{cases} \dot{q} = p, \\ \dot{p} = -\operatorname{sign} q + f(z), \\ \dot{z} = 1, \end{cases} \tag{3}$$

with the following initial conditions: $q(0) = x_0$, $p(0) = \dot{x}_0$, and $z(0) = 0$. As we can clearly see, the vector field $\mathcal{R}_1 = (p, -\operatorname{sign} q + f(z), 1)$ which corresponds to this system is divergence free div $\mathcal{R}_1 = 0$ (because of no friction). As is known, the flow maps associated to divergence-free vector fields are volume preserving. Thus, chaos in this system has a conservative nature.

Note that obtaining an analytical solution to perturbed (1) or (3) is a rather complicated problem and cannot be solved explicitly.

In Fig. 1, we show sensitivity of the solution to (1) to initial conditions which is typical for chaotic dynamics. Also in this figure we show the Fourier spectrum of the solution together with the Fourier spectrum of the corresponding excitation force. As can be seen, the spectrum of solution becomes continuous comparing to the excitation force which is explicitly discrete. These plots confirm the chaotic dynamics in the system at given parameters of the excitation force and initial conditions (parameters are shown in the caption).

In the same way, consider the dynamics of a particle with unit mass in a double-well non-smooth potential

$$V(x) = ||x| - a|$$

under external excitation $f(t)$. Parameter a determines positions of the fixed points (in this case we have two stable points $x = -a$ and $x = a$, and one unstable point $x = 0$).

The Lagrange function in this case has the form:

$$\mathcal{L}(x, \dot{x}, t) = \frac{\dot{x}^2}{2} - ||x| - a| + xf(t),$$

and the corresponding equation of motion reads

$$\ddot{x} + \text{sign}\,(|x| - a)\,\text{sign}\,x = f(t), \quad x(0) = x_0, \quad \dot{x}(0) = \dot{x}_0. \tag{4}$$

As it has been done in the previous subsection, (4) can be presented in the form of equivalent system. Within the same notations the corresponding system reads

$$\begin{cases} \dot{q} = p, \\ \dot{p} = -\text{sign}\,(|q| - a)\,\text{sign}\,q + f(z), \\ \dot{z} = 1, \end{cases} \tag{5}$$

with the same initial conditions: $q(0) = x_0$, $p(0) = \dot{x}_0$, and $z(0) = 0$. The vector field $\mathcal{R}_2 = (p, -\text{sign}\,(|q| - a)\,\text{sign}\,q + f(z), 1)$ in this case is also divergence free $\text{div}\mathcal{R}_2 = 0$. Thus, chaos in this system is also conservative.

Note that this double-well potential seems quite similar to a double-well potential within the Duffing model [48]. However, these systems are rather different because one of them is smooth continuous function, while another is a piece-wise linear function with three points ($x = -a$, $x = 0$ and $x = a$) at which the derivative is not determined. This fact brings some features to the nonlinear dynamics of this system and enriches dynamical regimes.

2.2 Poincaré Sections, Bifurcation Diagrams, and Lyapunov Characteristic Exponents

In this subsection, we show numerical results for Poincaré sections, bifurcation diagrams, and Lyapunov characteristic exponents in the case of a double-well non-smooth potential.

A useful way to analyze the chaotic regime of dynamical system is to obtain the Poincaré section for the corresponding equation. Rather than considering the phase portrait during all simulation time (this phase portrait is a continuous curve), the Poincaré section is the discrete set of points in the phase space which corresponds to cross section of the phase portrait by some plane at every period of the external excitation. In this sense, the Poincaré section gives a convenient opportunity to make a decision about dynamical regimes (periodic, quasi-periodic, or chaotic) established in the system.

In Fig. 2, we show the corresponding Poincaré sections for (4) at different values of the amplitude of the excitation force A (namely, $A = 3$ and $A = 4$). As can be seen from these figures, the structure of Poincaré sections becomes sufficiently complicated comparing to the case of smooth potentials (e.g., in the standard quadratic oscillatory potential where the Poincaré section is a single point which corresponds

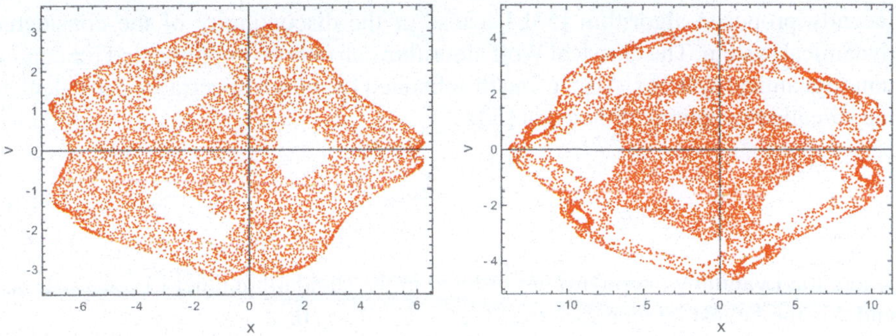

Fig. 2 Poincaré sections for (4) with $a = 1$ and initial conditions $x_0 = 0$, $\dot{x}_0 = 0$ and the following parameters of the excitation force: $A = 3$, $\omega = 2$, $\xi = 0$ (left panel); $A = 4$, $\omega = 2$, $\xi = 0$ (right panel)

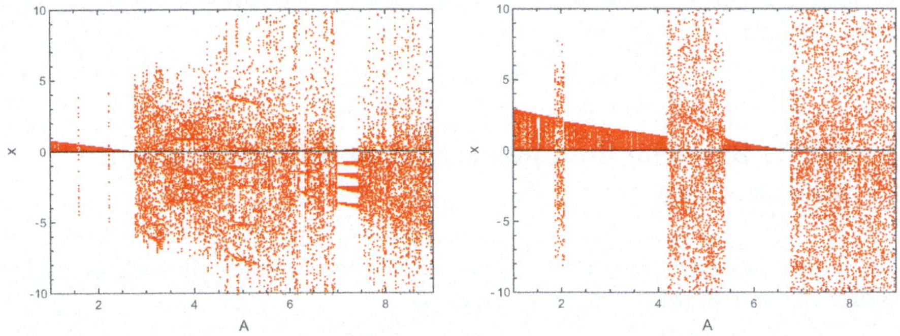

Fig. 3 Bifurcation diagrams for (4) with $a = 1$ (left panel) and $a = 2$ (right panel) depending on the amplitude of the external excitation A at $\omega = 2$ and $\xi = 0$. Initial conditions are $x_0 = 0$ and $\dot{x}_0 = 0$

to periodic motion, when the period has doubled it consists of two points, and so on). A shape and complexity of sections are also increased due to chaotic regime at the presented parameters. However, due to the symmetry of the corresponding double-well potential, sections are also symmetric.

In Fig. 3, we show also numerical results for bifurcation diagrams of (4) (depending on amplitude A) at various values of parameter a. As can be seen from the left panel in the range of A from $A \approx 3$ to $A \approx 7$, the system exhibits sufficiently chaotic dynamics (the corresponding Lyapunov characteristic exponents also prove this statement). When the parameter a increase, ($a = 2$) the corresponding ranges of chaotic motion become narrower, while transition to chaos becomes sharp.

As is well known, Lyapunov exponents provide a quantitative measure of chaotic motion by the divergence of nearby trajectories. In this paper, instead of the standard approach to numerical evaluation of the Lyapunov spectrum based on the Wolf algorithm (the Gram-Schmidt orthonormalization procedure) [49–52], we use the recently proposed algorithm [53] because of the discontinuity of the considered dynamical system. The classical Wolf algorithm can only be implemented for continuous dynamical systems. Also, using the obtained Lyapunov spectra we can calculate the so-called Lyapunov dimension [54]

$$D_L = j - \frac{\sum_{i=1}^{j} \lambda_i}{\lambda_{j+1}}, \tag{6}$$

where the Lyapunov exponents are ordered in the standard manner ($\lambda_1 \geqslant \ldots \geqslant \lambda_n$) and j is the largest integer such that $\lambda_1 + \cdots + \lambda_j > 0$.

The Lyapunov characteristic exponents for (4) with initial conditions $x_0 = 0, \dot{x}_0 = 0$ and initial phase $\xi = 0$ have the following quantities: $\{0.29522, 0.0, -0.29522\}$ when $A = 3$, $\omega = 2$, $a = 1$ and the corresponding Lyapunov dimension is 3.0; $\{0.21308, 0.0, -0.21308\}$ when $A = 4$, $\omega = 2$, $a = 1$ and the Lyapunov dimension is 3.0. These results prove that in the system under consideration the conservative chaos establishes.

3 PRNG Based on the Chaotic System with Non-smooth Nonlinearities

For further numerical simulation of the PRNG, we use the following equivalent system of the first-order differential equations:

$$\ddot{x}_1 = -\frac{\mathrm{d}U}{\mathrm{d}x_1} + A\cos(\omega t). \Leftrightarrow \begin{cases} \dot{x}_1 = x_2, \\ \dot{x}_2 = -\dfrac{\mathrm{d}U}{\mathrm{d}x_1} + A\cos\omega t, \\ \dot{x}_3 = 1. \end{cases} \tag{7}$$

A proposed model of chaotic oscillator with non-smooth double-well potential is described by the following system:

$$\begin{matrix} \ddot{x}_1 = -\dfrac{\mathrm{d}U}{\mathrm{d}x_1} + A\cos\omega t, \\ U(x_1) = \left||x_1| - a\right|. \end{matrix} \Leftrightarrow \begin{cases} \dot{x}_1 = x_2, \\ \dot{x}_2 = -\mathrm{sign}(|x_1| - a)\mathrm{sign}\, x_1 + A\cos\omega t, \\ \dot{x}_3 = 1. \end{cases} \tag{8}$$

A double-well non-smooth potential within the proposed model comparing to the well-known chaotic Duffing potential is shown in Fig. 4. Comparing these nonlinearities we have to remember that the Duffing potential is a smooth continuous function while the proposed potential is a piece-wise linear function with three points at which the derivative is not determined. While at first sight these systems seem similar, but in fundamental level they are rather different and this fact brings some features to the nonlinear dynamics of this system and enriches dynamical regimes.

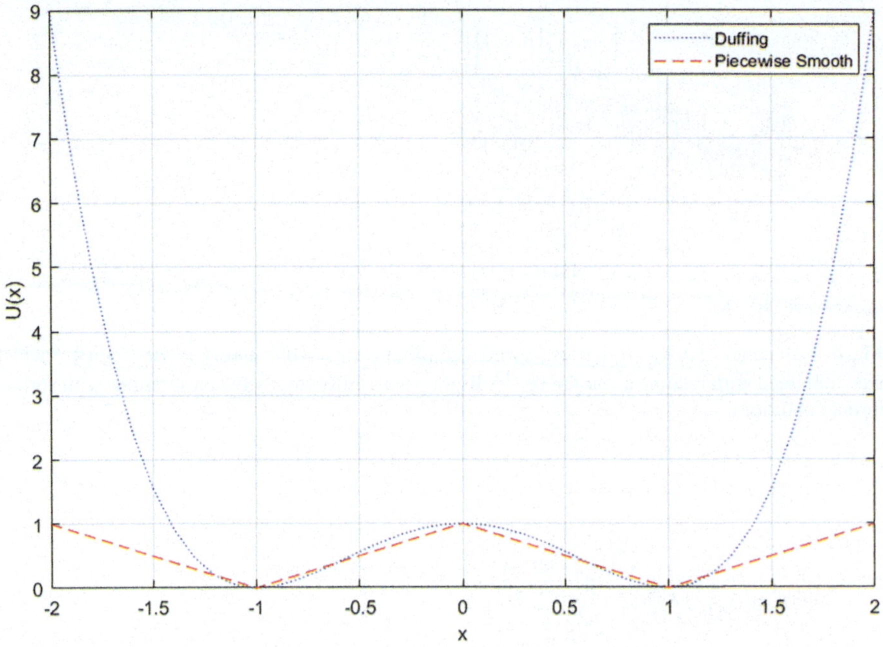

Fig. 4 A non-smooth double-well potential comparing to the Duffing oscillatory potential

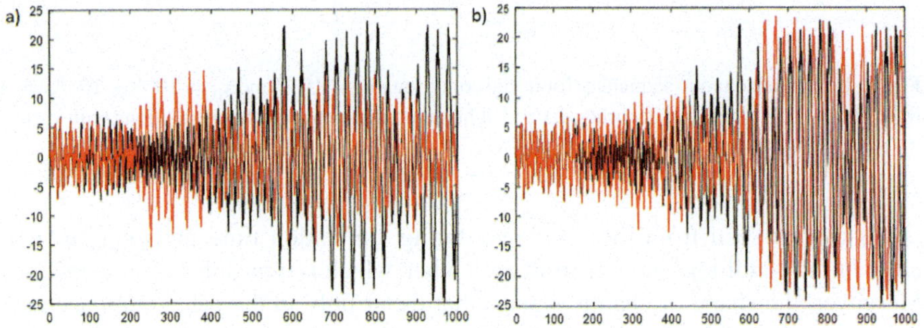

Fig. 5 Chaotic sequences for initial conditions $x_0 = 10^{-6}$ and $x'_0 = 10^{-6} + 10^{-15}$, which were obtained with various relative accuracy: 10^{-6} (left panel) and 10^{-12} (right panel)

Additionally to the initial conditions, the calculation accuracy also plays an important role in the generation of chaotic signals. To analyze the sensitivity of a solution to the proposed chaotic oscillator system, two sequences were calculated that differ in the initial conditions by 10^{-15} for various relative accuracy of calculations. The solution was implemented using the standard 4th-order Runge-Kutta method with absolute accuracy of 10^{-16}. Results of numerical simulations are shown in Figs. 5, 6, and 7.

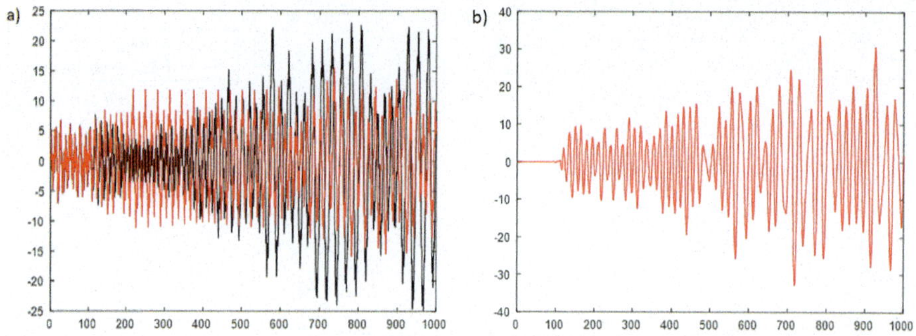

Fig. 6 Left panel: chaotic sequences for initial conditions $x_0 = 10^{-6}$ and $x_0' = 10^{-6} + 10^{-6}$ which were obtained with relative accuracy 10^{-12}; Right panel: difference between solutions with various initial conditions

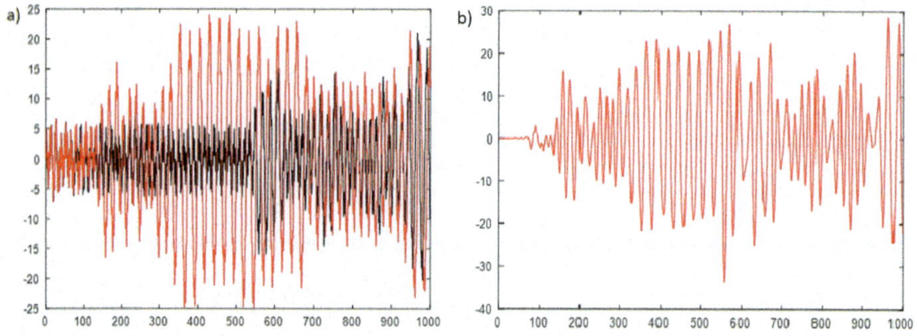

Fig. 7 Left panel: chaotic sequences for initial conditions $x_0 = 10^{-6}$ and $x_0' = 10^{-6} + 10^{-6}$ which were obtained with relative accuracy 10^{-6}; Right panel: difference between solutions with various initial conditions

As can be seen from these figures, during simulation time the solutions with close initial conditions diverge from each other, which is typical for chaotic regimes. Moreover, the shape of chaotic signals that differ only in the calculation accuracy differs significantly, which is caused by accumulation of calculation errors.

To implement the sequence of samples of chaotic signal generating by the proposed conservative chaotic oscillator, a finite-difference form of (8) can be used. Fixing initial conditions $x_1(0) = x_0$, $\dot{x}_1(0) = v_0$ and time step Δt, we have the following explicit finite-difference mapping:

$$\begin{aligned}
&x_1(1) = x_0,\\
&x_1(2) = x_0 + v_0 \Delta t,\\
&x_1(i) = 2x_1(i-1) - x_1(i-2) -\\
&-\Delta t^2 \left\{ \text{sign}\left[|x_1(i-1)| - a\right] \text{sign}\left[x_1(i-1)\right] + A\cos(i\omega \Delta t) \right\}, \quad i = 3 \ldots N.
\end{aligned} \tag{9}$$

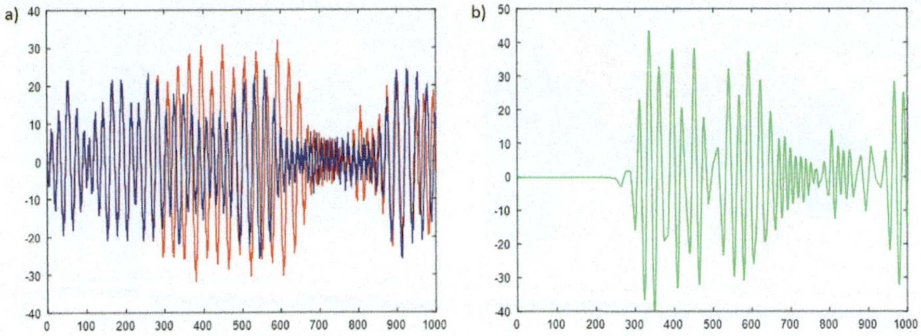

Fig. 8 Left panel: chaotic sequences for initial conditions $x_0 = 10^{-6}$ and $x_0' = 10^{-6} + 10^{-3}$; Right panel: difference between the obtained chaotic sequences

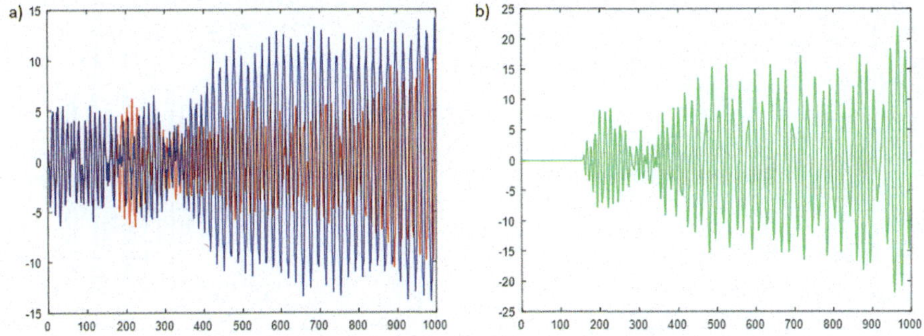

Fig. 9 Left panel: chaotic sequences for initial conditions $x_0 = x_0' = 10^{-6}$, $v_0 = 10^{-6}$ and $v_0' = 10^{-6} + 10^{-5}$; Right panel: difference between the obtained chaotic sequences

To analyze the sensitivity of the presented discrete realization of chaotic oscillator to initial conditions, we calculated two sequences with initial conditions differing by 10^{-3}. Numerical results are shown in Fig. 8. As can be seen from this figure, the system exhibits sensitivity to initial conditions which, as was mentioned above, is typical for chaotic regimes. Simulations were performed for parameters $a = 1.5$, $A = 10$, $\omega = 2$ with time step $\Delta t = 0.1$. While the time step is decreasing, the sensitivity to initial conditions is increasing. When the time step is $\Delta t = 0.01$ and the difference between initial conditions is $x_0 - x_0' \sim 10^{-7}$, the obtained sequences diverge noticeably starting at $t \sim 200$.

The sensitivity to a change in the initial velocity is higher than to a change in the initial position of the chaotic oscillator (these results are shown in Figs. 9 and 10). As previously, simulations were performed for parameters $a = 1.5$, $A = 10$, $\omega = 2$ with time step $\Delta t = 0.1$.

Correlation functions for the generated sequences are shown in Fig. 11. A fast decay of the auto-correlation function is typical for chaotic systems and a low level of the cross-correlation function of two sequences with close initial conditions is also observed. The obtained results confirm the chaotic nature of oscillations and the

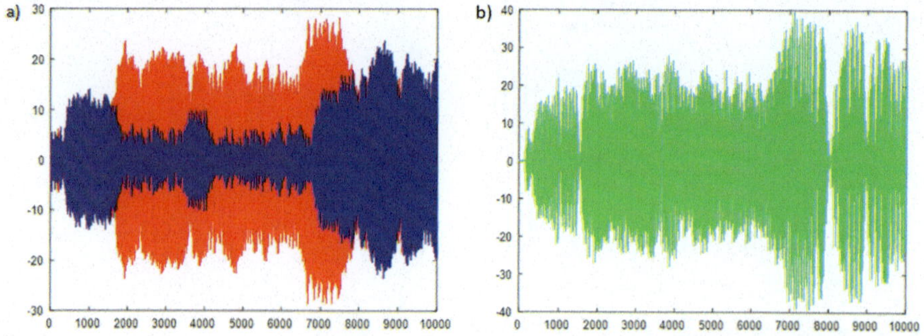

Fig. 10 Left panel: chaotic sequences of length 10^5 points for initial conditions $x_0 = x_0' = 10^{-6}$, $v_0 = 10^{-6}$ and $v_0' = 10^{-6} + 10^{-5}$; Right panel: difference between the obtained chaotic sequences

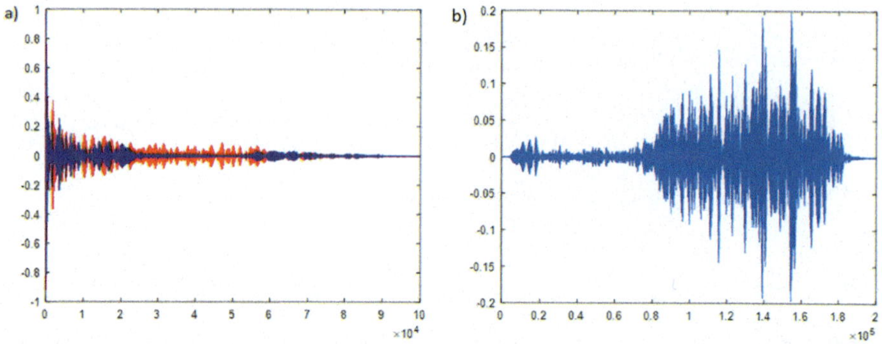

Fig. 11 Auto-correlation functions (left panel) and the cross-correlation function (right panel) of chaotic sequences corresponding to initial conditions $x_0 = x_0' = 10^{-6}$, $v_0 = 10^{-6}$ and $v_0' = 10^{-6} + 10^{-5}$

divergence of close trajectories with time, which are required for the generation of pseudorandom numbers.

4 Test on Randomness and Comparison with Other Chaotic-Based PRNGs

4.1 Test on Randomness

Using sequences of numbers which are obtained from the discrete model of conservative chaotic oscillator with non-smooth nonlinearity, a pseudorandom bit sequence is generated. In the process of post-processing, we performed extraction of the least significant bits of the mantissa of a number from 5 to 36. The generation speed rate is

Table 1 Results of NIST tests for the proposed PRNG (9)

Test	η	Result
Frequency test	0.991	Success
Frequency test within a block	0.992	Success
Runs test	0.991	Success
Tests for the longest run of ones in a block	0.992	Success
Binary matrix rank	0.986	Success
Discrete fourier transform test	0.989	Success
Non-overlapping template matching test	0.988	Success
Overlapping template matching test	0.990	Success
Maurer's universal statistical test	0.999	Success
Lempel-Ziv compression test	0.996	Success
Linear complexity test	0.990	Success
Serial test	0.985	Success
Approximate entropy test	0.988	Success
Cumulative sums test (forward)	0.991	Success
Cumulative sums test (reverse)	0.988	Success
Random excursion test (Average value)	0.985	Success
Random excursion variants test (Average value)	0.986	Success

around 17 Mbps. Obtained bit sequences were successfully passed statistical NIST tests [55]. Tests were performed using $N = 10^3$ sequences of 10^4 bits.[2] Each test defines a proportion η as the ratio of sequences passed test successfully $\tilde{n}$ to the total number of sequences N (i.e., $\eta = \frac{\tilde{n}}{N}$). Results of main NIST tests are shown in Table 1.

As can be seen from Table 1, the proposed PRNG successfully passed main NIST statistical tests since proportion η of sequences that successfully passed each test is within the interval determined by the relation:

$$(1-\alpha) \pm 3\sqrt{\frac{\alpha}{N}},$$

[2] Following recommendations by the NIST, for the Binary Matrix Rank Test, Linear Complexity Test, Lempel-Ziv Compression Test, Random Excursion Test, and Random Excursion Variants Test, we used $N = 10^3$ sequences of 10^6 bits.

Table 2 Characteristics of various PRNGs

Type of PRNG based on	LLE	Security claim	Speed, Mbps
Two chaotic maps ($T = 9$, $n = 1024$) [29]	N/A	2^{120}	1.99
Three chaotic maps ($T = 5$, $n = 2048$) [30]	N/A	2^{111}	5.36
Discrete chaotic mapping with low security ($n = 10$) [33]	N/A	2^{2}	6.65
Discrete chaotic mapping with medium security ($n = 10$) [33]	N/A	2^{22}	3.36
Discrete chaotic mapping with high security ($n = 10$) [33]	N/A	2^{41}	0.34
Modified discrete chaotic mapping ($n = 10$) [34]	~5	2^{42}	24.45
Modified discrete chaotic mapping ($n = 15$) [34]	~5	2^{80}	31.27
Modified discrete chaotic mapping ($n = 20$) [34]	~5	2^{122}	34.18
Pseudorandomly enhanced logistic map [31]	~10	2^{128}	1.7
Proposed PRNG	~0.3	$\mathbf{2^{128}}$	**17.42**

where $\alpha = 0.01$ is the probability of a false rejection of a random sequence and N is the number of sequences. As it can be seen, for $N = 10^3$ sequences this interval is $[0.9805, 0.9995]$.

The degree of randomness of a sequence with a uniform distribution can be additionally increased by the mixing randomization procedure as is described in [46, 56]: elements of the sequence x_i are loaded into the buffer array V of length $k \sim 100$; an element with number $j = \left[k \cdot \dfrac{x_n}{\max(x_i)} \right]$, $n = k + 1$ is retrieved; buffer element V_j is replaced by x_{n+2}; then the work with the buffer is repeated for $n = k + 3$, and so on.

4.2 Comparison of Various PRNGs

Following recommendations of the NIST and ECRYPT, the cipher of PRNG must generate more than 2^{128} various secret keys avoiding attacks by the direct search method using computing powers that may be available until 2040 AD [57, 58]. Characteristics of speed and strong cryptography (in the form of attack complexity) for various PRNGs that have no effective hacking algorithms are shown in Table 2. In Table 2, the speed characteristic of the proposed PRNG was estimated using MATLAB simulation and the following hardware: Intel I7 8700 3.2 GHz CPU, 64 GB RAM, and Windows 10 64-bit OS.

A secret key (or the security claim which is connected to attack complexity of the PRNG) of the proposed PRNG consists of parameters of the discrete oscillator model: $\Delta t \in [0.1, 0.4]$ with step 10^{-8}, $a \in [1, 21]$ with step 10^{-3}, $A \in [1, 11]$ with step 10^{-6}, $w \in [1, 11]$ with step 10^{-6}, initial values $x_0 \in [-2a, 2a]$ with step 10^{-4}, and $v_0 \in [-2, 2]$ with step 10^{-7}. These minimal step sizes correspond to divergence (relative to each other) of two chaotic sequences with close initial conditions and parameters. Thus, the key space of this PRNG is estimated by $3.6 \times 10^{38} \approx 2^{128}$.

As can be seen from Table 2, the proposed PRNG exhibits relatively high-speed characteristics[3] and attack complexity, which is required in main NIST and ECRYPT recommendations [57, 58], while the largest Lyapunov exponent (LLE) is relatively low compared to other chaotic-based PRNGs. Note that the proposed PRNG is based on the low-dimensional conservative chaotic system. As is known, for encryption purposes, low-dimensional chaotic systems are more attractive than high dimensional [31]. Moreover, such a low-dimensional chaotic system can be easily implemented in any programmable electronic device even under limited hardware resources. This makes it possible to consider the proposed PRNG as a promising and efficient source of pseudorandom numbers which can be applied in the field of modern chaotic-based digital communications and cryptography.

5 Conclusions

In this chapter, we have proposed a novel PRNG which is based on the model of conservative chaotic oscillator with non-smooth nonlinearity. Main characteristics of this system such as Poincaré sections, bifurcation diagrams, Lyapunov characteristic exponents (together with the Lyapunov dimension), auto-correlation, and cross-correlation functions were obtained and discussed during numerical simulations. Based on the considered model, a novel PRNG was proposed and the main characteristics of this generator were simulated and compared to another well-known chaotic-based PRNGs. Note that the considered chaotic system admits wide variation of parameters (at least 6 parameters can be changed in a wide range to obtain

[3] In the case of a high-speed FPGA-based chaotic TRNG which is described in [32], the output rate can be raised until 58.76 Mbps.

independent random sequences). Moreover, because of conservative nature, this systems exhibits a high sensitivity to changes of these parameters (especially to changes in initial conditions and time step) which allows to make a fine-tuning of the corresponding PRNG during its algorithmic implementation at the simulation level. The obtained results (especially the speed characteristic which is around 18 Mbps, and security claim which is around 2^{128} and satisfies main requirements of NIST and ECRYPT) make it possible to conclude that the proposed PRNG might have promising applications in the field of modern fast operating chaotic-based digital communications and cryptography.

Acknowledgements This research was supported by the Russian Science Foundation (RSF) grant number 23-29-00696.

References

1. E.N. Lorenz, J. Atmos. Sci. **20**(2), 130 (1963). https://doi.org/10.1175/1520-0469(1963)020<0130:DNF>2.0.CO;2
2. O.E. Rössler, Phys. Lett. A **57**(5), 397 (1976). https://doi.org/10.1016/0375-9601(76)90101-8
3. O.E. Rössler, Phys. Lett. A **71**(2), 155 (1979). https://doi.org/10.1016/0375-9601(79)90150-6
4. L.O. Chua, IEEE Trans. Circuits Syst. **27**(11), 1059 (1980). https://doi.org/10.1109/TCS.1980.1084745
5. T. Matsumoto, IEEE Trans. Circuits Syst. **31**(12), 1055 (1984). https://doi.org/10.1109/TCS.1984.1085459
6. A.S. Andreatos, A.P. Leros, J. Eng. Sci. Technol. Rev. **6**(4), 90 (2013). https://doi.org/10.25103/jestr.064.11
7. A. Arneodo, P.H. Coullet, C. Tresser, Commun. Math. Phys. **79**, 573 (1981). https://doi.org/10.1007/BF01209312
8. A. Arneodo, P.H. Coullet, E.A. Spiegel, Phys. Lett. A **92**(8), 369 (1982). https://doi.org/10.1016/0375-9601(82)90455-8
9. A. Arneodo, P.H. Coullet, E.A. Spiegel, Geophys. Astrophys. Fluid Dyn. **31**(1–2), 1 (1985). https://doi.org/10.1080/03091928508219264
10. A. Arneodo, P.H. Coullet, E.A. Spiegel, C. Tresser, Physica D: Nonlinear Phenomena **14**(3), 327 (1985). https://doi.org/10.1016/0167-2789(85)90093-4
11. S. Vaidyanathan, Advanced computing, in *CCSIT, Communications in Computer and Information Science*. ed. by N. Meghanathan, B.K. Kaushik, D. Nagamalai (2011), pp. 98–107
12. E.N. Lorenz, Tellus A **36A**(2), 98 (1984). https://doi.org/10.1111/j.1600-0870.1984.tb00230.x
13. H. Wang, Y. Yu, G. Wen, J. Appl. Math. **2014**(11), 296279 (2014). https://doi.org/10.1155/2014/296279
14. J.C. Sprott, Phys. Rev. E **50**, R647 (1994). https://doi.org/10.1103/PhysRevE.50.R647
15. J.R. Piper, J.C. Sprott, IEEE Trans. Circuits Syst. II: Express Briefs **57**(9), 730 (2010). https://doi.org/10.1109/TCSII.2010.2058493
16. J.C. Sprott, S. Jafari, V.T. Pham, Z.S. Hosseini, Phys. Lett. A **379**(36), 2030 (2015). https://doi.org/10.1016/j.physleta.2015.06.039
17. S. Cang, A. Wu, R. Zhang, Z. Wang, Z. Chen, Int. J. Bifurc. Chaos **28**(07), 1850087 (2018). https://doi.org/10.1142/S0218127418500876

18. G. Qi, J. Hu, Z. Wang, Appl. Math. Model. **78**, 350 (2020). https://doi.org/10.1016/j.apm.2019.08.023
19. S. Cang, Y. Li, W. Xue, Z. Wang, Z. Chen, Nonlinear Dyn. **99**(2), 1699 (2020). https://doi.org/10.1007/s11071-019-05385-9
20. N. Wang, G. Zhang, H. Bao, Nonlinear Dyn. **99**, 3197 (2020). https://doi.org/10.1007/s11071-020-05465-1
21. P. Meleshenko, M. Semenov, A. Klinskikh, Nonlinear Dyn. **101**, 2523 (2020). https://doi.org/10.1007/s11071-020-05956-1
22. P. Meleshenko, J. Vib. Eng. Technol. **10**, 2203 (2022). https://doi.org/10.1007/s42417-022-00514-y
23. M. Kunze, *Non-smooth dynamical systems*, *Lecture Notes in Mathematics*, vol. 1744 (Springer Science & Business Media, 2000). https://doi.org/10.1007/BFb0103843
24. R. Leine, D. Van Campen, B. Van de Vrande, Nonlinear Dyn. **23**(2), 105 (2000). https://doi.org/10.1023/A:1008384928636
25. M. Bernardo, C. Budd, A.R. Champneys, P. Kowalczyk, Piecewise-smooth dynamical systems: theory and applications, in *Applied Mathematical Sciences*, vol. 163 (Springer Science & Business Media, 2008). https://doi.org/10.1007/978-1-84628-708-4
26. O. Makarenkov, J.S.W. Lamb, Physica D: Nonlinear Phenomena **241**(22), 1826 (2012). https://doi.org/10.1016/j.physd.2012.08.002
27. J.S. Pei, J.P. Wright, F. Gay-Balmaz, J.L. Beck, M.D. Todd, Nonlinear Dyn. **95**(2), 1165 (2019). https://doi.org/10.1007/s11071-018-4622-2
28. S. Li, X. Ma, X. Bian, S.K. Lai, W. Zhang, Nonlinear Dyn. **99**(2), 1621 (2020). https://doi.org/10.1007/s11071-019-05380-0
29. M. François, T. Grosges, D. Barchiesi, R. Erra, Informatica **24**(2), 181 (2013). https://doi.org/10.15388/Informatica.2013.391
30. M. François, T. Grosges, D. Barchiesi, R. Erra, Commun. Nonlinear Sci. Numer. Simul. **19**(4), 887 (2014). https://doi.org/10.1016/j.cnsns.2013.08.032
31. M. Murillo-Escobar, C. Cruz-Hernández, L. Cardoza-Avendaño, R. Méndez-Ramírez, Nonlinear Dyn. **87**(1), 407 (2017). https://doi.org/10.1007/s11071-016-3051-3
32. I. Koyuncu, A.T. Özcerit, Comput. Electr. Eng. **58**, 203 (2017). https://doi.org/10.1016/j.compeleceng.2016.07.005
33. D. Lambić, M. Nikolić, Nonlinear Dyn. **90**(1), 223 (2017). https://doi.org/10.1007/s11071-017-3656-1
34. D. Lambić, M. Nikolić, Filomat **33**(8), 2257 (2019). https://doi.org/10.2298/FIL1908257L
35. B. Karakaya, A. Gülten, M. Frasca, Chaos. Solitons Fractals **119**, 143 (2019). https://doi.org/10.1016/j.chaos.2018.12.021
36. J. Zheng, H. Hu, H. Ming, X. Liu, Chaos, Solitons Fractals **138**, 109863 (2020). https://doi.org/10.1016/j.chaos.2020.109863
37. A.V. Tutueva, E.G. Nepomuceno, A.I. Karimov, V.S. Andreev, D.N. Butusov, Chaos. Solitons Fractals **133**, 109615 (2020). https://doi.org/10.1016/j.chaos.2020.109615
38. X.Y. Wang, X. Qin, Nonlinear Dyn. **70**(2), 1589 (2012). https://doi.org/10.1007/s11071-012-0558-0
39. Y. Dodis, D. Pointcheval, S. Ruhault, D. Vergniaud, D. Wichs, in *Proceedings of the 2013 ACM SIGSAC conference on Computer and communications security* (2013), pp. 647–658. https://doi.org/10.1145/2508859.2516653
40. M.S. Turan, E. Barker, J. Kelsey, K.A. McKay, M.L. Baish, M. Boyle, NIST Spec. Publ. **800**(90B), 1 (2018)
41. M. Matsumoto, T. Nishimura, ACM Trans. Model. Comput. Simul. (TOMACS) **8**(1), 3 (1998). https://doi.org/10.1145/272991.272995
42. L. Blum, M. Blum, M. Shub, SIAM J. Comput. **15**(2), 364 (1986). https://doi.org/10.1137/0215025
43. S. Tezuka, *Linear Congruential Generators* (Springer, Boston, MA, 1995), pp. 57–82. https://doi.org/10.1007/978-1-4615-2317-8_3
44. G. Marsaglia, J. Stat. Softw. **8**(14), 1 (2003). https://doi.org/10.18637/jss.v008.i14

45. F. Panneton, P. L'ecuyer, ACM Trans. Model. Comput. Simul. **15**(4), 346 (2005). https://doi.org/10.1145/1113316.1113319
46. D.E. Knuth, *The Art of Computer Programming: Seminumerical Algorithms*, 3rd edn., vol. 2 (Addison-Wesley Longman Publishing Co., Inc., 1997)
47. L. Landau, E. Lifshitz, *Course of Theoretical Physics, Mechanics*, 3rd edn., vol. 1(Butterworth-Heinemann, 1976)
48. I. Kovačić, M.J. Brennan (eds.), *The Duffing Equation: Nonlinear Oscillators and their Behaviour* (Wiley, New York, 2011)
49. A. Wolf, J.B. Swift, H.L. Swinney, J.A. Vastano, Physica D: Nonlinear Phenomena **16**(3), 285 (1985). https://doi.org/10.1016/0167-2789(85)90011-9
50. I. Goldhirsch, P.L. Sulem, S.A. Orszag, Physica D: Nonlinear Phenomena **27**(3), 311 (1987). https://doi.org/10.1016/0167-2789(87)90034-0
51. J. Awrejcewicz, A.V. Krysko, N.P. Erofeev, V. Dobriyan, M.A. Barulina, V.A. Krysko, Entropy **20**(3), 175 (2018). https://doi.org/10.3390/e20030175
52. R. Fiedler, H. Hetzler, MATEC Web Conf. **241**, 01009 (2018). https://doi.org/10.1051/matecconf/201824101009
53. M. Semenov, S. Borzunov, P. Meleshenko, Nonlinear Dyn. **109**, 1805 (2022). https://doi.org/10.1007/s11071-022-07492-6
54. J.L. Kaplan, J.A. Yorke, in *Functional Differential Equations and Approximation of Fixed Points*, ed. by H.O. Peitgen, H.O. Walther (Springer Berlin Heidelberg, 1979), pp. 204–227. https://doi.org/10.1007/BFb0064319
55. J. Soto, in *Proceedings of the 22nd National information systems security conference*, vol. 10 (NIST Gaithersburg, MD, 1999), p. 12
56. C. Bays, S. Durham, ACM Trans. Math. Softw. **2**(1), 59 (1976). https://doi.org/10.1145/355666.355670
57. Yearly report on algorithms and keysizes (2009–2010), D. SPA. 13 Rev. 1.0. Tech. rep., ICT-2007-216676 (2010). https://www.ecrypt.eu.org/ecrypt2/documents/D.SPA.13.pdf/
58. A.W. Dent, Inf. Secur. Tech. Rep. **15**(1), 21 (2010). https://doi.org/10.1016/j.istr.2010.10.006

Hysteretic Beam and Temperature Effect

Mikhail E. Semenov, Evgeny A. Karpov, and Peter A. Meleshenko

Abstract In this work, we focus on the problem of the dynamics of vibrations of the hysteresis beam taking into account the temperature effects depending on the external load: shock, periodic and seismic. The classical Euler-Bernoulli model is considered as an initial stage for a beam model. Also, we introduced the distributed hysteresis properties to consider the hysteresis beam. The well-known Bouc-Wen model is used as a hysteresis carrier. The temperature simulation is implemented within the one-dimensional heat conduction equation. To connect the beam oscillation equation and the thermal conductivity equation, we used a dissipation function which physically corresponds to transformation of the oscillation energy into heat. The results of numerical simulation for classical and hysteresis beams are presented and discussed. Particularly, the obtained results show that the beam with hysteresis properties exhibits sufficiently different dynamical behaviour compared to the classical model. The temperature distribution also exhibits specific difference between the classical and hysteresis beams.

1 Introduction

Modelling of structural systems is currently the most important stage of design engineering. It helps to determine the reaction of such systems to external loads of various types. It is also important to study the dependence of the behaviour of structural systems on external factors including temperature, beam wind, solar radiation, precipitation, etc. The modelling can help to determine the strength of structural systems,

M. E. Semenov (✉)
Geophysical Survey of Russia Academy of Sciences, Lenina av. 189, 249035 Obninsk, Russia
e-mail: mkl150@mail.ru

Mathematics Department, Voronezh State Technical University, XX- letiya Oktyabrya st. 84, 394006 Voronezh, Russia

M. E. Semenov · E. A. Karpov · P. A. Meleshenko
Digital Technologies Department, Voronezh State University, Universitetskaya sq. 1, 394006 Voronezh, Russia
e-mail: melechp@yandex.ru

M. Belhaq (ed.), *Advances in Nonlinear Dynamics and Control of Mechanical and Physical Systems*, Springer Proceedings in Physics 301,
https://doi.org/10.1007/978-981-99-7958-5_19

which is of utmost importance in seismic hazard zones. Beams are key elements affecting the dynamic features of structural systems. It is known that the strength of a beam depends primarily on the properties of its material.

As a result of load or thermal expansion, the inner forces within the material start to resist to regain their initial shape. When the deformation is small, the process can be described as a linear one. However, under real conditions, deformations can rarely be considered small. If, when the load is lifted, the material reaches a new equilibrium different from the initial one, the process should be considered as hysteresis. Hysteresis occurs as a result of dislocations [1] in the material caused by crystal lattice defects. We should note that the dislocation density affects the strength of the material. Dislocations can move [2] and restructure when affected by loads and temperature variations. The energy lost during each cycle of restructuring or annihilation is eventually converted into inner energy of the material, i.e. internal friction (the material is heated). Therefore, equations describing the dynamics of engineering structures should take into account the temperature effects.

A specific feature of hysteresis is that the output signal depends on the history of previous states of the system as well as on the input. Hysteresis can be found in a number of fields, including physics [3–6], electrical engineering, mechanics [7], thermodynamics, chemistry, biology, etc. [8, 9]. An example of hysteresis in physics is hysteresis in magnetic materials. In electrical engineering, hysteresis is found primarily in relay systems, where the reaction to the input is not immediate and depends on the history of the states of the system [10]. Hysteretic properties are used to design devices facilitating energy dissipation [11]. The theory of hysteretic models was greatly contributed to by M. A. Krasnoselsky who introduced the idea of basic hysteretic elements, hysterons [12].

1.1 Equation of Beam Vibrations and Thermal Conductivity

The behaviour of beams under mechanical load is usually described by one of the following theories (Fig. 1): the Euler-Bernoulli beam theory or the Timoshenko-Ehrenfest beam theory [13]. They differ in that the second theory takes into account a wider range of factors affecting the behaviour of beams, while the first theory presents a simpler model describing less complicated situations. The model of beam vibrations described in this article is based on the Euler-Bernoulli theory, namely on the assumption that the beam is pinned at both ends and has zero initial conditions:

$$u_{tt} + a^2 u_{xxxx} = g(x, t), \quad t > 0, x \in [0, L], \tag{1}$$

where $u = u(x, t)$ is the beam deflection, indices denote partials, $g(x, t)$ is the function characterising the external load, $a^2 = EI$ is the product of the modulus of elasticity and the moment of inertia, $t > 0$ is the time of modelling, and L is the length of the beam.

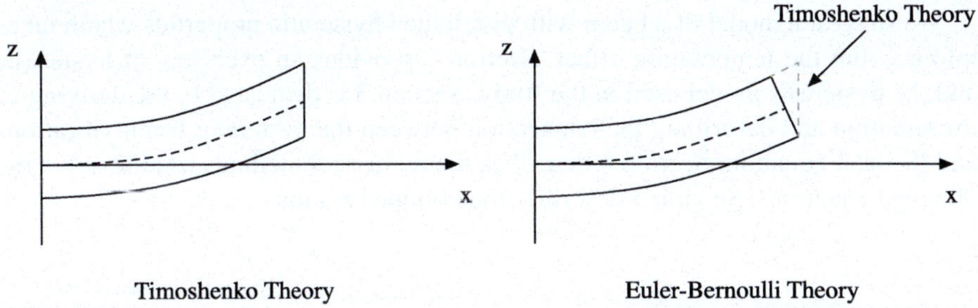

Fig. 1 Graphical representation of the fundamental difference between the Timoshenko theory and the Euler-Bernoulli theory

Beam dynamics is usually studied based on the external factors. Thus, the behaviour of beams under impulse, seismic, periodic, constant, and other types of loads is described in [14–17]. The main task of such studies is to investigate the stability of beams under various conditions. Modelling helps to determine the type of material and the shape of beams that should be used under particular types of load. Beam models can also be used as basic elements of bridges. A number of studies [18–21] focus on the behaviour of beams under moving loads. They consider both simple models, which take into account longitudinal vibrations only, and models taking into account transverse vibrations. Some studies focus on the reduction of beam vibrations. Works [22, 23] suggest special boundary conditions which facilitate the damping of external loads.

Another important task of design engineering is to study temperature distribution in solids [22, 24, 25]. It helps to determine the parts that are more susceptible to temperature, the time of thermal equilibrium, etc. The distribution of heat in a body is usually modelled using a well-known heat equation:

$$\omega_t - b^2\omega_{xx} = \phi(x, t) \tag{2}$$

where $\omega = \omega(x, t)$ is the temperature of the body at the point with coordinates (x, t), b^2 is the coefficient of thermal conductivity, and $\phi(x, t)$ is the heat sink function.

The equation is a basic heat and mass transfer equation and is widely used in physics, geology, meteorology, and other fields of study. Heat equation describes the way heat is distributed within the material and the dependence of the distribution on various parameters including temperature, thermal conductivity, density, and heat capacity. The equation can be used to solve a number of problems, namely to determine the temperature inside the body, to calculate the time required to reach the equilibrium, etc. Recent studies [26, 27] demonstrated the importance of studying the stability of heat equation. Another important problem is control over temperature conditions either by means of regulating the temperature at the ends of the modelled beam or by direct external regulation [28]. In our study, we used the heat equation to model the distribution of temperature within a beam.

We suggest a model of a beam with distributed hysteretic properties which takes into account the temperature effect. Section 2 provides an overview of hysteresis and the hysteresis model used in the study. Section 3 is dedicated to the deriving of the equation and describing the connection between the hysteretic beam vibrations and the heat equation. Section 4 describes the numerical methods used to solve the obtained equations. Section 5 describes the obtained results.

2 Hysteresis in Materials

As we have already mentioned, hysteresis occurs in a large number of physical systems. It means that when the external loads change, the system does not return to its original state. Instead, it continues changing until it reaches a new equilibrium. Another characteristic feature of hysteresis is that the system has a memory of its previous states. Hysteresis is modelled using two approaches: constructive and phenomenological.

The constructive approach involves designing a hysteresis model based on physical principles and mechanisms characteristic for the modelled system. Models obtained using this approach are applied in the modelling of materials under load. A basic model designed within the constructive approach is the play-operator [29]. It can be used to model the behaviour of materials under load with a certain degree of accuracy. However, this model does not take into account all the processes occurring within the material. A more complicated model within the constructive approach is the Prandtl-Ishlinskii model [12], which is based on the play-operator. This model is presented as a continuous flowchart and describes the behaviour of materials with greater accuracy.

The phenomenological approach can be used to describe hysteresis without taking into account the physical principles and mechanisms characteristic for the modelled system. Models obtained using this approach are often used to describe complex systems, when it is impossible to take into account the effect of external factors and all the physical processes within the system. In other words, a phenomenological model can be considered as a "black box", which reacts to an input signal with an output signal (Fig. 2). A well-known phenomenological model is the Bouc-Wen model developed by American scientists Bouc [30, 31] and Wen [32]. The model is based on the assumption that the hysteresis loop can be reduced to a series of

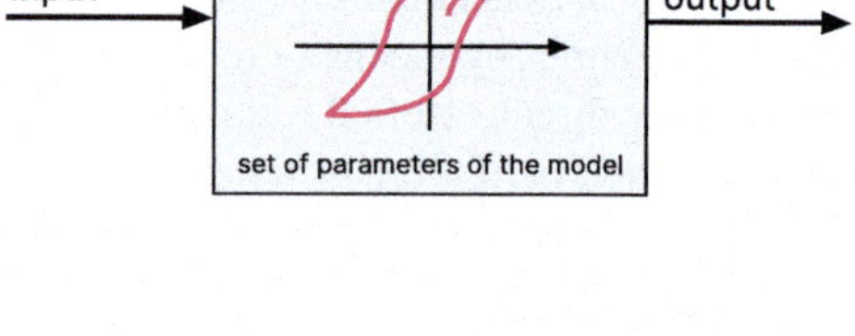

Fig. 2 Phenomenological approach to hysteresis modelling—Bouc-Wen model

harmonic vibrations. Each vibration is characterised by its own amplitude and phase. A complete hysteresis loop is obtained by summation of all harmonic vibrations. The model can be mathematically described as a first-order differential equation with four parameters which can regulate the shape and the size of the loop:

$$\dot{\Gamma} = \dot{x}\left\{A - \left[\beta \operatorname{sign}(\dot{x}\Gamma) + \gamma\right]|\Gamma|^n\right\} \tag{3}$$

where $x = x(t)$ is the input presented as a time-dependent function, $\Gamma = \Gamma(t)$ is the output presented as a time-dependent function (the response of the operator), and γ, n, β, and $A > 0$ are dimensionless parameters of the model. One of the problems is to identify the parameters of this model [33]. Since in some studies the only known parameter is the response of the operator to the external load, it is important to develop special algorithms for each specific task [34–37].

An advantage of the Bouc-Wen model is that it is flexible and can be adapted to various physical processes. Another advantage is that the model can be easily implemented using any mathematical software and has low computational cost, which makes it effective for engineering calculations. However, the Bouc-Wen model also has some drawbacks. For instance, it cannot be used to describe hysteretic properties in any frequency range or at any amplitude.

Both approaches have their strengths and drawbacks, and the choice depends on the problem that needs to be solved. Constructive models are usually more accurate than phenomenological ones. However, their design and software implementation can require more time and data. Therefore, due to the simplicity of implementation and application, in our study, we used the Bouc-Wen model.

3 Deriving a Hysteretic Beam Equation Taking into Account the Temperature Conditions

In this section, we obtain vibration equations of a beam taking into account the distributed hysteretic properties and temperature. The temperature conditions were modelled using classical heat equation (2). It is assumed that the material of the beam has hysteretic properties [38]. Then the dependence of normal stress on the deformation is nonlinear, i.e. it is described by the effect of the hysteresis operator on the deformation summand ε:

$$\sigma = \Gamma[\varepsilon]. \tag{4}$$

The bending moment with respect to the transverse vibrations axis, due to the normal stress in the cross-section Ω, is described as

$$M = \int_{\Omega} r\sigma dS = \int_{\Omega} r\Gamma[\varepsilon]dS \tag{5}$$

where r is the distance from the symmetry axis of the beam to the considered part of the beam's cross-section. Therefore, hysteretic properties are distributed along the whole beam, and at each point the material has hysteretic properties. Taking this into account, the equation of beam vibrations is presented as follows:

$$u_{tt} + \Gamma[u_{xx}]_{xx} = g(x, t), \tag{6}$$

where u is the beam deflection at point x at a moment t, Γ is the hysteretic operator (in our study, it is the Bouc-Wen operator (3), and $g(x, t)$ is the function characterising the external load.

Let us now consider the connection between the heat equation and the equation describing the hysteretic beam vibrations. It is understood that external factors provoke internal friction within the beam's material. Therefore, the energy of external vibrations is converted into the internal energy of the system (energy dissipation). In our study, we considered the conversion of energy into heat. Let's present the above-mentioned dependence using the dissipation function which describes the dissipative force as

$$F = \tau u_x^2 \tag{7}$$

where τ is the dissipation factor. The connection is implemented by inserting the dissipation function on the right-hand sides of the (2) and (6) as shown in Fig. 3.

Taking into account the fact that the energy is converted into heat (7), and considering the distributed hysteretic properties (6), zero initial conditions, the boundary conditions, and the heat (2), the final equation of the hysteretic beam taking into account the temperature is presented as follows:

$$\begin{cases} \omega_t - \omega_{xx} = \tau u_x^2, \\ u_{tt} + \Gamma[u_{xx}]_{xx} - \tau u_x^2 = g(x, t), \\ \omega(x, 0) = 1, \\ \omega(0, t) = \omega(L, t) = 1, \\ u(x, 0) = 0, \ u_t(x, 0) = 0, \\ u(0, t) = u_{xx}(0, t) = u(L, t) = u_{xx}(L, t) = 0, \\ t > 0, x \in [0, L]. \end{cases} \tag{8}$$

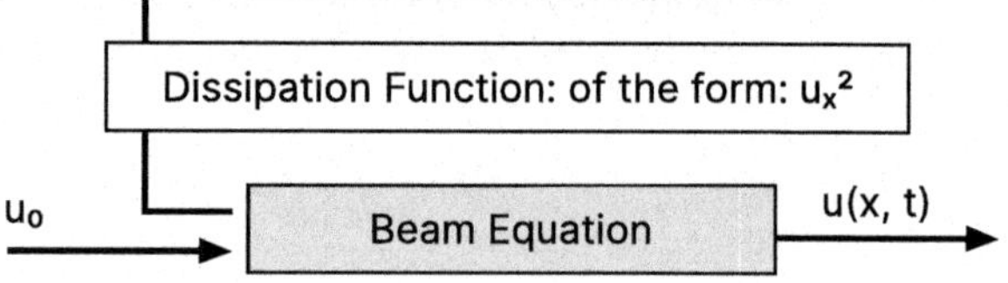

Fig. 3 Schema of the modelling coupled beam and heat equations

4 Numerical Solutions of the Equation

This section describes a solution of the vibration equation of the hysteretic beam taking into account temperature conditions (8). It is obvious that an analytical solution of the equation is non-trivial due to the summand describing the hysteretic properties. It is hard to find such a solution using common approaches. Therefore, in our study, we used a numerical solution. One of the well-known techniques used to solve partial differential equations involves the finite-difference methods. It involves discretisation of the spatial domain and the time interval into pre-set steps. Then, for each point of the discretised space, a solution of the initial equation is found. There are two schemes: implicit (the values of the grid on each temporal layer are based on the previous ones) and explicit (the values of the grid on each temporal layer are based on the neighbouring layers). In our study we used the explicit scheme (Fig. 4).

Let's consider a two-dimensional space domain $\mho = \{0 \le x \le \mathrm{L}, 0 \le t \le \mathrm{T}\}$ with a uniform grid. Let the grid steps be $h_x = L/N$ and $h_t = T/M$, where N and M are positive integers characterising the number of nods in the grid for the spatial and temporal coordinates respectively.[1] Let u_i^j denote the value of the function of the approximated solution at point $(x_i, t_j) = (h_x * i, h_t * j)$, where

Fig. 4 Region $\mho = \{0 \le x \le \mathrm{L}, 0 \le t \le \mathrm{T}\}$ with a M × N uniform grid. The white nodes are calculated by solving the equation. The black nodes are calculated based on the boundary conditions

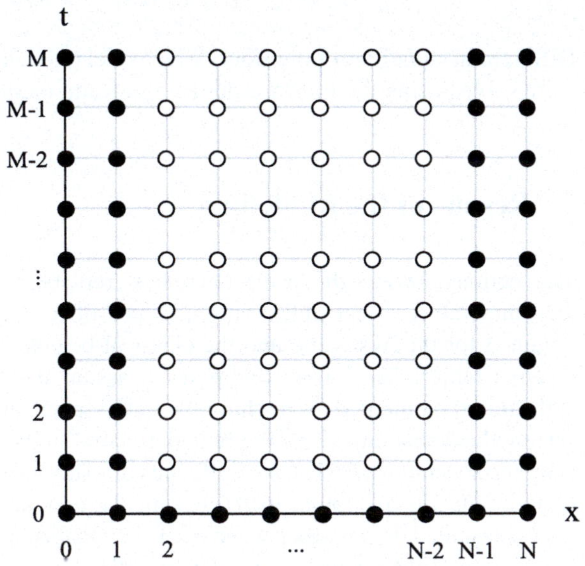

[1] We should note that the number of nodes in the grid for the temporal coordinate (value $M \gg N$) is significantly greater than that for the spatial coordinate in order to ensure the convergence of the explicit difference scheme.

$i \in (0, N-1)$, $j \in (0, M-1)$. By approximating the temporal derivative of the first order using an approximation operator $\Lambda_a f(\cdot)$:

$$\Lambda_t u = \frac{u_i^{j+1} - u_i^{j-1}}{2h_t}, \tag{9}$$

approximating the second-order temporal derivative:

$$\Lambda_{tt} u = \frac{u_i^{j-1} - 2u_i^j + u_i^{j+1}}{h_t^2}, \tag{10}$$

approximating the second-order spatial derivative:

$$\Lambda_{xx} u = \frac{u_{i-1}^j - 2u_i^j + u_{i+1}^j}{h_x^2}, \tag{11}$$

and approximating the second spatial derivative of the hysteresis operator:

$$\Lambda_{xx}\Gamma[u_{xx}] = \Lambda_{xx}\Gamma[\Lambda_{xx}u] = \frac{\Gamma[\Lambda_{xx}u]_i^{j-1} - 2\Gamma[\Lambda_{xx}u]_i^j + \Gamma[\Lambda_{xx}u]_i^{j+1}}{h_x^2}, \tag{12}$$

we obtain the following explicit difference scheme for (8):

$$\begin{cases} \Lambda_t \omega - \Lambda_{xx}\omega = (\tau \Lambda_x u)^2, \\ \Lambda_{tt} u + \Lambda_{xx}\Gamma[\Lambda_{xx}u] - (\tau\Lambda_x u)^2 = g \end{cases}. \tag{13}$$

Differential equations of the Bouc-Wen model (calculating the response to the input) were solved using the fourth-order Runge-Kutta method.

5 Results of the Modelling

The section describes the results of the hysteretic beam vibrations taking into account the temperature effect under various types of loads. It also compares the results obtained for the hysteretic and the classical beams. The modelling parameters are given in Table 1. The initial conditions are assumed to be zero (the beam is stationary at the initial moment, does not have the initial speed, and is fixed at both ends). When seismic load was modelled, the boundary conditions included a function characterising the seismic wave. The studied types of loads were periodic load (presented as a periodic function), impulse load (presented as a Dirac-like distribution), and seismic load (presented as a mother wavelet of the Daubechies wavelets). Let us begin the description of the obtained results by modelling the effect of periodic loads.

Table 1 Parameters of computational experiments

Parameter	Value
L—the length of the beam	1
T—modelling time	20
a^2—the coefficient in the beam equation	1
h_t—discretisation step for the temporal coordinate	0.0125
h_x—discretisation step for the spatial coordinate	0.05
A—the Bouc-Wen operator parameter	1
β—the Bouc-Wen operator parameter	0.2
γ—the Bouc-Wen operator parameter	0.5
n—the Bouc-Wen operator parameter	0.2
τ—the dissipation factor for the dissipation function	1

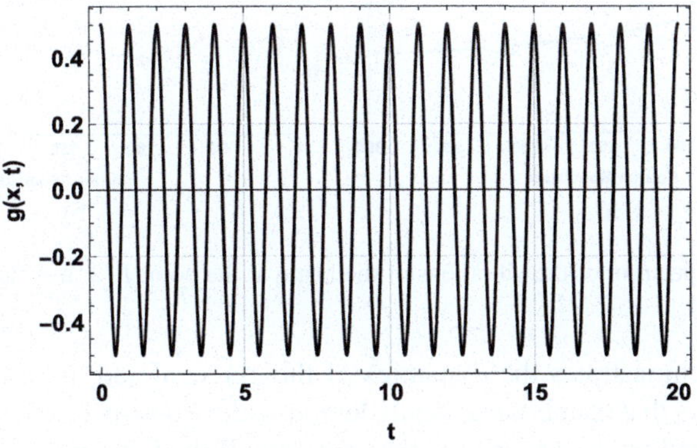

Fig. 5 Periodic load function graph depending on time at $L/2$ point

5.1 *Periodic Loads*

The section describes the behaviour of the beam under periodic loads. It is known that periodic loads occur due to monotonous movement along the beam (for instance, works conducted in the building and vehicle traffic). In our study, we used the following trigonometric function of one variable as a periodic load function (Fig. 5):

$$g(x, t) = \frac{1}{2}\cos(2\pi t); \tag{14}$$

the effect of the load depends on time only and is constant along the length of the beam.

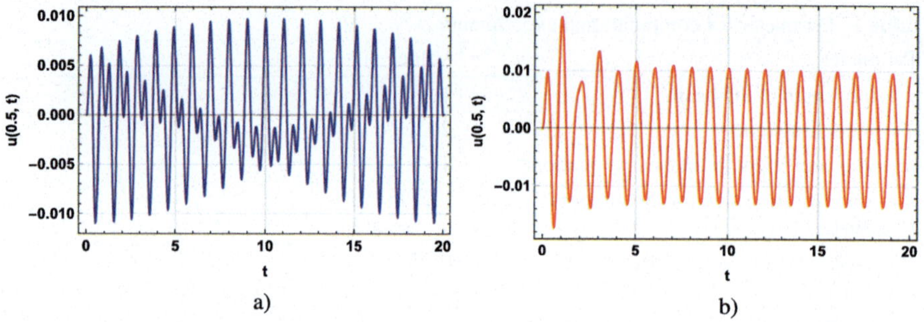

Fig. 6 Vibration graphs at the point $L/2$ depending on time: the blue curve denotes the classical beam, and the red curve denotes the Bouc-Wen model

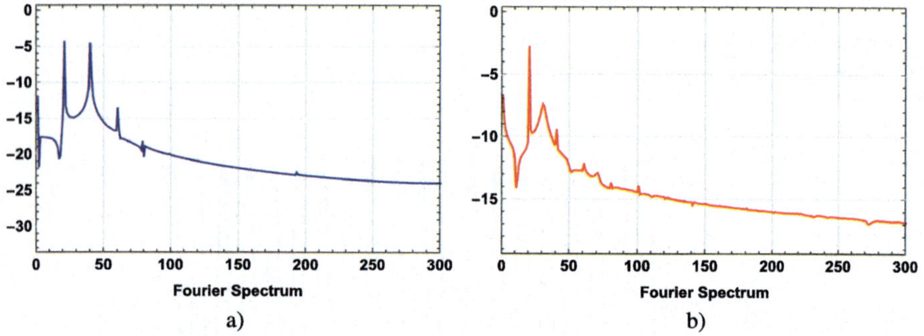

Fig. 7 Fourier spectrum of the vibrations of the beam at the point $L/2$: **a** – classical model; **b** – Bouc-Wen model

Figure 6 demonstrates the dynamics of the classical and the hysteretic beams. Figure 6a shows that the classical beam demonstrates quasi-periodic vibrations. This is also confirmed by the analysis of the spectrum of the Fourier transform (Fig. 7a), which demonstrated two prevailing frequencies. The hysteretic beam, on the contrary, demonstrated periodic vibrations (Fig. 6b). The analysis of the spectrum of the Fourier transform (Fig. 7b) demonstrated that the hysteretic beam helps to reduce quasi-periodic vibrations of the second frequency almost to zero.

Figure 9 demonstrates the distribution of temperature along the length of the beam over time. As we can see, the extreme temperatures are different for the classical (Fig. 9a) and the hysteretic (Fig. 9b) beams. The frequency of temperature changes is higher for the classical beam due to the quasi-periodic nature of vibrations. The graphs show that the maximum temperature is observed in the middle of the beam and gradually reduces towards the ends. This is accounted for by the fact that we used thermal equilibrium ratios (the temperature at the boundaries of the beam is assumed to be constant).

The analysis of the hysteresis curve (Fig. 8) demonstrated that the energy obtained from external vibrations is dissipated and the amount of the dissipated energy equals the area of the loop. Each cyclic change in the ratio of deformation and stress facili-

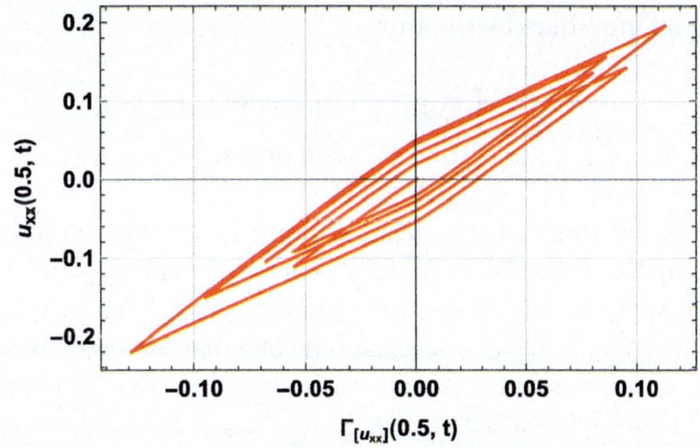

Fig. 8 Parametric graphs of hysteresis loops (time t as a parameter): input-output dependence of the input u_{xx} on the output $\Gamma[u_{xx}]$ at the point $L/2$

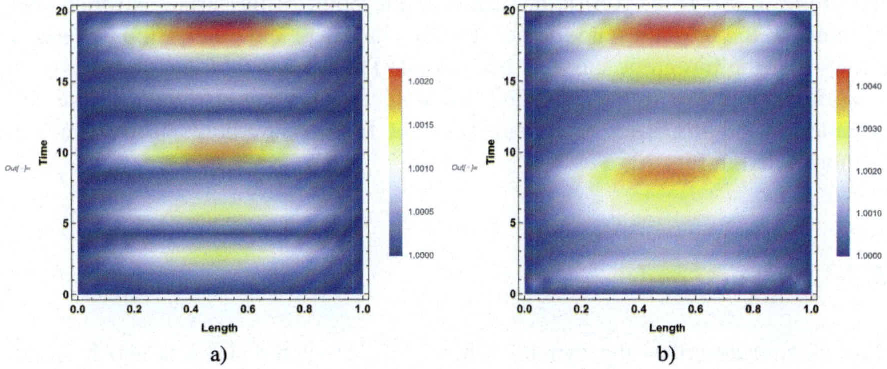

Fig. 9 Temperature distribution along the length of the beam at a certain period of time. The histogram of the temperature distribution is shown to the right of the corresponding graph. **a** is the classical model; **b** is the model with distributed hysteresis properties

tates the dissipation of energy within each cycle. For the classical beam, this dependence is linear (presented as a line with a certain slope angle). This is why the classical beam does not take into account the dissipation of energy during the deformation of the material.

5.2 *Impulse Loads*

This section analyses the behaviour of the beam under impulse loads. Impulse load is usually considered to be an explosion or a strike. In our study we considered the latter one. This type of load is characterised by a short-term (impulse) impact (for instance, when an object falls on the beam and a hammer strike). The impact is modelled using a Dirac-like distribution:

$$g(x,t) = \begin{cases} F_{impulse}, & \text{if } t = 0 \text{ and } x = 0 \\ 0, & \text{otherwise} \end{cases} \tag{15}$$

where $F_{impulse}$ is the force of the strike at point (x, t). We should note that here the parameters of the Bouc-Wen model (3) are different from those described in Table 1. Namely, we used the following parameters: $A = 0.7$, $\beta = 0.1$, and the others remained the same. We should also consider the fact that the load is momentary and to make sure that the equation remains convergent over time, the dissipation function (7) was used with the dissipation factor of $\tau = 10^{-4}$.

The load (15) is applied at a moment $t = 0.5$ to a point $x = 1.5$ with a force of $F_{impulse} = 300$. The result of the impact is presented in Fig. 10. Figure 10a shows that after the strike, the classical beam starts vibrating at a constant frequency. The hysteretic model (Fig. 10b) is deformed at the point of the strike and then seeks to return to the original state (demonstrating elasticity) and finally stops. Figure 11 shows that the hysteretic curve is not closed. This means that the structure of the material changed or it does not have any defects. This can also indicate the deterioration of the properties of the material. The temperature fields of the beams are demonstrated in Fig. 12.

5.3 Seismic Loads

This section describes the dynamic characteristics of the classical and hysteretic beams under seismic loads. A mother wavelet of the Daubechies wavelets was used as the load function because it resembles short-term seismic impact. It is known that seismic load affects the bearing of the beam. Therefore, zero boundary conditions are presented as follows (Fig. 13):

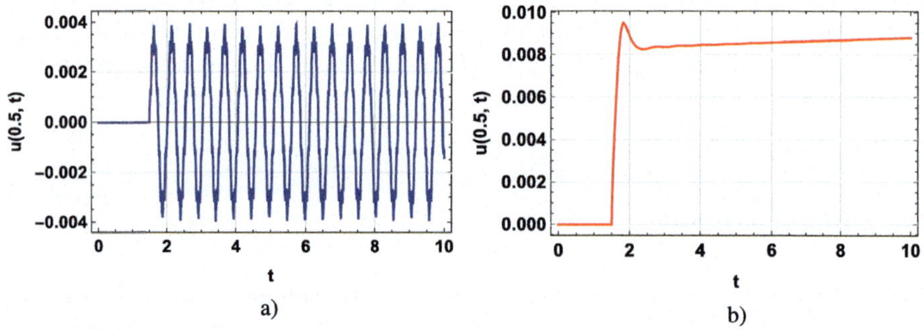

Fig. 10 Vibration graphs at the point $L/2$ depending on time: the blue curve denotes the classical beam, and the red curve denotes the Bouc-Wen model

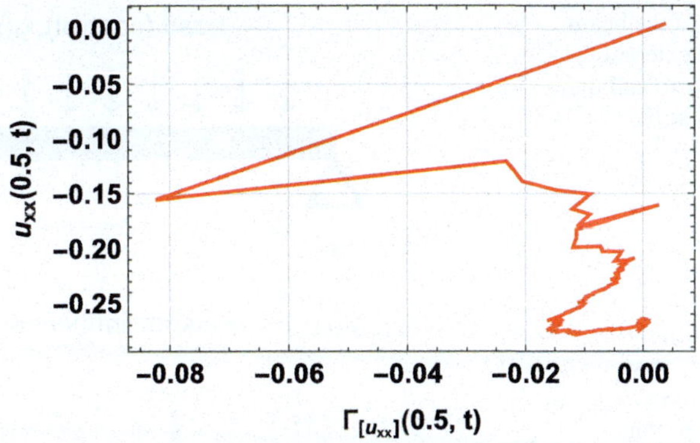

Fig. 11 Parametric graphs of hysteresis loops (time t as a parameter): input-output dependence of the input u_{xx} on the output $\Gamma[u_{xx}]$ at the point $L/2$

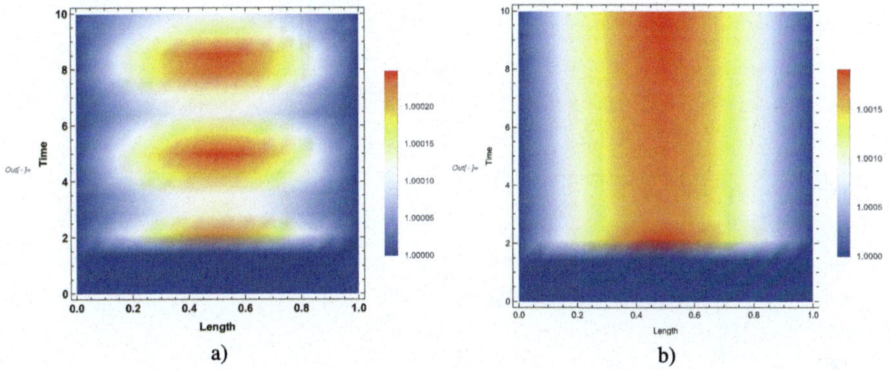

Fig. 12 Temperature distribution along the length of the beam at a certain period of time. The histogram of the temperature distribution is shown to the right of the corresponding graph. **a** is the classical model; **b** is the model with distributed hysteresis properties

$$u(0, t) = u(L, t) = S(t) \tag{16}$$

where $S(t)$ is the function characterising seismic load. In numerical experiments, we used a mother wavelet of the Daubechies wavelets with parameters $a = 6, b = 4$ (Fig. 14).

Figure 15 demonstrates the behaviour of the beams under seismic load applied to the middle. We can see that for the classical beam (Fig. 15a), the maximum amplitude is 1.5. It corresponds to the maximum seismic load (Fig. 14). The amplitude of the hysteretic beam is larger—2.1. We can see that the beams react differently to seismic load. The classical beam reproduces the structure of the seismic wave. The hysteretic beam has a different structure: initially the vibrations repeat the structure of the seismic wave (in the same way as the classical beam), but then they start to

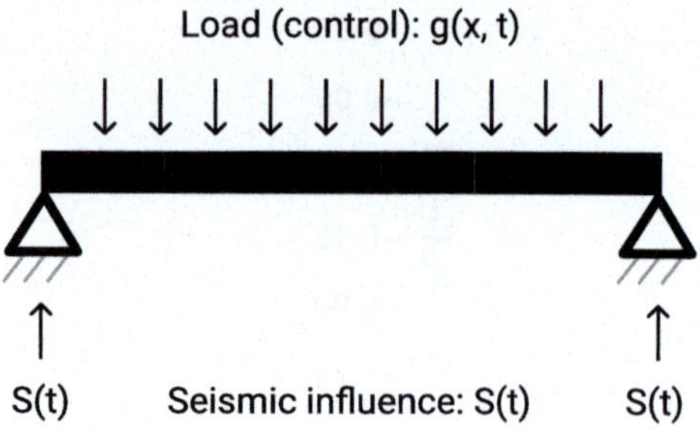

Fig. 13 Pinning of the beam depending on the external load $g(x, t)$ (control) and the seismic load $S(t)$ at the attachment points

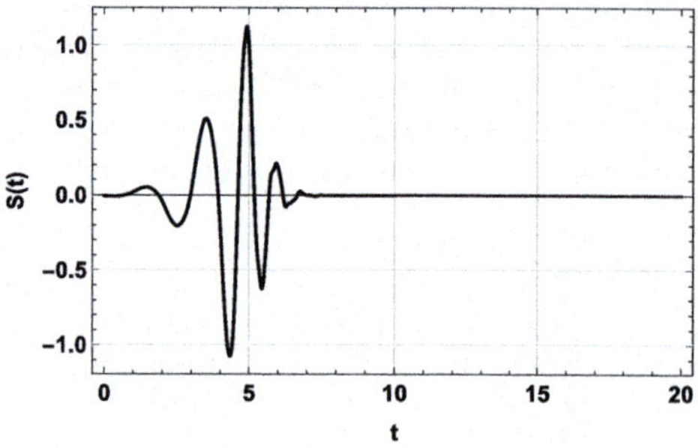

Fig. 14 Seismic load function (the Daubechies mother wavelet of the 10th order with parameters $a = 1$, $b = 9$) graph depending on the time

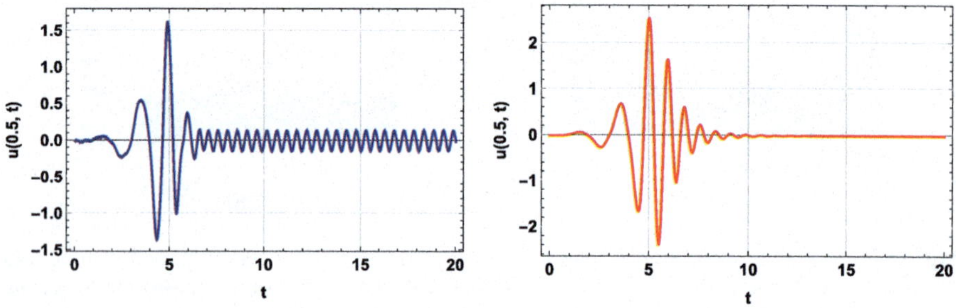

a) b)

Fig. 15 Vibration graphs at the point $L/2$ depending on time: the blue curve denotes the classical beam, the red curve denotes the Bouc-Wen model

slow down and finally stop. The classical beam, on the contrary, vibrates at a constant frequency.

The distribution of temperature along the beam is demonstrated in Fig. 17. The highest temperature is registered during the impact of the seismic wave and is concentrated in the middle of the beam. In Fig. 16, we show the hysteresis curve which corresponds to the hysteresis beam under seismic load. As was pointed out above, the area of the loop determines the dissipated energy. Comparing the loop in Fig. 8, the dissipation in this case is sufficiently higher.

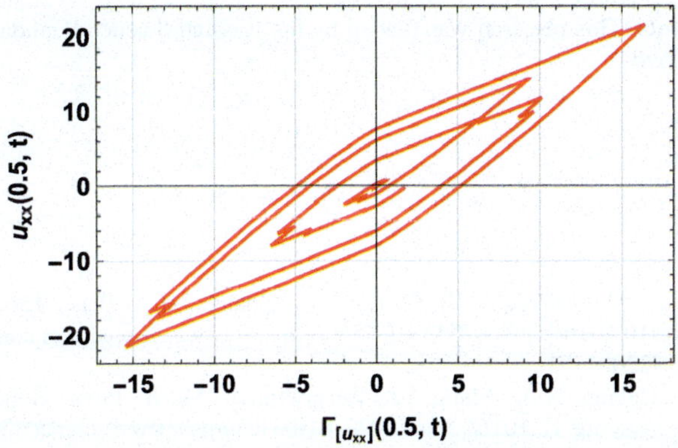

Fig. 16 Parametric graphs of hysteresis loops (time t as a parameter): input-output dependence of the input u_{xx} on the output $\Gamma[u_{xx}]$ at the point $L/2$

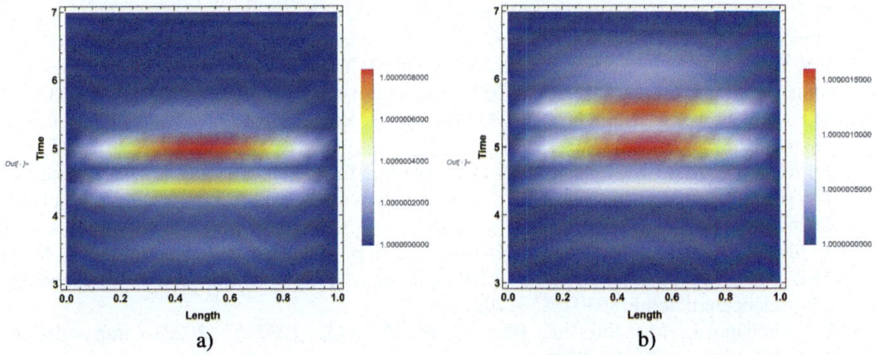

Fig. 17 Temperature distribution along the length of the beam at a certain period of time. The histogram of the temperature distribution is shown to the right of the corresponding graph. **a** is the classical model; **b** is the model with distributed hysteresis properties

Conclusions

The article considered a classical model of beam vibrations and a model taking into account distributed hysteretic properties and temperature under external load of various types: impulse load, periodic load, and seismic load. The hysteretic properties were modelled using a modification of a classical equation. The Bouc-Wen model was used as a hysteron. The connection between the heat equation and the beam vibration equation is established using the dissipation function. The numerical experiments demonstrated that the hysteretic beam behaves differently from the classical beam. Temperature distribution for the hysteretic beam is also different. The obtained results can be used to design and model structural systems for seismic hazard zones as well as structural systems exposed to periodic and impulse loads.

Acknowledgements This research was funded by the Russian Science Foundation (RNF) grant number 23-29-00696.

References

1. S. Lu, J. Zhao, M. Huang, Z. Li, G. Kang, X. Zhang, Int. J. Plast. **156**, 103356 (2022). https://doi.org/10.1016/j.ijplas.2022.103356, https://www.sciencedirect.com/science/article/pii/S0749641922001371
2. S. Xu, J.Y. Cheng, N.A. Mara, I.J. Beyerlein, J. Mech. Phys. Solids **168**, 105031 (2022). https://doi.org/10.1016/j.jmps.2022.105031, https://www.sciencedirect.com/science/article/pii/S0022509622002083
3. M.E. Semenov, E.A. Karpov, S.G. Tikhomirov, P.A. Meleshenko, M. Teplyakova, in *Vibration Engineering and Technology of Machinery*. ed. by J.M. Balthazar (Springer International Publishing, Cham, 2021), pp. 3–36
4. M.E. Semenov, S.V. Borzunov, P.A. Meleshenko, A.V. Lapin, Mathematics **10**(18) (2022). https://doi.org/10.3390/math10183262, https://www.mdpi.com/2227-7390/10/18/3262
5. S.V. Borzunov, M.E. Semenov, N.I. Sel'vesyuk, P.A. Meleshenko, A.M. Solovyov, Math. Model. Comput. Simul. **14**(2), 305 (2022). https://doi.org/10.1134/S207004822202003X
6. M.E. Semenov, A.M. Solovyov, P.A. Meleshenko, J. Vib. Control **27**(1–2), 43 (2021). https://doi.org/10.1177/1077546320923436
7. A. Meibodi, N.A. Alexander, Soil Dyn. Earthq. Eng. **135**, 106160 (2020). https://doi.org/10.1016/j.soildyn.2020.106160, https://www.sciencedirect.com/science/article/pii/S0267726119309728
8. J. Jiao, Y. Chang, S. Kobayashi, X. Liu, G. Li, B. Wu, C. He, J. Magn. Magn. Mater. **554**, 169310 (2022). https://doi.org/10.1016/j.jmmm.2022.169310, https://www.sciencedirect.com/science/article/pii/S0304885322002578
9. E. Jurčišinová, M. Jurčišin, Phys. Lett. A **443**, 128225 (2022). https://doi.org/10.1016/j.physleta.2022.128225, https://www.sciencedirect.com/science/article/pii/S0375960122003073
10. A. Solovyov, M. Semenov, P. Meleshenko, O. Reshetova, M. Popov, E. Kabulova, Procedia Eng. **201**, 578 (2017). https://doi.org/10.1016/j.proeng.2017.09.634, https://www.sciencedirect.com/science/article/pii/S187770581734167X. 3rd International Conference "Information Technology and Nanotechnology", ITNT-2017, Samara, Russia
11. M.E. Semenov, P.A. Meleshenko, S.V. Borzunov, O.O. Reshetova, A.I. Barsukov, Micromachines **14**(2) (2023). https://doi.org/10.3390/mi14020310, https://www.mdpi.com/2072-666X/14/2/310
12. M. Krasnosel'skii, M. Niezgodka, A. Pokrovskii, *Systems with Hysteresis* (Springer, Berlin, 2012). https://books.google.ru/books?id=3tH6CAAAQBAJ
13. S. Timoshenko, *History of Strength of Materials: With a Brief Account of the History of Theory of Elasticity and Theory of Structures*. Dover Civil and Mechanical Engineering Series (Dover Publications, 1983). https://books.google.ru/books?id=tkScQmyhsb8C
14. F. Giunta, A. Cicirello, Eng. Struct. **179**, 66 (2019). https://doi.org/10.1016/j.engstruct.2018.10.036, https://www.sciencedirect.com/science/article/pii/S014102961831071X
15. S. Malik, D.K. Singh, G. Bansal, V. Paliwal, A.R. Manral, Mater. Today: Proc. **46**, 10725 (2021). https://doi.org/10.1016/j.matpr.2021.01.548, https://www.sciencedirect.com/science/article/pii/S2214785321006398. International Conference on Technological Advancements in Materials Science and Manufacturing
16. Y. Miao, Y. Shi, H. Luo, R. Gao, Appl. Math. Model. **54**, 21 (2018). https://doi.org/10.1016/j.apm.2017.09.040, https://www.sciencedirect.com/science/article/pii/S0307904X17305930 doi:https://doi.org/10.1016/j.apm.2017.09.040.https://www.sciencedirect.com/science/

article/pii/S0307904X17305930 10.1016/j.apm.2017.09.040. https://www.sciencedirect.com/science/article/pii/S0307904X17305930
17. S. Borjigin, C.W. Kim, K.C. Chang, K. Sugiura, Eng. Struct. **176**, 500 (2018). https://doi.org/10.1016/j.engstruct.2018.09.014, https://www.sciencedirect.com/science/article/pii/S0141029618304760
18. M. Ghadiri, S.H.S. Hosseini, Compos. Part B: Eng. **173**, 106928 (2019). https://doi.org/10.1016/j.compositesb.2019.106928, https://www.sciencedirect.com/science/article/pii/S1359836819309928
19. G. Qiao, S. Rahmatalla, J. Sound Vib. **491**, 115771 (2021). https://doi.org/10.1016/j.jsv.2020.115771, https://www.sciencedirect.com/science/article/pii/S0022460X20306003
20. X. Zhang, D. Thompson, X. Sheng, J. Sound Vib. **481**, 115432 (2020). https://doi.org/10.1016/j.jsv.2020.115432, https://www.sciencedirect.com/science/article/pii/S0022460X20302649
21. T. Mazilu, Procedia Eng.**199**, 2561 (2017). https://doi.org/10.1016/j.proeng.2017.09.327, https://www.sciencedirect.com/science/article/pii/S1877705817338122. X International Conference on Structural Dynamics, EURODYN 2017
22. M. Krstic, J.M. Wang, in *Proceedings of the 2011 American Control Conference* (2011), pp. 2356–2361. https://doi.org/10.1109/ACC.2011.5990607
23. B. Painter, G. Ferrari, M. Amabili, Nonlinear Dyn. **108**(4), 2903 (2022). https://doi.org/10.1007/s11071-022-07458-8
24. A. Krysko, J. Awrejcewicz, I. Kutepov, V. Krysko, Int. J. Non-Linear Mech. **94**, 207 (2017). https://doi.org/10.1016/j.ijnonlinmec.2016.12.004, https://www.sciencedirect.com/science/article/pii/S0020746216303869. A Conspectus of Nonlinear Mechanics: A Tribute to the Oeuvres of Professors G. Rega and F. Vestroni
25. A. Warminska, E. Manoach, J. Warminski, S. Samborski, Contin. Mech. Thermodyn. **27**(4), 719 (2015). https://doi.org/10.1007/s00161-014-0381-6
26. X. Zong, J.F. Zhang, J. Zheng, Z. Liu, G. Ma, Appl. Math. Comput. **437**, 127511 (2023). https://doi.org/10.1016/j.amc.2022.127511, https://www.sciencedirect.com/science/article/pii/S0096300322005859
27. L. Wang, P. Lei, Q. Wu, Syst. Control Lett. **171**, 105417 (2023). https://doi.org/10.1016/j.sysconle.2022.105417, https://www.sciencedirect.com/science/article/pii/S0167691122001943
28. S. Ervedoza, K. Le Balc'h, M. Tucsnak, J. Funct. Anal. **283**(10), 109666 (2022). https://doi.org/10.1016/j.jfa.2022.109666, https://www.sciencedirect.com/science/article/pii/S0022123622002865
29. A. Visintin, in *The Science of Hysteresis*, ed. by G. Bertotti, I.D. Mayergoyz (Academic Press, Oxford, 2006), pp. 1–123. https://doi.org/10.1016/B978-012480874-4/50004-X, https://www.sciencedirect.com/science/article/pii/B978012480874450004X
30. R. Bouc, *Proceedings of the Fourth Conference on Nonlinear Oscillations* (Prague, 1967). https://cir.nii.ac.jp/crid/1570291225556669696
31. R. Bouc, *Modèle mathématique d'hystérésis. Application aux systèmes à un degré de liberté* (R. Bouc, 1969). https://books.google.ru/books?id=lAdmQwAACAAJ
32. Y.K. Lin, G.Q. Cai, Nonlinear dynamics, in *Engineering Systems*. ed. by W. Schiehlen (Springer, Berlin, 1990), pp. 189–196
33. A. Charalampakis, V. Koumousis, in *Parameters of Bouc-Wen Model Revisited* (2010)
34. M. Lin, B. Sun, C. Cheng, B. Zhao, Z. Peng, G. Meng, J. Sound Vib. **538**, 117242 (2022). https://doi.org/10.1016/j.jsv.2022.117242, https://www.sciencedirect.com/science/article/pii/S0022460X22004291
35. N. Karabutov, A. Shmyrin, IFAC-PapersOnLine **53**(2), 1225 (2020). https://doi.org/10.1016/j.ifacol.2020.12.1339, https://www.sciencedirect.com/science/article/pii/S2405896320317456. 21st IFAC World Congress
36. N.N. Son, C. Van Kien, H.P.H. Anh, Eng. Appl. Artif. Intell. **87**, 103317 (2020). https://doi.org/10.1016/j.engappai.2019.103317, https://www.sciencedirect.com/science/article/pii/S0952197619302702

37. X.B. Nguyen, T. Komatsuzaki, H.T. Truong, Int. Mech. Sci. **213**, 106848 (2022). https://doi.org/10.1016/j.ijmecsci.2021.106848, https://www.sciencedirect.com/science/article/pii/S0020740321005701
38. P. Krejci, Int. J. Non-Linear Mech. **37**(8), 1337 (2002). https://doi.org/10.1016/S0020-7462(02)00022-7, https://www.sciencedirect.com/science/article/pii/S0020746202000227 10.1016/S0020-7462(02)00022-7. https://www.sciencedirect.com/science/article/pii/S0020746202000227

Torsion-Bending Coupling in Nonlinear Model Order Reduction

Lucas G. C. Silveira and Morten H. Hansen

Abstract This study investigates the performance of two nonlinear model order reduction techniques, a quadratic manifold and a linear basis augmented with modal derivatives, in predicting the coupling of torsion and bending in a mechanical structure. We demonstrate that the quadratic manifold, which assumes a linear variation of the modal derivatives, overlooks the presence of modal hybridisations. Conversely, the linear basis augmented with modal derivatives successfully captures both the vertical and large angular displacements, even in the presence of modal veerings. We further introduce a simple yet practical selection algorithm for optimising the use of modal derivatives, although this approach reveals limitations in terms of its assumptions and error definition. This investigation highlights the potential of the linear basis method augmented with modal derivatives and prompts further research into improved selection strategies and error definitions, contributing to the advancement of more precise and efficient model order reductions.

1 Introduction

In many structural designs, the torsion-bending coupling phenomenon is important for optimising external loads like aerodynamic forces. It's crucial for a reduced-order model to accurately predict the twist angle of the deformed structure, such as a wind turbine blade. However, linear modal truncations do not capture the variation of the modal basis, and quadratic manifolds are unable to capture nonlinear variations of the modal basis. This paper demonstrates that a linear modal basis supplemented with modal derivatives can successfully predict the twist angle, even when confronted with nonlinear variation in the modal basis like veered mode shapes. This is achieved by

L. G. C. Silveira (✉) · M. H. Hansen
Department of Mechanical and Electrical Engineering, University of Southern Denmark, Alsion 2, 6400, Sønderborg, Denmark
e-mail: silveiralucas@sdu.dk

M. H. Hansen
e-mail: mortenhhansen@sdu.dk

M. Belhaq (ed.), *Advances in Nonlinear Dynamics and Control of Mechanical and Physical Systems*, Springer Proceedings in Physics 301,
https://doi.org/10.1007/978-981-99-7958-5_20

treating the contributions of linear modes and modal derivatives as separate degrees of freedom.

When conducting dynamic simulations for large-scale models, structural components are often represented using simplified models, such as modal truncations or beam elements, due to the prohibitive computational demands associated with the direct utilisation of high-fidelity models. Reduced-order models are approximate models that leverage the reduction of high-fidelity models to retain their essential characteristics, while significantly curtailing computational expense. When used in dynamic analysis, they have the potential to enhance simulation fidelity and narrow the fidelity gap between models utilised by load simulation engineers and structural component designers.

The usage of model order reduction methods based on the linear projection onto modal coordinates, notably the Craig-Bampton method [4], has been in practice since the 1960s. However, the presence of nonlinear effects requires the extension of these classical methods. This need led to the proposal of modal derivatives by Idelsohn and Cardona [5]. Jain [6] and Rutzmozer [9] recommended representing the modal basis via a second-order Taylor expansion, with quadratic terms provided by the modal derivatives. It was demonstrated by Jain [6] that the quadratic manifold is a subset of the more general augmented linear basis with modal derivatives "where the amplitudes of the modal derivatives are taken as predefined functions of the amplitudes of the mode shapes". This enhanced linear basis offers increased flexibility in determining the contributions from both modes and modal derivatives. Silveira [10] showed that when veered modes constitute a substantial part of a structure's deflection, the quadratic manifold might be inadequate in accurately capturing the system dynamics. Modal veering poses an inherent limitation to quadratic manifolds but not to modal derivative augmented linear basis.

2 Theoretical Framework

2.1 Modal Derivatives

The dynamic behaviour of a structure, as determined by a large displacement finite element code that employs the Total Lagrange formulation [2], is described by a second-order ordinary differential equation system of the form:

$$\mathbf{M}\ddot{\mathbf{u}} + \mathbf{C}\left(\mathbf{u}\right)\dot{\mathbf{u}} + \mathbf{f}\left(\mathbf{u}\right) = \mathbf{g}\left(t\right) \tag{1}$$

where $\mathbf{u}$ denotes the generalised degrees of freedom of the system, $\mathbf{f}$ represents the generalised nonlinear internal elastic force vector in the reference configuration, $\mathbf{g}$ signifies the forcing vector due to external loads, and $\mathbf{C}$ is the damping matrix. The equations of motion in the Total Lagrange formulation are derived with respect to the undeformed or reference configuration, thereby leading to the mass matrix **M** being generally constant [2].

In contrast to linear systems, the modal properties of a nonlinear system depend on its current configuration and can change as the state of the system changes. For instance, in the presence of large deformations or rotations, the system's stiffness matrix is state-dependent. This means the modal properties can change drastically as the system undergoes large deformations. Modal derivatives quantify this change in relation to state parameters.

In the context of geometric nonlinear structural systems, modal derivatives usually describe the variation in vibration modes as a function of the system's deflection state, particularly in the direction of a specific mode. More specifically, a modal derivative represents how much a vibration mode changes as we perturb the system's deflection state in the direction of that particular mode.

The modal derivatives can be obtained by solving a closed set of equations derived from the system's equations of motion and the normalisation condition for its mode shapes:

$$\begin{bmatrix} \left(\mathbf{K}|_{eq} - \omega_i^2|_{eq}\,\mathbf{M}\right) & \mathbf{M}\,\boldsymbol{\phi}_i|_{eq} \\ \left(\mathbf{M}\,\boldsymbol{\phi}_i|_{eq}\right)^T & 0 \end{bmatrix} \begin{bmatrix} \left.\frac{\partial \boldsymbol{\phi}_i}{\partial q_j}\right|_{eq} \\ \left.\frac{\partial \omega_i^2}{\partial q_j}\right|_{eq} \end{bmatrix} = \begin{bmatrix} -\left.\frac{\partial \mathbf{K}}{\partial q_j}\right|_{eq} \boldsymbol{\phi}_i|_{eq} \\ 0 \end{bmatrix} \tag{2}$$

More detailed information about the derivation and calculation of modal derivatives can be found in [10].

2.2 *Quadratic Manifold*

When dealing with small displacements in structural dynamics, it is generally assumed that both the stiffness matrix and modal properties remain constant. This allows the displacements to be represented by a linear superposition of mode shapes

$$u_i\left(\mathbf{q}\right) = \Phi_{il} q_l \tag{3}$$

Here, the modal amplitudes **q**, often referred to as *modal coordinates*, represent the magnitudes of the mode shapes in the system. However, when dealing with finite displacements, the linear relationship in (3) does not lead to good results. To improve the approximation of system dynamics in the presence of finite displacements, we can consider a second-order Taylor expansion:

$$u_i\left(\mathbf{q}\right) = \Phi_{il} q_l + \frac{1}{2}\Theta_{ilm} q_m q_l \tag{4}$$

where $\mathbf{\Theta}$ is the symmetric component of the modal derivatives; the antisymmetric part does not contribute to the deflection because of the symmetry of the product

$q_m q_l$. As detailed in [6, 9, 10], inserting the quadratic manifold assumption in (1), the structure equation of motion can be written in terms of the modal coordinates using a Galerkin projection as

$$\mathbf{P}_\Gamma^T \mathbf{M} \mathbf{P}_\Gamma \ddot{\mathbf{q}} + \mathbf{P}_\Gamma^T (\mathbf{M}\boldsymbol{\Theta}\dot{\mathbf{q}}) \dot{\mathbf{q}} + \mathbf{P}_\Gamma^T \mathbf{C} \mathbf{P}_\Gamma \dot{\mathbf{q}} + \mathbf{P}_\Gamma^T \mathbf{f} = \mathbf{P}_\Gamma^T \mathbf{g} \tag{5}$$

where the tangent (i.e. state-dependent) projection $\mathbf{P}_\Gamma$ is given by

$$\mathbf{P}_\Gamma = \boldsymbol{\Phi} + \boldsymbol{\Theta}\mathbf{q} \tag{6}$$

A critical underlying assumption for the quadratic manifold, as presented in (4), is that the mode shapes vary linearly with modal coordinates. Therefore, the validity and applicability of this model greatly depend on how well this assumption holds for the specific system under consideration.

2.3 Linear Basis Augmented with Modal Derivatives

A different, but more general, approach is to augment the linear basis by including normalised modal derivatives

$$\mathbf{R} = \left[\frac{\boldsymbol{\phi}_1}{\|\boldsymbol{\phi}_1\|}, \frac{\boldsymbol{\phi}_2}{\|\boldsymbol{\phi}_2\|}, \ldots, \frac{\boldsymbol{\phi}_n}{\|\boldsymbol{\phi}_n\|}, \frac{\boldsymbol{\Theta}_{11}}{\|\boldsymbol{\Theta}_{11}\|}, \frac{\boldsymbol{\Theta}_{12}}{\|\boldsymbol{\Theta}_{12}\|}, \ldots, \frac{\boldsymbol{\Theta}_{nn}}{\|\boldsymbol{\Theta}_{nn}\|} \right] \tag{7}$$

The quadratic manifold is in effect a subset of the subspace defined by the projection basis $\mathbf{R}$ [6]. The linear basis augmented with modal derivatives in effect substitutes the product $q_m q_l$ in (4) by additional degrees of freedom.

The modal derivatives are in general not orthogonal to the mode shapes. The singular value decomposition provides a way of obtaining an orthonormal basis $\mathbf{U}$ equivalent to $\mathbf{R}$:

$$\mathbf{R} = \mathbf{U}\boldsymbol{\Sigma}\mathbf{V}^T \tag{8}$$

$$\mathbf{u} = \mathbf{U}\mathbf{q}_U \tag{9}$$

Using $\mathbf{U}$ as the orthogonal basis of a Galerkin projection by inserting (9) into (1) and premultiplying it by $\mathbf{U}^T$, one obtains

$$\overline{\mathbf{M}}\mathbf{q} + \overline{\mathbf{C}}\mathbf{q} + \overline{\mathbf{f}}(\mathbf{q}) = \overline{\mathbf{g}}(t) \tag{10}$$

where the subscript of $\mathbf{q}_U$ has been omitted to improve readability. The reduced mass matrix $\overline{\mathbf{M}}$, damping matrix $\overline{\mathbf{C}}$, elastic forces vector $\overline{\mathbf{f}}$, and external forces vector $\overline{\mathbf{g}}$ are given by

$$\overline{\mathbf{M}} = \mathbf{U}^T\mathbf{M}\mathbf{U}, \quad \overline{\mathbf{C}} = \mathbf{U}^T\mathbf{C}\mathbf{U}, \qquad \overline{\mathbf{f}} = \mathbf{U}^T\mathbf{f}, \qquad \overline{\mathbf{g}} = \mathbf{U}^T\mathbf{g} \tag{11}$$

2.3.1 Sequential Backward Selection of Modal Derivatives

A linear basis augmented with modal derivatives provides a convenient approach to reducing computation time. However, the introduction of modal derivatives quickly increases the size of the basis. If n denote the number of mode shapes, the resulting number of modal derivatives is $(n^2 + n)/2$. This rapid increase in basis size calls for an efficient method to optimise the basis.

It is noteworthy that not all modal derivatives are equally useful; some contribute more to the computational accuracy than others. In this context, a sequential backward selection procedure provides an economical means of reducing the basis size. This method, based on a stepwise regression approach, eliminates, one at a time, the modal derivatives which impact the least on the projection error, thus systematically shrinking the basis size while preserving its essential features.

Assuming that the static solution can aid in optimising the reduction basis proves to be a reasonable and effective strategy. Minimising the error in static deflection provides a simple criterion for selecting which modal derivatives to include in the reduced model. This optimisation approach contributes to enhancing the computational efficiency of dynamic simulations, while ensuring a high level of accuracy. The procedure for the sequential backward selection of modal derivatives is outlined in Algorithm 1.

Algorithm 1 Backward Elimination

```
procedure BACKWARDELIMINATION(R, sol)
    while error ≤ criteria do
        for j in 1 to n_MD remaining do
            Construct R_temp by deleting MD_j from R
            Perform SVD on R_temp to get V
            Project sol on V and back to get proj_sol
            Calculate error [j] = norm (proj_sol − sol) /norm (sol)
        end for
        Remove MD corresponding to min (error) from R
        Increment i by 1
    end while
    return R, errors
end procedure
```

It's worth noting that the selection of features is greedy in the sense that once a feature is removed, it can't be added back even if it might be beneficial in the presence of other features. So while this function should work well if the error associated with each feature is relatively independent, it might perform poorly if there is a high degree of interaction between features.

Furthermore, if the error definition is not well-suited to the specific system under consideration, it could lead to inaccurate modal derivative selection, and, consequently, inaccurate predictions.

2.4 Reduction of Internal Forces

The model order reduction methods detailed in Sects. 2.2 and 2.3 are contingent on pre-calculated evaluations of internal forces **f**, and tangent stiffness **K**. Since these are nonlinear functions of the generalised displacement vector **u**, their evaluation by the high-fidelity FEM model becomes computationally expensive, undermining the efficiency of these methods.

An alternative approach, proposed by Mignolet [8], circumvents this problem by directly approximating the reduced elastic forces $\overline{\mathbf{f}}$ in (10) using a third-order polynomial:

$$\overline{f}_i \approx K^{(0)}_{il}\, q_l + \frac{1}{2} K^{(1)}_{ilm}\, q_m q_l + \frac{1}{6} K^{(2)}_{ilmn}\, q_n q_m q_l \tag{12}$$

where $\mathbf{K}^{(0)}$, $\mathbf{K}^{(1)}$, and $\mathbf{K}^{(2)}$ are second-, third-, and fourth-order tensors symmetric over all indices. They can be preliminarily computed via numerical differentiation of the stiffness matrix or through enforced displacement methods [7].

To further improve the computational efficiency, Silveira [10] suggested taking advantage of the symmetries in these tensors to expedite the computation of $\overline{f}_i$ during dynamic simulations. Specifically, a re-expression of (12) is proposed:

$$\overline{f}_i \approx K^{(0)}_{il}\, q_l + A_{ij} b_j + B_{ij} c_j \tag{13}$$

where b_j and c_j are half-vectorisations of $q_k q_j$ and $q_l q_k q_j$, respectively. This replacement capitalises on tensor symmetries and consequently provides a gain in performance by reducing the number of computations.

Given that the quadratic manifold is a subset of the augmented linear basis [6], the approximation described above can be used in conjunction with the model order reduction described in Sect. 2.2 with no accuracy loss through the methodology detailed in [10].

3 Numerical Examples

The accuracy of the two previously described nonlinear model order reduction methods in representing bending-torsion coupling is examined through the numerical example shown in Fig. 1. This figure illustrates an L-shaped cantilever beam, 2m in length. The cross-sectional dimensions of this beam are characterised by a width of

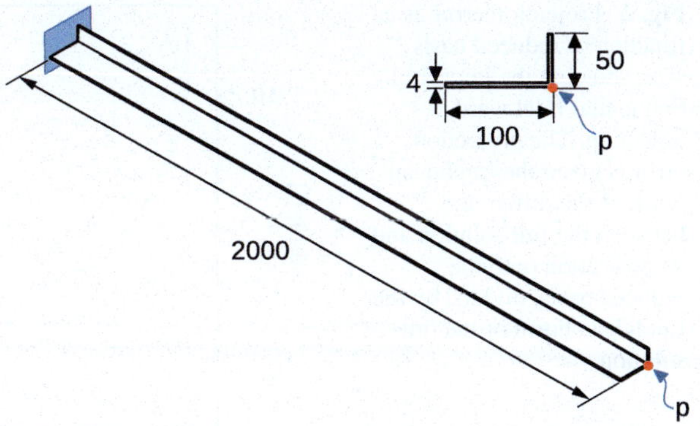

Fig. 1 Schematic representation of the L-shaped cantilever beam used in the numerical example, with a length of 2 m, a width of 100 mm, a height of 50 mm, and a thickness of 4 mm

100 mm, a height of 50 mm, and a thickness of 4 mm. The beam's material properties are defined by Young's modulus E of 70 GPa, Poisson's ratio ν of 0.3, and a density ρ of 2700 kg/m^3. The beam is discretised using a three-dimensional regular mesh. The converged mesh is made up of 4161 nodes, arranged into 696 hexahedral elements; each element is comprised of 20 nodes and employs quadratic shape functions. The left extremity of the beam is clamped and a constant vertical load distribution $\mathbf{t}(x) = 1.5 \cdot 10^5 x$ N/m^2 is applied on its right end surface.

A three-dimensional nonlinear finite element solution was developed in Python, using the Total Lagrange formulation [2]. The corresponding finite element method (FEM) code, inclusive of documentation and symbolic scripts utilised for the mathematical derivation, alongside automatic code generation, is accessible in the relevant text repositories (refer to the data availability declaration). In addition to this, simulations were executed via the commercially available FEM software, COMSOL. These served as a benchmark to validate the fidelity of the Python implementation.

The previously outlined reduced-order models were implemented in Python and subsequently integrated with the Python FEM code. A set of 17 modes, along with their corresponding 153 modal derivatives, were utilised in formulating the quadratic manifold and the augmented linear basis. Time simulations were executed employing the Newmark time integration algorithm, with parameters set to $\alpha = 0.5$, $\beta = 0.25$, and relative tolerance of $1.0e{-5}$. The chosen time step for these simulations was $\Delta t = 1 \cdot 10^{-4}$ s. The adequacy of the time step was confirmed via a convergence study applied to the full solution. The specific numerical scripts employed in this section are made available in the corresponding text repositories (refer to the data availability declaration).

The sequential backward selection method, as outlined in Sect. 2.3.1, was employed to discern the modal derivatives that most significantly influence the accuracy of the reduced-order model. The static solution derived from the full Finite Element Method (FEM) model was initially projected onto the reduced basis and then re-projected back. The projection error is calculated using the formula:

$$error = \left\| \mathbf{u}_{proj.} - \mathbf{u}_{full} \right\| \;/\; \left\| \mathbf{u}_{full} \right\| \tag{14}$$

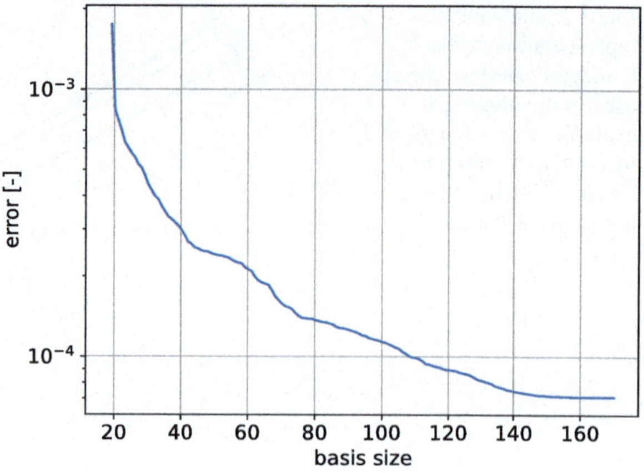

Fig. 2 Projection error as a function of reduced basis size, showing the impact of Sequential Backward Selection. The projection error is given the Euclidean norm of the difference between the full solution and its projection onto the reduced basis, divided by the Euclidean norm of the full solution

This process allowed for the systematic removal of the least impactful modal derivatives one by one, effectively reducing the basis size while retaining its key features. It is important to note that the backward selection procedure exclusively influenced the count of modal derivatives; the number of mode shapes remained constant throughout the process. Figure 2 shows the projection error as a function of the basis size. A reduced basis of size 120 (i.e. 17 mode shapes and 103 modal derivatives) was selected and compared with the original reduced basis of size 170.

Figure 3a shows the results for the displacement of point p in the L-shaped cantilever beam as a function of time predicted using the different models. In it, the abbreviation ALB refers to augmented linear basis, QM to the quadratic manifold, and FKR to the reduction of the internal force vector and tangent stiffness matrix. The abbreviation "sel." refers to the reduced basis size obtained from the backward selection. Figure 3b shows the difference between the value predicted by the linear and augmented linear basis model order reduction models from the full FEM model.

Figure 4a shows the results for the twist angle around point p. The linear MOR illustrated by the green curve refers to a linear modal truncation composed of the first 17 mode shapes. Figure 4b shows the difference between the value predicted by the linear and augmented linear basis model order reduction models from the full FEM model.

From Fig. 3, it can be observed that all reduced-order models except the quadratic manifold show a good agreement with the full model. Even the linear modal truncation is able to capture the vertical displacement of point p.

However, from Fig. 4 it can be observed that the linear reduction and the quadratic manifold fail to predict the angular deflection. Around 250 ms, the linear reduction predicts a negative twist while the full model predicts a positive one. The linear basis augmented with modal derivatives is able to predict the twist angle even at deflections as high as 8°. The linear basis augmented with selected modal derivatives still accurately predicts the twist angle.

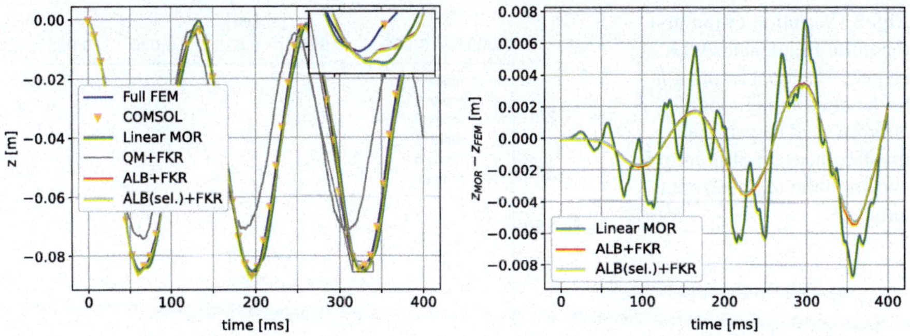

(a) Vertical displacement (b) Vertical displacement error

Fig. 3 Vertical displacements of node p in the solution history of the numerical example of a L-shaped cantilever beam shown in Fig. 1 (**a**) and the difference between the value predicted by different MOR models and the full FEM model (**b**). The abbreviation ALB refers to augmented linear basis, QM to the quadratic manifold, FKR to the reduction of the internal forces, and sel. to the reduced basis size obtained from the backward selection

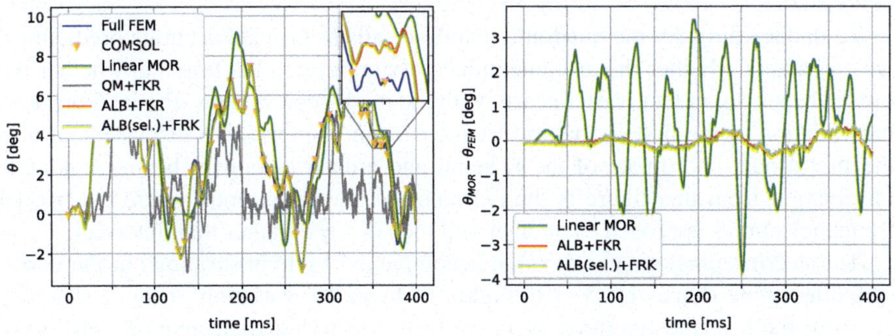

(a) Twist angle (b) Twist angle error

Fig. 4 Angular displacements of node p in the solution history of the numerical example of a L-shaped cantilever beam shown in Fig. 1. The abbreviation ALB refers to augmented linear basis, QM to the quadratic manifold, FKR to the reduction of the internal forces, and sel. to the reduced basis size obtained from the backward selection

3.1 Discussion

The vertical displacements of the system under consideration are relatively small, and even a linear reduction model is capable of accurately predicting these vertical displacements. This is a significant observation because it demonstrates the efficacy of a simpler, linear model in capturing the behaviour of the system.

In contrast, the quadratic manifold model, which is based on more complex assumptions, fails to predict the vertical displacement effectively. A key assumption underlying the quadratic manifold, as outlined in (4), is that the mode shapes vary linearly with the modal coordinates. The validity of this model, therefore, largely

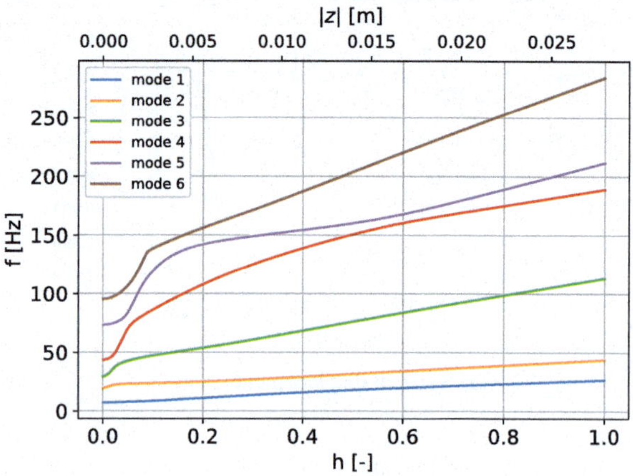

Fig. 5 Variation of the first 6 natural frequencies with respect to a displacement h in the dominating modal coordinate. The vertical displacement $|z|$ of point p is shown in the top horizontal axis

hinges on the extent to which this assumption is applicable to the specific system being studied.

To understand why the quadratic manifold fails to provide accurate predictions, we can assess whether the aforementioned assumption holds true, i.e. whether the mode shapes do, in fact, vary linearly with the modal coordinates. Figure 5 presents the variation of the first six natural frequencies in response to a displacement h in the first modal coordinate of the L beam and arched cantilever beam. A notable observation from this figure is the existence of several regions where two modal frequency curves approach each other and then diverge, rather than intersect.

The veering phenomenon is often accompanied by hybridisations in the corresponding mode shapes [1, 3, 10]. Figure 6 shows the variation of mode shapes 3 and 4 of the L-shape cantilever beam with respect to a displacement h equal to 0, 0.025, and 0.035 in the first modal coordinate. Mode shape 3 is initially characterised predominantly by a torsion around the x-axis. Mode shape 4 is initially characterised predominantly by a bending along x-axis. In the hybridisation zone, they exhibit characteristics of both of the original mode shapes. After the hybridisation zone, the main characteristics of the mode shapes are swapped. For a better visualisation, the 3D geometries of the veered mode shapes depicted on the figure are also made available on the text repository (refer to the data availability declaration).

This insight is crucial because the quadratic manifold assumes that the mode shapes vary linearly with the modal coordinates, an assumption that does not account for the occurrence of modal hybridisations. Therefore, the quadratic manifold's failure to predict the vertical displacement could be attributed to its inability to accurately represent these hybridisations. This underscores the importance of carefully scrutinising the underlying assumptions of any model and matching them to the specific characteristics of the system under study.

As shown in Fig. 4, despite the relatively small vertical displacements, the angular deflection of point p around the x-axis is fairly substantial, with a magnitude of approximately 10°. It's noteworthy that the linear reduction model, while effective

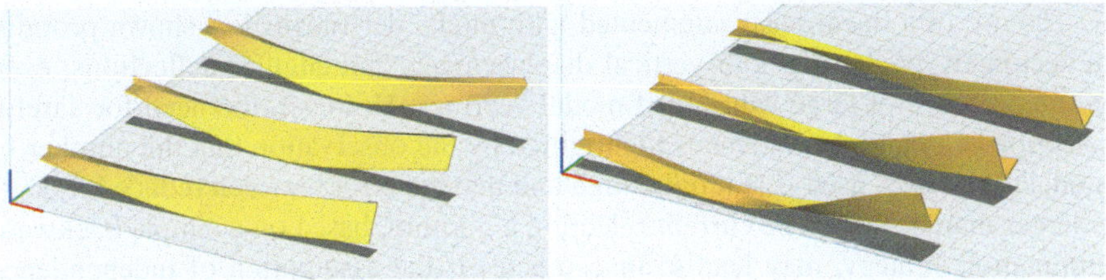

(a) Mode 3, h=0, 0.025, 0.035 (b) Mode 4, h=0, 0.025, 0.035

Fig. 6 Veering of orthonormal mode shapes 3 and 4 of the L-shaped beam with respect to a displacement $h = 0, 0.025, 0.035$, from front to back respectively, in the first modal coordinate $\boldsymbol{\phi}_1$

for predicting the smaller vertical displacements, fails to accurately forecast these larger angular deflections.

When the linear basis is augmented with modal derivatives, it offers a more accurate prediction of the angular deflection, even with its relatively large values. This method exhibits robust predictions despite the presence of modal veerings. The reason behind this accuracy can be attributed to the fact that the contribution of modal derivatives is given by additional degrees of freedom. As such, the linear basis augmented with modal derivatives is not affected by the modal hybridisations. This indicates that the method's effectiveness and applicability extend beyond simple linear movements, offering greater versatility in predicting complex structural behaviours.

In addition to the aforementioned points, it is notable that the linear basis augmented with selected modal derivatives still yields accurate predictions for the angular displacements despite incorporating fewer modal derivatives. This observation suggests that some modal derivatives carry more significant information, at least within the present context. This underscores the importance of carefully selecting the modal derivatives to be included in the model. It's worth noting that the number of modal derivatives necessary for good accuracy is still relatively large. This suggests that further improvements in the selection of modal derivatives could be beneficial.

4 Conclusions

The paper has explored the applicability of linear and quadratic manifolds, as well as a linear basis augmented with modal derivatives for predicting vertical displacements and angular deflections. Our results reveal that while linear reduction accurately predicts vertical displacements, it struggles to accurately predict large angular deflections. The quadratic manifold, which assumes that mode shapes vary linearly with modal coordinates, fails to predict vertical displacements accurately. A significant insight gained from this study is that this failure is likely due to modal hybridisations, which are not accounted for in the quadratic manifold.

The use of a linear basis augmented with modal derivatives has shown promise in accurately predicting both vertical displacements and angular deflections, even in the presence of large values and modal veerings. However, the need for careful selection of modal derivatives is highlighted by the observation that the number of modal derivatives used is still quite high and that the error between reduced models is larger than desired. The current selection algorithm, based on a simple backward elimination strategy, may lead to inaccuracies if the assumption of independence between columns does not hold. Furthermore, the selection of modal basis and modal derivative basis is dependent on the load case and different selections may be needed for different load cases.

Moving forward, future work should focus on refining the modal derivative selection process, particularly considering the correlation between columns and the definition of error. This paper paves the way for such future investigations, highlighting the potential of modal derivatives in enhancing the prediction accuracy of reduced-order models. Ultimately, the goal is to develop a robust and efficient reduced-order model that can accurately predict the behaviour of complex systems under various conditions.

Declarations

Data availability. The Total Lagrange finite element code, implemented in Python, used in this text for the full simulations can be found in the GitHub repository https://github.com/silveira-lucas/Nonlinear_FEM. The Nonlinear Model Order Reduction scripts, implemented in Python, used in this work can be found in the GitHub repository https://github.com/silveira-lucas/TBCNMOR_2023.

References

1. E. Balmès, High modal density, curve veering, localization: a different perspective on the structural response. J. Sound Vib. **161**(2), 358–363 (1993). https://doi.org/10.1006/jsvi.1993.1078, https://www.sciencedirect.com/science/article/pii/S0022460X83710783
2. T. Belytschko, W. Liu, B. Moran, K. Elkhodary, *Nonlinear Finite Elements for Continua and Structures* (Wiley, 2013). https://books.google.dk/books?id=e_w8AgAAQBAJ
3. F. Benedettini, D. Zulli, R. Alaggio, Frequency-veering and mode hybridization in arch bridges, in *IMAC-XXVII, Conference and Exposition on Structural Dynamics* (2009). https://hdl.handle.net/11697/39428
4. R.R. Craig, M.C.C. Bampton, Coupling of substructures for dynamic analyses. AIAA J. **6**(7), 1313–1319 (1968). https://doi.org/10.2514/3.4741
5. S.R. Idelsohn, A. Cardona, A reduction method for nonlinear structural dynamic analysis. Comput. Methods Appl. Mech. Eng. **49**(3), 253–279 (1985). https://doi.org/10.1016/0045-7825(85)90125-2
6. S. Jain, P. Tiso, J. Rutzmoser, D. Rixen, A quadratic manifold for model order reduction of nonlinear structural dynamics. Comput. Struct. **188** (2017). https://doi.org/10.1016/j.compstruc.2017.04.005
7. M. Karamooz Mahdiabadi, P. Tiso, A. Brandt, D.J. Rixen, A non-intrusive model-order reduction of geometrically nonlinear structural dynamics using modal derivatives. Mech. Syst. Sig-

nal Process. **147**, 107126 (2021). https://doi.org/10.1016/j.ymssp.2020.107126, https://www.sciencedirect.com/science/article/pii/S0888327020305124

8. M.P. Mignolet, A. Przekop, S.A. Rizzi, S.M. Spottswood, A review of indirect/non-intrusive reduced order modeling of nonlinear geometric structures. J. Sound Vib. **332**(10), 2437–2460 (2013). https://doi.org/10.1016/j.jsv.2012.10.017, https://www.sciencedirect.com/science/article/pii/S0022460X12008188
9. J. Rutzmoser, D. Rixen, P. Tiso, S. Jain, Generalization of quadratic manifolds for reduced order modeling of nonlinear structural dynamics. Comput. Struct. **192** (2016). https://doi.org/10.1016/j.compstruc.2017.06.003
10. L.G.C. Silveira, M.H. Hansen, Modal veering effect on quadratic manifold model order reduction accuracy (2023). preprint

Bayesian Estimation of Laser Linewidth From Delayed Self-Heterodyne Measurements

Lutz Mertenskötter and Markus Kantner

Abstract We present a statistical inference approach to estimate the frequency noise characteristics of ultra-narrow linewidth lasers from delayed self-heterodyne beat note measurements using Bayesian inference. Particular emphasis is on the estimation of the intrinsic (Lorentzian) laser linewidth. The approach is based on a statistical model of the measurement process, taking into account the effects of the interferometer as well as the detector noise. Our method therefore yields accurate results even when the intrinsic linewidth plateau is obscured by detector noise. The regression is performed on periodogram data in the frequency domain using a Markov chain Monte Carlo method. By using explicit knowledge about the statistical distribution of the observed data, the method yields good results already from a single time series and does not rely on averaging over many realizations, since the information in the available data is evaluated very thoroughly. The approach is demonstrated for simulated time series data from a stochastic laser rate equation model with $1/f$-type non-Markovian noise.

1 Introduction

Ultra-narrow linewidth lasers are critical components of many applications in modern science and technology, ranging from precision metrology, e.g., gravitational wave interferometers [1] and optical atomic clocks [2], to coherent optical communication systems [3] and ion-trap quantum computers [4]. To perform well in these technologies, the laser requires a high degree of spectral coherence, i.e., a well-defined phase, and/or a sharply defined frequency and thus low-frequency noise. The (intrinsic) laser linewidth is quantified by the width of the optical power spectrum, which in real systems is usually broadened by additional $1/f$-like technical noise (also *flicker*

L. Mertenskötter (✉) · M. Kantner
Weierstrass Institute for Applied Analysis and Stochastics, Mohrenstr. 39, 10117 Berlin, Germany
e-mail: mertenskoetter@wias-berlin.de

M. Kantner
e-mail: kantner@wias-berlin.de

M. Belhaq (ed.), *Advances in Nonlinear Dynamics and Control of Mechanical and Physical Systems*, Springer Proceedings in Physics 301,
https://doi.org/10.1007/978-981-99-7958-5_21

noise) [5–7]. As the width of the optical power spectrum is dominantly determined by the frequency noise, the corresponding frequency noise power spectral density (FN-PSD) provides an almost complete characterization of the spectral quality of the laser, where the effects of non-Markovian noise can be well separated from that of white noise. The latter manifests itself as a plateau in the high-frequency part of the FN-PSD, from which the intrinsic (Lorentzian) laser linewidth [8, 9] can be deduced; that is of major interest for most of the aforementioned applications. In ultra-narrow linewidth lasers, the determination of the white noise plateau can be challenging, since it often sets in only at very high frequencies and is obscured by $1/f$-type noise (at low frequencies) or by detector noise (at high frequencies).

The standard technique for the experimental measurement of the laser linewidth is the delayed self-heterodyne (DSH) method [10, 11], which involves measuring the beat signal between the optical field with a delayed and frequency-shifted copy of itself. This method is attractive because it can provide a direct measurement of the linewidth without the need for an external frequency standard or active frequency stabilization. Evaluation of the DSH measurement data is however non-trivial, as both the footprint of the interferometer as well as the detector noise must be removed in order to obtain an artifact-free reconstruction of the FN-PSD [12].

In this paper, we present a Bayesian estimation approach to infer on the laser's FN-PSD from time series data. Our method is based on statistical modeling of the DSH measurement process and allows to extract accurate estimates of key parameters such as the intrinsic linewidth, even when the white noise plateau is obscured by detector noise. The method is demonstrated for simulated time series data based on a stochastic rate equation model for a single-mode semiconductor laser.

2 Delayed Self-Heterodyne Method

2.1 Experimental Setup

The DSH method is a standard technique for measuring the laser linewidth [11], particularly of ultra-narrow linewidth lasers. It involves splitting of the laser beam into two paths using an acousto-optic modulator (AOM), where one of the two beams is also frequency-shifted; see Fig. 1. The frequency-shifted beam is then delayed using a long fiber before being finally superimposed with the other beam at a photodetector. The optical field received by the detector $E(t) = \mathrm{Re}(\mathcal{E}(t))$ reads

$$\mathcal{E}(t) = \sqrt{P(t)}\,\mathrm{e}^{i(\omega_0 t+\phi(t))} + \sqrt{P(t-\tau_d)}\,\mathrm{e}^{i((\omega_0+\Delta\omega)(t-\tau_d)+\phi(t-\tau_d))} + \xi_E(t), \quad (1)$$

where ω_0 is the nominal continuous wave (CW) frequency, τ_d is the interferometer delay, $\phi(t)$ is the (randomly fluctuating) optical phase, $P(t)$ is the (fluctuating) amplitude and $\Delta\omega$ is the frequency shift induced by the AOM. Moreover, $\xi_E(t)$ describes additive detector noise. Conventional photodetectors capture only the slow

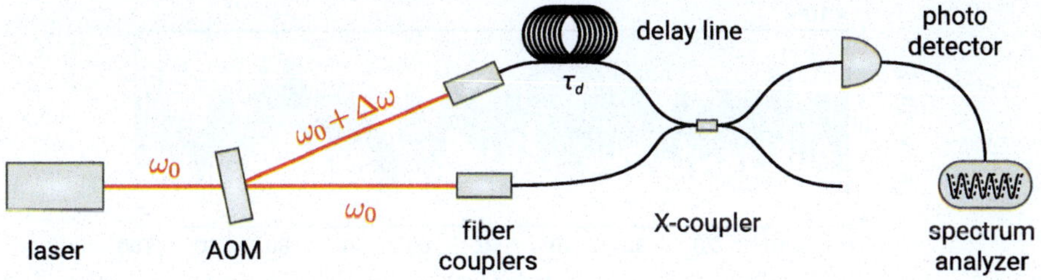

Fig. 1 Experimental setup of the DSH method, where an acousto-optic modulator (AOM) separates the incoming laser beam into two beams. One part of the signal is delayed by a long fiber and frequency-shifted. The two beams are superimposed at a photodetector, which captures only the slow beat note signal. The picture is reprinted with permission from [12]

beat note in the intensity signal $I(t) \propto |E(t)|^2$. The spectrum analyzer then further downshifts the beat frequency and removes the DC component of the signal. From this signal and its quadrature component (generated by a Hilbert transform), one can then extract the phase jitter and finally conclude on the fluctuations of the instantaneous laser frequency [12].

2.2 *Periodogram and Power Spectral Density*

The power spectral density (PSD) $S_{z,z}(\omega)$ of a stationary stochastic process $z(t)$ is given by the Fourier transform of its auto-correlation function $C_{z,z}(\tau) = \langle z(t)\, z(t+\tau)\rangle$ (Wiener–Khinchin theorem). Given only a sample of the trajectory (in discrete time), the PSD is typically estimated from the periodogram [13, 14], which is given by the absolute square of the (discrete) Fourier transform of the time series

$$\widehat{S}_{z,z}(\omega) = |\mathcal{F}[z(t)](\omega)|^2 . \tag{2}$$

The PSD then follows as the expectation value of the periodogram (ensemble mean)

$$S_{z,z}(\omega) = \langle \widehat{S}_{z,z}(\omega)\rangle . \tag{3}$$

Our goal is to estimate the FN-PSD of the free-running laser, which will be denoted by $S_{x,x}(\omega)$ in the following. As the frequency follows from the optical phase by differentiation with respect to time $\omega(t) = \dot{\phi}(t)$, their PSDs are connected by

$$S_{x,x}(\omega) = \omega^2 S_{\phi,\phi}(\omega) . \tag{4}$$

Note that the FN-PSD $S_{x,x}(\omega)$ is not directly observed in the experiment, but must be reconstructed from the phase jitter $\phi(t) - \phi(t - \tau_d)$ that can be deduced from the slow beat note of the intensity time series corresponding to (1).

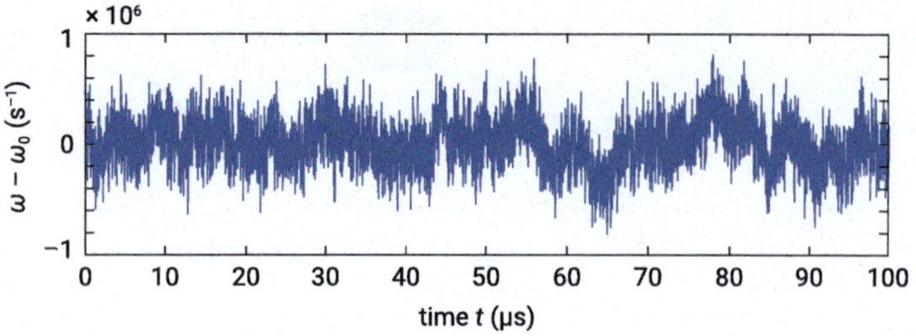

Fig. 2 Simulated time series of frequency fluctuations $\omega(t) = \dot{\phi}(t)$ around the nominal continuous wave frequency ω_0. The instantaneous laser frequency exhibits characteristic drifts, which are commonly observed in semiconductor lasers. The time series has been simulated using the SDE model (14), where frequency drifts are induced by the $1/f$-type colored noise contributions. The trajectory in the plot shows a moving average (over 50 ns) to improve the visibility of the effect

In the following, we describe the measured signal of the DSH experiment (in terms of frequency fluctuations) by

$$z(t) = (h * x)(t) + \xi(t), \tag{5}$$

where $z(t)$ is the observed time series, the convolution kernel $h(t) = \delta(t) - \delta(t - \tau_d)$ is the transfer function of the interferometer, $x(t)$ is the hidden time series of the instantaneous frequency fluctuations of the laser (*i.e.*, $x(t) \mathrel{\hat{=}} \omega(t) - \omega_0 = \dot{\phi}(t)$) and $\xi(t)$ is (colored) additive measurement noise (not correlated with the hidden signal). A sample time series of frequency fluctuations $x(t)$, which exhibits characteristic frequency drifts as commonly observed in semiconductor lasers, is shown in Fig. 2. Our goal is to characterize the statistical properties of the fluctuating time series $x(t)$. Fourier transform of (5) yields a relation between the PSDs of the observed and the hidden signal

$$S_{z,z}(\omega) = |H(\omega)|^2 S_{x,x}(\omega) + S_{\xi,\xi}(\omega), \tag{6}$$

where the Fourier transformed transfer function reads

$$H(\omega) = 1 - \exp(i\omega\tau_d). \tag{7}$$

The PSDs of the hidden signal and the detector noise are assumed to obey the following functional forms [12]:

$$S_{x,x}(\omega) = \frac{C}{|\omega|^{\nu}} + S_{\infty}, \qquad S_{\xi,\xi}(\omega) = \sigma^2\omega^2. \tag{8}$$

The model equation for $S_{x,x}(\omega)$ is a phenomenological relation [5, 6] that includes both $1/f^{\nu}$-type noise (described by C and ν) and a white noise plateau, where S_{∞} quantifies the intrinsic linewidth. The detector noise PSD $S_{\xi,\xi}(\omega)$ follows from the

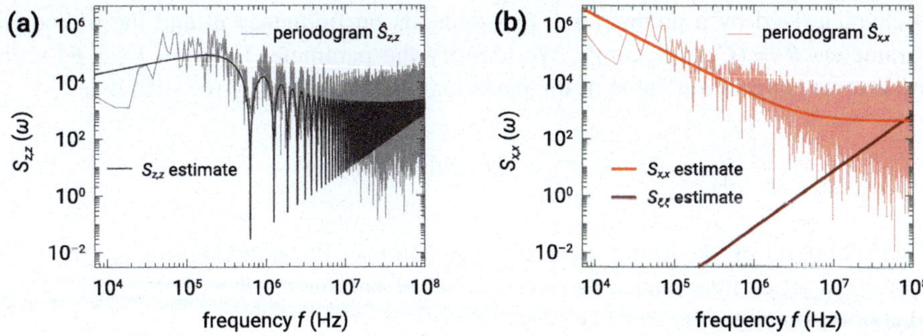

Fig. 3 **a** Periodogram $\widehat{S}_{z,z}(\omega)$ of an observed time series $z(t)$ along with the power spectral density $S_{z,z}(\omega)$ estimated using the MCMC method. The signal features sharp dropouts at the roots of the transfer function (7) at frequencies $f_n = n/\tau_d$, $n \in \mathbb{Z}$. **b** Estimated FN-PSD $S_{x,x}(\omega)$ and periodogram $\widehat{S}_{x,x}(\omega)$ of the hidden time series $x(t)$. The detector noise PSD $S_{\xi,\xi}(\omega)$ has a quadratic frequency dependency. Both PSDs are obtained from statistical inference on the observed periodogram $\widehat{S}_{z,z}(\omega)$

assumption of spectrally white phase fluctuation measurement noise, which must be multiplied by ω^2 to arrive at the corresponding measurement noise for the frequency fluctuations; cf. (4). Figure 3 shows a double-logarithmic plot of the PSDs considered here.

3 Parameter Inference

For the estimation of the parameters characterizing the FN-PSD of the laser, we perform a Bayesian regression on the PSD $S_{z,z}(\omega)$ of the detected signal using the transfer function (7) and the model relations (8), which have a highly nonlinear frequency dependency. In the regression procedure, we regard the periodogram $\widehat{S}_{z,z}(\omega)$ (available at a set of discrete frequencies) as observed data. In order to conduct a maximum likelihood estimation using frequency domain data, knowledge about the expected statistical distribution of the periodogram $\widehat{S}_{z,z}(\omega)$ is required.

The frequency fluctuations $x(t)$ and the measurement noise $\xi(t)$ are assumed to normally distributed, which is in excellent agreement with experimental data. The detected time series $z(t)$ observed at discrete instances of time is thus a multi-variate Gaussian characterized by its covariance matrix. By random variable transformation [13], we then find the periodogram of the measured time series to be exponentially distributed

$$\widehat{S}_{z,z}(\omega) \sim \mathrm{Exp}(\lambda(\omega, \boldsymbol{\theta})), \tag{9}$$

where the probability distribution function (PDF)

$$p\left(\widehat{S}_{z,z}(\omega), \lambda(\omega, \boldsymbol{\theta})\right) = \lambda(\omega, \boldsymbol{\theta})\, \mathrm{e}^{-\lambda(\omega,\boldsymbol{\theta})\, \widehat{S}_{z,z}(\omega)} \tag{10}$$

is characterized by a parameter λ that depends on frequency ω and the unknown parameters $\boldsymbol{\theta} = (C, \nu, S_{\infty}, \sigma)^T$. We identify the parameter function $\lambda(\omega, \boldsymbol{\theta})$ with the inverse expectation value of the periodogram data given by (6) such that

$$\frac{1}{\lambda(\omega, \boldsymbol{\theta})} = S_{z,z}(\omega, \boldsymbol{\theta}) = |H(\omega)|^2 S_{x,x}(\omega, \boldsymbol{\theta}) + S_{\xi,\xi}(\omega, \boldsymbol{\theta}). \tag{11}$$

The likelihood of observing a certain realization of the periodogram $\widehat{S}_{z,z}(\omega_k)$ (at a discrete set of frequencies ω_k) given a set of parameters $\boldsymbol{\theta}$ is then given by the likelihood function

$$L(\boldsymbol{\theta}) = \prod_k p\left(\widehat{S}_{z,z}(\omega_k), \lambda(\omega_k, \boldsymbol{\theta})\right), \tag{12}$$

where the function p is given by (10). Note that the underlying joint probability distribution factorizes completely, as the periodogram data $\widehat{S}_{z,z}(\omega)$ at different frequencies are statistically independent.

Bayesian regression on the parameters $\boldsymbol{\theta}$ is now performed by maximizing the likelihood function (12) using a Markov chain Monte Carlo (MCMC) approach [15]. We employ the Metropolis–Hastings algorithm to sample a Markov chain of parameter sets $\boldsymbol{\theta}^{(j-1)} \to \boldsymbol{\theta}^{(j)} \to \cdots$ that is distributed according to $L(\boldsymbol{\theta})$. The maximum of this distribution is then located at the parameter set $\boldsymbol{\theta}^*$ which is most likely to underlie the observed data.

We would like to stress that in the present scenario, the regression on spectral data (i.e., on the periodogram $\widehat{S}_{z,z}(\omega)$) as described above is advantageous compared to the direct estimation on time domain data. The reason for this is that the construction of the likelihood function for the observed time series is either conceptually challenging due to the long interferometer delay or would require a high-dimensional Markovian embedding. Furthermore, the system exhibits long-range time correlations due to the presence of $1/f$-type noise, which would require very long time series to be considered in the regression. These issues render the estimation procedure in the time domain computationally expensive, but can be easily overcome by transforming the problem into the frequency domain.

4 Results

We demonstrate the estimation method described above for simulated data using the stochastic laser rate equations given in Appendix A. The DSH measurement is simulated as described in [12] with simulation code that is available online [16].

Figure 4 shows histograms of the sampled Markov chains $\boldsymbol{\theta}^{(j)}$, which are estimators of the marginal distributions of the joint PDF of the parameters that is proportional to $L(\boldsymbol{\theta})$. A notable feature of the MCMC method is that it provides not only an estimate of the most probable set of parameters but rather their entire distribution, allowing to assess the uncertainty of the estimates; see Table 1. In addition to the

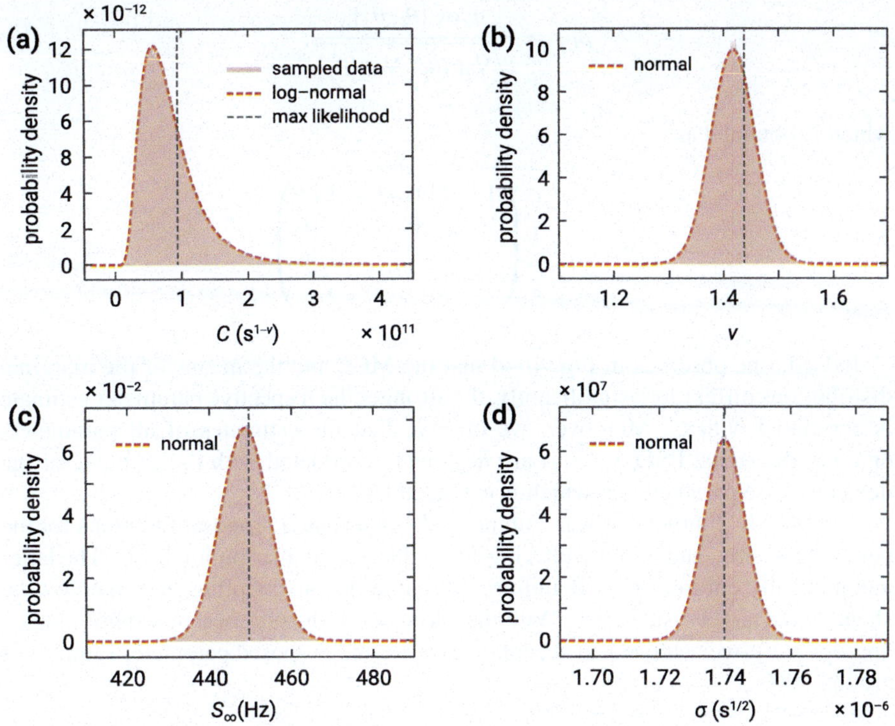

Fig. 4 Histograms of estimated parameters characterizing the FN-PSD obtained using the MCMC method. The sampled distributions are shown along with normally (Gaussian) and log-normally distributed probability density distributions. The dashed black line indicates the position of the maximum likelihood estimate. The estimation results are summarized in Table 1

Table 1 Mean values and standard deviations (std) of the marginal distributions and the maximum likelihood estimator (MLE) of the parameters $\boldsymbol{\theta}$ obtained using the MCMC method

Parameter (unit)	Mean ± std	MLE
C ($s^{1-\nu}$)	$(8.6 \pm 4.8) \cdot 10^{10}$	$9.6 \cdot 10^{10}$
ν	1.42 ± 0.04	1.44
S_∞ (Hz)	448.8 ± 5.9	449.6
σ ($s^{1/2}$)	$1.74 \cdot 10^{-6} \pm 6.4 \cdot 10^{-9}$	$1.74 \cdot 10^{-6}$

histogram, the plot also indicates the position of the maximum of $L(\boldsymbol{\theta})$ as a dashed line. The corresponding parameter set $\boldsymbol{\theta}^*$ is denoted as the maximum likelihood estimator (MLE), *i.e.*, the maximum of the joint PDF. It should be noted that these values partially differ significantly from the means of the marginal distributions. The reason for this is that the different parameters are strongly correlated, and the marginal distribution of C has a maximum far from its mean. The mutual correlation of all parameter estimates is quantified by the matrix of Pearson correlation coefficients

$$\rho_{i,j} = \frac{\text{Cov}\left(\theta_i, \theta_j\right)}{\sqrt{\text{Var}\left(\theta_i\right)\,\text{Var}\left(\theta_j\right)}}, \tag{13}$$

which is obtained as

$$\rho = \begin{array}{c} \\ C \\ \nu \\ S_\infty \\ \sigma \end{array} \begin{array}{c} \begin{array}{cccc} C & \nu & S_\infty & \sigma \end{array} \\ \begin{pmatrix} 1 & 0.93 & 0.41 & -0.14 \\ & 1 & 0.47 & -0.16 \\ & & 1 & -0.46 \\ & & & 1 \end{pmatrix} \end{array}.$$

Indeed, one observes in Fig. 4b–d that the MLE and the means of the marginal distributions differ more significantly, the stronger the respective parameter estimate is correlated with C. Moreover, we observe that the estimates of all parameters entering the signal PSD $S_{x,x}(\omega)$ are negatively correlated with the estimate of the detector noise parameter σ, which is expected from (6).

The MCMC routine was run with normally distributed proposal functions for the parameters ν, S_∞ and σ and with a log-normal proposal distribution for C. The latter automatically enforces $C > 0$ and provides an efficient sampling as it matches the shape of the target distribution. With the variances of the proposal distribution tuned appropriately, we achieved an acceptance rate of the proposed parameter samples of around 30%.

5 Conclusions

The application of Bayesian inference methods to data from DSH laser linewidth measurements allows for an accurate extraction of the parameters characterizing the FN-PSD along with its uncertainties. Based on statistical modeling of the underlying measurement process, the method enables a reliable characterization of the spectral coherence of the laser even when the intrinsic linewidth plateau is obscured by detector noise. Furthermore, by invoking explicit assumptions about the statistical distribution of the measured data, the method extracts the information encoded in the available data very extensively, so that no averaging over many realizations or long times is required.

Acknowledgements This work was funded by the German Research Foundation (DFG) under Germany's Excellence Strategy—EXC2046: MATH+ (project AA2-13).

Appendix

A Stochastic Laser Rate Equations

We describe a set of Itô-type stochastic differential equations (SDEs) modeling the fluctuation dynamics of a generic single-mode semiconductor laser. In the presence of noise, the evolution of the photon number P, the optical phase ϕ and the carrier number N in the active region is described by

$$\begin{aligned} \mathrm{d}P &= \left(-\gamma\left(P - P_{\mathrm{th}}\right) + \Gamma v_g g\left(P, N\right) P + \Gamma v_g g_{\mathrm{sp}}\left(P, N\right) + \sigma_P\left(P\right) \mathcal{F}_P\left(t\right)\right) \mathrm{d}t \qquad (14a) \\ &\quad + \sqrt{\gamma\left(1 + P_{\mathrm{th}}\right) P}\, \mathrm{d}W_{\mathrm{out}}^{P} + \sqrt{\gamma P_{\mathrm{th}}\left(1 + P\right)}\, \mathrm{d}W_{\mathrm{in}}^{P} + \sqrt{\Gamma v_g g_{\mathrm{sp}}\left(P, N\right)}\, \mathrm{d}W_{\mathrm{sp}}^{P} \\ &\quad + \sqrt{\Gamma v_g g_{\mathrm{sp}}\left(P, N\right) P}\, \mathrm{d}W_{\mathrm{st-em}}^{P} + \sqrt{\Gamma v_g g_{\mathrm{abs}}\left(P, N\right) P}\, \mathrm{d}W_{\mathrm{st-abs}}^{P}, \\ \mathrm{d}\phi &= \left(\Omega_0 + \frac{\alpha_H}{2} \Gamma v_g g\left(P, N\right) + \frac{\sigma_P\left(P\right)}{2P} \mathcal{F}_\phi\left(t\right)\right) \mathrm{d}t \qquad (14b) \\ &\quad + \frac{1}{2P}\left(\sqrt{\gamma\left(1 + P_{\mathrm{th}}\right) P}\, \mathrm{d}W_{\mathrm{out}}^{\phi} + \sqrt{\gamma P_{\mathrm{th}}\left(1 + P\right)}\, \mathrm{d}W_{\mathrm{in}}^{\phi} + \sqrt{\Gamma v_g g_{\mathrm{sp}}\left(P, N\right)}\, \mathrm{d}W_{\mathrm{sp}}^{\phi}\right. \\ &\qquad \left. + \sqrt{\Gamma v_g g_{\mathrm{sp}}\left(P, N\right) P}\, \mathrm{d}W_{\mathrm{st-em}}^{\phi} + \sqrt{\Gamma v_g g_{\mathrm{abs}}\left(P, N\right) P}\, \mathrm{d}W_{\mathrm{st-abs}}^{\phi}\right), \\ \mathrm{d}N &= \left(\frac{\eta I}{q} - R\left(N\right) - \Gamma v_g g\left(P, N\right) P - \Gamma v_g g_{\mathrm{sp}}\left(P, N\right) + \sigma_N\left(N\right) \mathcal{F}_N\left(t\right)\right) \mathrm{d}t \qquad (14c) \\ &\quad + \sqrt{\frac{\eta I}{q}}\, \mathrm{d}W_I + \sqrt{R\left(N\right)}\, \mathrm{d}W_R - \sqrt{\Gamma v_g g_{\mathrm{sp}}\left(P, N\right) P}\, \mathrm{d}W_{\mathrm{st-em}}^{P} \\ &\quad - \sqrt{\Gamma v_g g_{\mathrm{abs}}\left(P, N\right) P}\, \mathrm{d}W_{\mathrm{st-abs}}^{P} - \sqrt{\Gamma v_g g_{\mathrm{sp}}\left(P, N\right)}\, \mathrm{d}W_{\mathrm{sp}}^{P}. \end{aligned}$$

Here, γ is the optical loss rate, P_{th} is the thermal photon number (Bose–Einstein factor), Γ is the optical confinement factor, v_g is the group velocity, Ω_0 is a detuning from the nominal CW frequency, α_H is the linewidth enhancement factor (Henry factor), η is the injection efficiency, I is the pump current and q is the elementary charge. The net-gain is modeled as

$$g\left(P, N\right) = \frac{g_0}{1 + \varepsilon P} \log\left(\frac{N}{N_{\mathrm{tr}}}\right),$$

where g_0 is the gain coefficient, ε is the inverse saturation photon number (modeling gain compression) and N_{tr} is the carrier number at transparency. Following [9], the rate of spontaneous emission into the lasing mode is modeled as

$$g_{\mathrm{sp}}\left(P, N\right) = \frac{1}{2} \frac{g_0}{1 + \varepsilon P} \log\left(1 + \left(\frac{N}{N_{\mathrm{tr}}}\right)^2\right),$$

which does not require any additional parameters and shows the correct asymptotics at low and high carrier numbers. The rate of stimulated absorption entering the noise amplitudes follows as $g_{\mathrm{abs}}(P, N) = g_{\mathrm{sp}}(P, N) - g(P, N)$. Finally, non-radiative recombination and spontaneous emission into waste modes is described by

$$R(N) = AN + \frac{B}{V}N^2 + \frac{C}{V^2}N^3.$$

We refer to [12] for a list of parameter values used in the simulation.

The model equations (14) include white and colored noise contributions. Here, $\mathrm{d}W \sim \mathrm{Normal}(0, \mathrm{d}t)$ denotes the increment of the standard Wiener processes modeling Gaussian white noise. Wiener processes with different sub- and superscripts are statistically independent. Colored noise sources $\mathcal{F}_{P,\phi,N}(t)$ are constructed as superpositions of Ornstein–Uhlenbeck processes, where the parameters are calibrated to result in power spectral densities showing a $1/f^{\nu}$-type frequency dependency. See [12] for details.

References

1. B.P. Abbott, R. Abbott, R. Adhikari et al., LIGO: the laser interferometer gravitational-wave observatory. Rep. Prog. Phys. **72**(7), 076901 (2009). https://doi.org/10.1088/0034-4885/72/7/076901
2. A.D. Ludlow, M.M. Boyd, J. Ye, E. Peik, P.O. Schmidt, Optical atomic clocks. Rev. Modern Phys. **87**(2), 637–701 (2015). https://doi.org/10.1103/RevModPhys.87.637
3. K. Kikuchi, Fundamentals of coherent optical fiber communications. J. Lightwave Technol. **34**(1), 157–179 (2016). https://doi.org/10.1109/JLT.2015.2463719
4. N. Akerman, N. Navon, S. Kotler, Y. Glickman, R. Ozeri, Universal gate-set for trapped-ion qubits using a narrow linewidth diode laser. New. J. Phys. **17**(11), 113060 (2015). https://doi.org/10.1088/1367-2630/17/11/113060
5. K. Kikuchi, Effect of 1/f-type FM noise on semiconductor-laser linewidth residual in high-power limit. IEEE J. Quant. Electron. **25**(4), 684–688 (1989). https://doi.org/10.1109/3.17331
6. L.B. Mercer, 1/f frequency noise effects on self-heterodyne linewidth measurements. J. Lightwave Technol. **9**(4), 485–493 (1991). https://doi.org/10.1109/50.76663
7. G.M. Stéphan, T.T. Tam, S. Blin, P. Besnard, M. Têtu, Laser line shape and spectral density of frequency noise. Phys. Rev. A **71**(4), 043809 (2005). https://doi.org/10.1103/PhysRevA.71.043809
8. C. Henry, Phase noise in semiconductor lasers. J. Lightwave Technol. **4**(3), 298–311 (1986). https://doi.org/10.1109/JLT.1986.1074721
9. H. Wenzel, M. Kantner, M. Radziunas, U. Bandelow, Semiconductor laser linewidth theory revisited. Appl. Sci. **11**(13), 6004 (2021). https://doi.org/10.3390/app11136004
10. T. Okoshi, K. Kikuchi, A. Nakayama, Novel method for high resolution measurement of laser output spectrum. Electron. Lett. **16**(16), 630 (1980). https://doi.org/10.1049/el:19800437
11. M. Schiemangk, S. Spießberger, A. Wicht, G. Erbert, G. Tränkle, A. Peters, Accurate frequency noise measurement of free-running lasers. Appl. Opt. **53**(30), 7138 (2014). https://doi.org/10.1364/AO.53.007138
12. M. Kantner, L. Mertenskötter, Accurate evaluation of self-heterodyne laser linewidth measurements using Wiener filters. Opt. Express **31**(10), 15994–16009 (2023). https://doi.org/10.1364/OE.485866

13. M.B. Priestley, *Spectral Analysis and Time Series* (Academic, London, 1982)
14. H. Madsen, *Time Series Analysis* (Chapman and Hall, New York, 2007). https://doi.org/10.1201/9781420059687
15. C.A. Naesseth, F. Lindsten, T.B. Schön, Elements of sequential Monte Carlo. Found. Trends Mach. Learn. **12**(3), 307–392 (2019). https://doi.org/10.1561/2200000074
16. M. Kantner, L. Mertenskötter, *Laser Noise (GitHub repository)* (2023). https://github.com/kantner/LaserNoise

The manufacturer's authorised representative in the EU is Springer Nature Customer Service Centre GmbH, Europaplatz 3, 69115 Heidelberg, Germany. If you have any concerns regarding our products, please contact ProductSafety@springernature.com

Printed and bound by CPI Group (UK) Ltd, Croydon, CR0 4YY
01/10/2026
02225900-0007